WATER HYACINTH

Environmental Challenges, Management and Utilization

WATER HYACINTH

Environmental Challenges, Management and Utilization

Editors

Shao Hua Yan
Jiangsu Academy of Agricultural Sciences
Nanjing, Jiangsu Province
China

Jun Yao Guo
Agricultural Resources and Environment
Jiangsu Academy of Agricultural Sciences
Nanjing, Jiangsu Province
China

CRC Press
Taylor & Francis Group
Boca Raton London New York

CRC Press is an imprint of the
Taylor & Francis Group, an **informa** business
A SCIENCE PUBLISHERS BOOK

Cover illustration reproduced by kind courtesy of Lin Shang

CRC Press
Taylor & Francis Group
6000 Broken Sound Parkway NW, Suite 300
Boca Raton, FL 33487-2742

Printed on acid-free paper
Version Date: 20170110

International Standard Book Number-13: 978-1-4987-4389-1 (Hardback)

Library of Congress Cataloging-in-Publication Data

Names: Yan, Shao Hua, editor. | Guo, Jun Yao, editor.
Title: Water hyacinth : environmental challenges, management, and utilization / editors, Shao Hua Yan, Jiangsu Academy of Agricultural Sciences, Nanjing, Jiangsu Province, China, Jun Yao Guo, Agricultural Resources and Environment, Jiangsu Academy of Agricultural Sciences, Nanjing, Jiangsu Province, China.
Description: Boca Raton, FL : CRC Press, 2017. | Includes bibliographical references and index.
Identifiers: LCCN 2016056214| ISBN 9781498743891 (hardback : alk. paper) | ISBN 9781498743907 (e-book : alk. paper)
Subjects: LCSH: Water hyacinth. | Water hyacinth--Control.
Classification: LCC SB615.W3 W36 2017 | DDC 628.9--dc23
LC record available at https://lccn.loc.gov/2016056214

Visit the Taylor & Francis Web site at
http://www.taylorandfrancis.com

and the CRC Press Web site at
http://www.crcpress.com

Preface

Water hyacinth (*Eichhornia crassipes*) is a large vascular floating aquatic plant adapted to a variety of environments; it has high yields and vigorous reproductive capacity. Because of these characteristics, water hyacinth is considered to be among the top 10 invasive weeds.

Its fast growth and prolificacy have caused severe ecological disasters and have had serious impacts on the socio-economic development worldwide. In rivers, lakes and water reservoirs, water hyacinth has affected transportation, fishery and tourism and has clogged pipes and canals, seriously impacting urban and industrial water supply and irrigation. On the death of water hyacinth, the biomass decay may lead to mosquito breeding and deterioration of water quality and aquatic life. By covering the water surface, water hyacinth blocks sunlight and decreases contact with air, which may result in decreased dissolved oxygen concentration and lower temperature in water. Phytoplankton, zooplankton, submerged plants and aquatic animals are affected by the water hyacinth mat, thereby impacting the food chain and aquatic biodiversity.

In contrast, because of the above-mentioned characteristics, water hyacinth is an excellent agent for controlling water pollution. Due to its strong capacity to absorb nutrients and pollutants from eutrophic/polluted waters, great adaptability, relatively easy harvesting and low maintenance costs compared to submerged or emergent plant species, water hyacinth is an excellent candidate for remediation of eutrophic waters and the biological treatment of polluted waters.

Under the experimental conditions, water hyacinth can decrease the eutrophication level of waters by taking up 2.5 g of nitrogen (N) and 0.2 g of phosphorus (P) per square meter per day. Water hyacinth has been used in various ecological engineering projects aimed at restoration of contaminated lakes. In Lake Dianchi (24°45' N 102°36' E, Kunming, China) in 2011, a pilot project using water hyacinth was implemented in an area of 5.25 square kilometers (representing 44% of the total area of the inner lake). The inner lake has a volume of 100 million cubic meters of water, with high average annual concentrations of ammonia (6.25 mg N L^{-1}), total nitrogen (13.1 mg N L^{-1}) and total phosphorus (0.8 mg P L^{-1}) in the influent streams. Six months after water hyacinth was planted, the concentrations of ammonia and total nitrogen were reduced to 0.34 and 3.3 mg N L^{-1}, respectively, and that of total P to 0.25 mg P L^{-1} in the water discharged from Xiyuan tunnel, the outlet of the inner lake.

In eutrophic fresh waters, water hyacinth can grow up to 600–900 tonnes of fresh biomass (35–54 tonnes of dry biomass) per hectare in temperate climate zone (Zheng et al. 2008). Water hyacinth biomass is suitable for production of bio-energy, feed and fertilizers.

Water hyacinth has a biogas production potential of 336 mL per gram Total Solids (TS), with gas containing 59% methane. Hence, 1 hectare of water hyacinth can produce 8550 cubic meters of methane fuel that can support 3.25 households (US single home type) with an averaged energy budget for a year. Water hyacinth can produce higher volume of pure gas per unit area than various agricultural residues (e.g., 2.2 times more than rice, 1.7 times more than corn and 2.5 times more than rapeseed).

Crude protein content of water hyacinth is around 180 to 210 g kg^{-1} dry weight, makes it a good candidate for producing silage for cattle, sheep, geese and other animals. The feed quality and feeding efficiencies can rival high-quality traditional forage.

Fresh water hyacinth or residues from biogas production (solid- and liquid-state fermentation) can be utilized in composting process to make organic fertilizer. The slurry after gas production from one hectare of water hyacinth can yield 34 tonnes of organic fertilizer to support 1.2 hectares of grain crop regarding the nitrogen, phosphorus and potassium demands.

Despite extreme efficiency of water hyacinth in purifying eutrophic or polluted water and growing biomass for bio-energy, feed and organic fertilizers, one should be cautious about water hyacinth being a really invasive species in aquatic environments and potentially a killer of aquatic life if not managed appropriately. For this reason, the first and utmost criterion is to control water hyacinth by either confining it to specific locations for growth or developing suitable harvest capacities for managing its population in natural water bodies in areas where the climate is suitable. In the same manner, water hyacinth planted for the purpose of remediation must be timely harvested and disposed of or utilized properly to avoid rotting that can cause even worse ecological problems.

To control water hyacinth natural populations as well as to use it for phytoremediation and/or as a source of biomass, we face many challenges.

The most economic way of harnessing water hyacinth would be to utilize it as a biomass resource. However, the first challenge is up to 95% of water in its biomass, which impedes harvesting, transport and disposal due to high costs involved. A key would be dehydration reliant on the new technologies, including development and production of special equipment. In addition, integration of technologies, management, and policy making is required to achieve highly efficient and low-cost production of water hyacinth biomass. The second challenge is to develop the commercially-viable and high-quality value-added products from dehydrated biomass of water hyacinth, such as the formulated feed for herbivores, purified biogas, organic fertilizers and growth media (bedding materials) for crops as well as horticulture nurseries. The third challenge is cooperation amongst scientists, policy makers, managers and citizens given that water bodies are public properties and their management is in common interest. For this reason, it may be necessary to develop policies regulating a compensation structure (similar to that related to off-setting carbon dioxide emissions) to support relevant industries.

The present book referenced and summarized water hyacinth research papers and monographs starting in 1817, combined with a compilation of technologies and critical thinking and practical experience from the authors' research teams. The monograph is divided into 12 chapters including Introduction (Chapter 1), Biology of Water Hyacinth (Chapter 2), Direct and Strong Influence of Water Hyacinth on Aquatic Communities in Natural Waters (Chapter 3), Impacts of Water Hyacinth on Ecosystem Services (Chapter 4), Mechanisms and Implications of Nitrogen Removal (Chapter 5), Mechanism and Implications of Phosphorus Removal (Chapter 6), Impact of Water Hyacinth on Removal of Heavy Metals and Organic Pollutants (Chapter 7), Ecological Engineering Using Water Hyacinth in Phytoremediation (Chapter 8), Impact of Water Hyacinth on Aquatic Environment in Phytoremediation of Eutrophic Lakes (Chapter 9), Utilization of Biomass for Energy and Fertilizer (Chapter 10), Utilization of Water Hyacinth Biomass for Animal Feed (Chapter 11), and Economic Assessment and Further Research (Chapter 12).

The book proposes a theoretical concept for sustainable development (social, economic and environmental) by systematically integrating environmental management with remediation of eutrophic waters using water hyacinth and with utilization of biomass resources produced in such remediation. The research focuses on pollution control, harvesting, processing and utilization of water hyacinth biomass using case studies of developing and integrating innovative technologies to bring about restoration projects of practical significance. The book also describes recent developments in producing purified biogas, organic fertilizers and feeds from the dehydrated biomass of water hyacinth. Further research and development needs are discussed in-depth, particularly regarding application prospects. We hope the book will contribute to (i) promoting ecological, social and economic applications of water hyacinth, (ii) producing valuable goods and (iii) utilizing technological innovations to make water hyacinth beneficial to mankind in the near future.

Acknowledgement

Authors would like to give our heartfelt thanks to Professor Zed Rengel from School of Earth and Environment, University of Western Australia (UWA) for his suggestions and advice. We also acknowledge the grant support from National Key Technology Research and Development Program (No. 2009BAC63B01).

S. H. Yan

16 June 2016

Overview

Eichhornia crassipes has three name variants: water hyacinth, water-hyacinth or waterhyacinth (Solms et al. 2009), the first one will be used throughout this book. Water hyacinth is a large floating vascular macrophyte and belongs to family Pontederiaceae. It is an intensively investigated and a highly Invasive Alien Species (IAS) with the large global impacts on ecological and socio-economic systems. The early record of this species can be traced back to 1816 in Brazil (The International Plant Names Index 2005). Water hyacinth was introduced as an ornamental plant to North America in the late 1800s (Chabot 2009), to Africa in early 1900s (Lindsey and Hirt 2000), and now has become naturalized in Central America, Africa, Asia, Australia and New Zealand (Ramey 2001). Water hyacinth occurs in at least 62 countries, causing extremely serious problems in the regions between 40° N and 45° S. It has been declared the most damaging Floating Aquatic Weed (FAW) worldwide, with large sums of money being spent on control attempts (Howard and Harley 1998). Water hyacinth distribution is mainly limited by its winter-survival temperature (around 7°C), although a short period of exposure to 5°C may be tolerated by the macrophytes (Owens and Madsen 1995). Therefore, the naturalized habitats are not located only within the 7°C water isothermal line, but they extend into colder areas as well (Fig. 1) (UNEP and GEAS 2013). Water hyacinth can start growing at 10°C (Gettys et al. 2009) and tolerates temperatures as high as 43°C (Howard and Harley 1998).

Noticeable ecosystem damage has been frequently reported around the world, especially in large warm lakes such as Lake Victoria (Kateregga and Sterner 2007), and in lakes and water reservoirs in Mexico, where a total coverage of 40,000 hectares was reported in 1993 (López et al. 1996). This invasive alien species was introduced into China as an ornamental plant in early 1900s (Duan et al. 2003) and is now considered a major invasive weed species in more than 19 provinces, especially in the southern parts of the country (Hong et al. 2005). A survey in Fujian province in 2003 revealed 130,000 locations with the presence of water hyacinth, covering about 270 square kilometers (Huang et al. 2004).

The widespread occurrence of water hyacinth is mainly due to combination of its unique biological characteristics, global warming and exacerbated eutrophication of surface waters. Because of the increased spread and

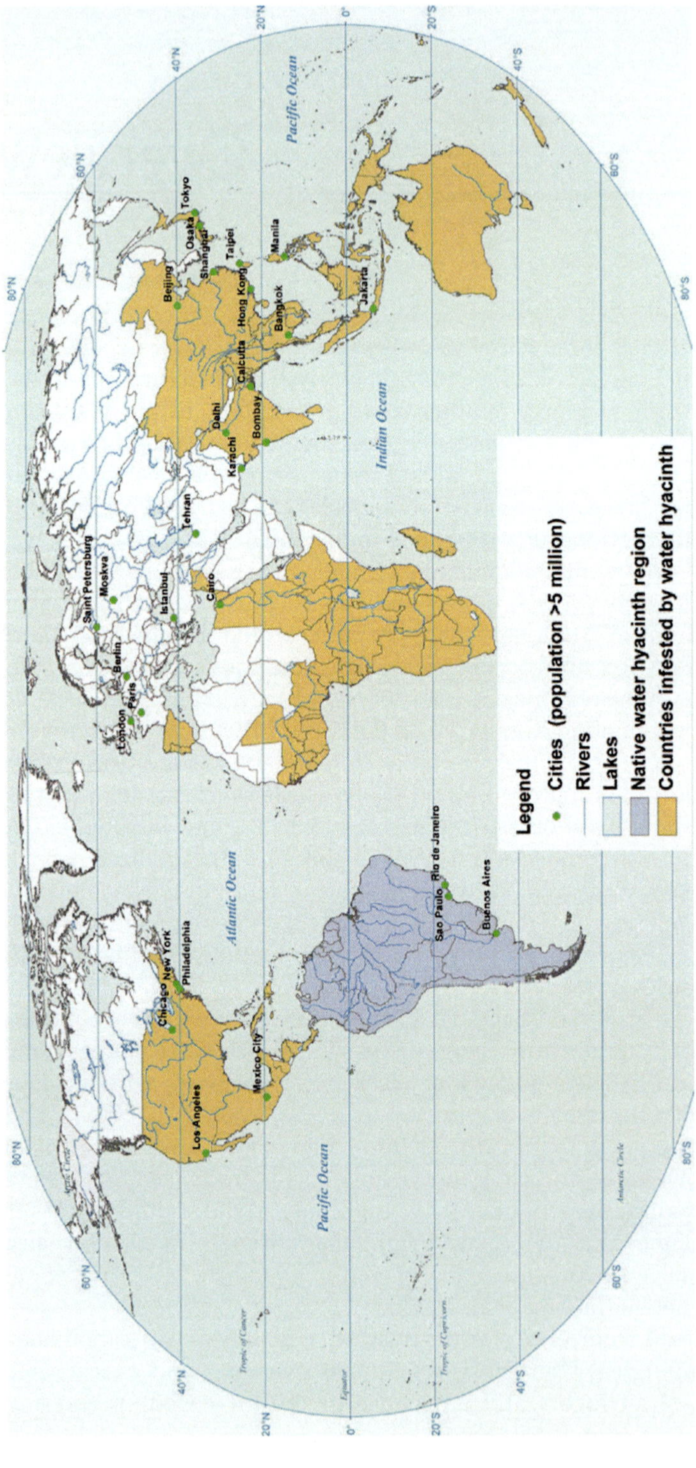

Fig. 1. Global distribution of water hyacinth (Map redrawn by UNEP/DEWA from Téllez et al. 2008) [(UNEP and GEAS 2013), picture reuse permission license number 3557980460437 from Elsevier].

aggravating effects of water hyacinth, there is intensive worldwide research, especially on management strategies and utilization of its huge biomass, including using this species to remediate eutrophic waters.

References cited

Chabot, J. F. 2009. *Foreign invaders: non-native species and their effect on North America's ecosystems.* New York, USA: Full Blast Productions, New York.

Duan, H., S. Qiang, H. Wu and J. Lin. 2003. Water hyacinth [*Eichhornia crassipes* (Mart.) Solms.]. *Weed Sciences* 2: 39–40 (In Chinese).

Gettys, L. A., W. T. Haller and M. Bellaud. 2009. *Biology and control of aquatic plants: a best management practices handbook.* 2nd ed. Marietta GA, USA: Aquatic Ecosystem Restoration Foundation.

Hong, C., Y. Wei, Y. Jia, X. Yang and H. Weng. 2005. The advances of water hyacinth for control and overall utilization studies. *Bulletin of Science and Technology* 21(4): 491–496 (In Chinese with English Abstract).

Howard, G. W. and K. L. S. Harley. 1998. How do floating aquatic weeds affect wetland conservation and development? How can these effects be minimised? *Wetlands Ecology and Management* 5(3): 215–225.

Huang, B., Y. Guo, J. Song and Li Zhisheng. 2004. Status and future of the preventing countermeasure to water hyacinth abroad. *Wuyi Science Journal* 20: 149–155 (In Chinese with English Abstract).

Kateregga, E. and T. Sterner. 2007. Indicators for an invasive species: water hyacinths in Lake Victoria. *Ecological Indicators* 7(2): 362–370.

Lindsey, K. and H. M. Hirt. 2000. *Use water hyacinth—a practical handbook of uses for water hyacinth from across the world.* Winnenden, Germany: Anamed International.

López, E. G., R. H. Delgadillo and M. M. Jiménez. 1996. Water hyacinth problems in Mexico and practised methods for control. In *Strategies for Water hyacinth Control - Report of a Panel of Experts Meeting*, ed. R. Charudattan, R. Labrada, T. D. Center, and C. Kelly-Begazo, 125–135 pp. Rome, Italy: Food and Agricultural Organization of the United Nations.

Owens, C. S. and J. D. Madsen. 1995. Low temperature limits of waterhyacinth. *Journal of Aquatic Plant Management* 33: 63–68.

Ramey, V. 2001. Water-hyacinth. *Florida Department of Environmental Protection.* Florida: Florida Department of Environment Protection. http://plants.ifas.ufl.edu/node/141.

Solms, M., J. A. Coetzee, M. P. Hill, M. H. Julien and H. A. Cordo. 2009. *Eichhornia crassipes* (Mart.) Solms-Laub. (Pontederiaceae). In: *Biological Control of Tropical Weeds using Arthropods*, ed. R. Muniappan, G. V. P. Reddy, and A. Raman, 183–210. Cambridge, UK: Cambridge University Press.

The International Plant Names Index. 2005. *Eichhornia crassipes* Mart.[DB/OL]. *Nova Genera et Species Plantarum, 1816.* http://www.ipni.org/ipni/idPlantNameSearch.do?id=134676-3&back_page=%2Fipni%2FeditSimplePlantNameSearch.do%3Ffind_wholeName%3DPontederia%2Bcrassipes%26output_format%3Dnormal.

UNEP and GEAS. 2013. Water hyacinth – can its aggressive invasion be controlled? *Environmental Development* 7: 139–154.

Zheng, J., Z. Chan, L. Chen, P. Zhu and J. Shen. 2008. Feasibility studies on N and P removal using water hyacinth in Taihu Lake region. *Jiangsu Agricultural Science* 3: 247–250 (In Chinese).

Contents

Part Three—Application of Water Hyacinth in Phytoremediation

Part Four—Utilization of Water Hyacinth Biomass as Natural Resource

CHAPTER 1

Introduction

S. H. Yan[1] and *J. Y. Guo*[2,*]

The world has known water hyacinth since 1816 when it was first recorded. Over the centuries, the plant has changed its status from a beautiful ornamental plant to a noxious weed or invasive alien species that negatively impacted natural aquatic systems and socio-economic development. However, in some situations, positive effects of water hyacinth were reported as a good macrophyte for phytoremediation.

1.1 Water hyacinth in the past and present

1.1.1 Distribution of water hyacinth in the past and present—impact of global warming

It is well known that a greenhouse effect has resulted in an elevation of average surface temperature on Earth (global warming) (Forster et al. 2007); increasing water temperature may allow water hyacinth expansion further north in the northern hemisphere and further south in the southern hemisphere.

Global warming represents global climate change evidenced by a century-scale rise in temperature on Earth. Evidence of global warming comes from multiple independent climate indicators in the biosphere, including changes in temperature of the Earth's surface, atmosphere and oceans, and changes in ice and snow cover. "Scientists from all over the world have independently verified this evidence many times" (Hartmann et al. 2013). Moreover, the observed rate of warming (about 0.13 ± 0.03°C per decade) over the period 1955–2005 was almost double that for the period 1906–1955 (about 0.07 ± 0.02°C per decade) (Trenberth et al. 2007), which implies that global warming has

[1] 50 Zhong Ling Street, Nanjing, China.
Email: shyan@jaas.ac.cn
[2] 5 Armagh Way, Ottawa, Canada.
* Corresponding author: guoj1210@hotmail.com

a trend of acceleration. This trend may continue even if the concentration of greenhouse gases in the atmosphere would be stabilized, with the temperature of Earth's surface increasing another half a degree by the end of 21st century (Meehl et al. 2005). Although a rise of the global mean surface temperature averaged only 0.12°C per decade during the period 1951–2012 (Hartmann et al. 2013), the impact on the aquatic systems is important because water is the major heat sink on the planet due to a higher heat capacity than air and most types of rocks.

It can be safely assumed that invasion of the global fresh water systems by water hyacinth will not stop but rather accelerate due to global warming. In addition to expanding the water hyacinth range, global warming will also extend its growing season and increase its growth rate.

1.1.2 Growth of water hyacinth impacted by global warming

Growth of water hyacinth follows a logistic model (Wilson et al. 2005) and is influenced by many factors, including initial standing population (biomass), available space, water nutrient concentration, weather conditions such as temperature and light and other limiting factors such as toxic chemicals, salinity and biological agents. The logistic model is described by the equation:

$$M_t = \frac{kM_0}{M_0 + (k - M_0)e^{-rt}} \qquad (1.1.2\text{-}1)$$

Where M_0 is initial standing population (g m^{-2}); M_t is the population (g m^{-2}) at time t (day); r is growth rate (g g^{-1} d^{-1}); and K is carrying capacity (g m^{-2}). When the initial population M_0 is far less then carrying capacity K, the model can be simplified to:

$$M_t = M_0 e^{rt} \qquad (1.1.2\text{-}2)$$

The growth rate r varied from 0.005 to 0.118 according to the environmental and growth conditions, averaging 0.062 (Wilson et al. 2005).

Although many factors may influence the growth of water hyacinth, temperature and concentration of nitrogen and phosphorus in water are very important in most natural fresh waters. Nutrient concentrations in water bodies vary in various trophic states. Total Phosphorus (TP) may become a limiting factor for growth of water hyacinth in oligotrophic and mesotrophic state ($TP \leq 0.024$ mg P L^{-1}) (Carlson and Simpson 1996), whereas nitrogen, phosphorus and temperature may be interacting factors in eutrophic state ($0.17 <$ total nitrogen (TN) ≤ 0.77 mg N L^{-1}, $0.024 < TP \leq 0.07$ mg P L^{-1}) (Porcella et al. 1979). In hypereutrophic state ($TN > 0.77$ mg N L^{-1}, $TP > 0.07$ mg P L^{-1}), temperature is the only factor dominating the growth of water hyacinth (Wang et al. 2002).

The influence of temperature on the growth pattern of most plants can be generalized based on Q_{10} (temperature coefficient) theory in the absence of other limiting factors. The Q_{10} principle can also be applied to water hyacinth.

The Q_{10} temperature coefficient represents an increase in rate of plant growth with each increase of 10°C under normal conditions. The value of Q_{10} can be calculated using the following formula:

$$Q_{10} = \left(\frac{R_2}{R_1}\right)^{10/(T_2-T_1)} \tag{1.1.2-3}$$

Where R_2 and R_1 are growth rates at temperatures T_2 and T_1.

For most plants, the value of Q_{10} usually is 2 to 3 (Atwell et al. 1999), which indicates that intrinsic growth rate (r) will be doubled or tripled with the increase of every 10°C within certain limits. Given proliferative growth behavior and adaptation to high temperature of 43°C, the Q_{10} value of water hyacinth may be reasonably assumed as 3. To estimate the impact of global warming on the growth of water hyacinth, equations (1.1.2-2 and 1.1.2-3) can be used.

By rearranging the formula for Q_{10} with natural logarithm and assuming the value 3 for water hyacinth, the impact of global warming of 0.12°C temperature increase over a decade may result in an increase in water hyacinth growth rate of 1.33%. If the averaged basic growth rate (r) of water hyacinth is assumed to be 0.0620 (g g^{-1} day^{-1}), a 1.33% increase would make it 0.0628. Even though this might appear to be a very small increase, it can result in a huge additional amount of biomass considering the exponential function describing temperature effects on plant growth. For example, at the growth rate of 0.0628, the biomass would be doubled every 11.0 days compared with 11.2 days with 0.0620.

Global warming and the promoted growth of water hyacinth are clearly linked. Also, water hyacinth growth is a process of carbon (CO_2) sequestration associated with energy storage. Based on composition of the biomass of water hyacinth, every kilogram of dry biomass of water hyacinth sequesters 0.28 kg of carbon (C), which is equivalent to the amount of carbon dioxide (CO_2) released from burning 0.42 L of gasoline.

1.1.3 Growth of water hyacinth affected by eutrophication

Eutrophication is a natural process and has been accelerated by deforestation, surface runoffs and leaching from farmland, and increased volumes of effluent from municipal wastewater processing plants due to urbanization and industrial development. Accelerated eutrophication is a global problem in rivers, lakes, water reservoirs and estuaries; the percentage of eutrophic lakes and reservoirs was about 28% in Africa, 54% in Asia-Pacific and 41–53% in South America, North America and Europe (UNEP 1994). In the process of accelerated eutrophication, the average concentrations of nitrogen and phosphorus in streams and rivers have doubled globally from the base lines of 0.63 mg N L^{-1} and 0.015 mg P L^{-1} and have increased even 10–50 times in western Europe and North America due to anthropogenic influence since

1900s (Meybeck 1982). In Lake Dianchi (24°45' N 102°36' E, Yunnan Province, China) the average concentration of total nitrogen increased from 7.5 mg N L^{-1} in 1988 (Institute of Environmental Science 1992) to 16.8 mg N L^{-1} in 2009 (Zhang et al. 2014), and the average concentration of total phosphorus increased from 0.23 mg P L^{-1} in 1988 to 1.46 mg P L^{-1} in 2009 (Institute of Environmental Science 1992).

The excessive nutrients (particularly nitrogen and phosphorus) in eutrophic waters can greatly promote growth of algae and aquatic vegetation. When water hyacinth is present, it would respond to increased nutrient concentrations, quickly, outgrowing algae and other aquatic vegetation, and becoming dominant in aquatic ecosystems. In tropical fresh water bodies in oligotrophic and mesotrophic states, nitrogen and phosphorus are often limiting factors for aquatic vegetation. In subtropical and temperate climate zone, these phenomena were also evidenced in the lakes of New Zealand and Australia and most of the Central European lakes (Mitchell 1974).

The response of water hyacinth to increasing nutrient concentrations is complex and influenced by other limiting factors, especially temperature. Wilson et al. (2005) suggested a hyperbola function to estimate the response of water hyacinth to the N and P increments and introduced a concept of half-saturation coefficients of nitrogen (h_n) and phosphorus (h_p) and further suggested that if $h_n/h_p \geq C_n/C_p$ (C_n = nitrogen concentration in water expressed in mg N L^{-1}, C_p = phosphorus concentration in water expressed in mg P L^{-1}), the growth of water hyacinth may mainly respond to nitrogen (or to phosphorus if $h_n/h_p < C_n/C_p$). The half-saturation coefficients of nitrogen (h_n) and phosphorus (h_p) varied according to the experiment and measurement conditions; estimated half-saturation coefficient of nitrogen (h_n) ranged from 0.01 to 1.10 mg L^{-1} and was arbitrarily set to 0.2 mg L^{-1} based on most studies (Wilson et al. 2005). The half-saturation coefficient of phosphorus (h_p) ranged from 0.01 to 0.1 mg·L^{-1} and was arbitrarily set to 1/7 of half-saturation coefficient of nitrogen (h_n) based on various studies (Wilson et al. 2005). These suggested half-saturation coefficients are in accordance with nutrient concentrations in water bodies with a trophic index 50–70 in eutrophic state classification (Carlson and Simpson 1996, Wang et al. 2002).

The impact of nutrient increment on the growth of water hyacinth can be assessed by measuring nitrogen concentration in mature leaves and roots of water hyacinth. For example, the nitrogen concentration averaged 15.3 g kg^{-1} dry weight for mature roots and 20.5 g kg^{-1} dry weight for mature leaves (Liu et al. 2010). Calculation shows that every additional kilogram of nitrogen in water will promote growth of 700 kg of fresh weight of water hyacinth (assuming a denitrification loss of nitrogen at 34%) (Gao et al. 2012).

In summary, growth of water hyacinth is promoted by accelerated eutrophication in aquatic environments, fueled by global warming and enhanced by water hyacinth having great reproductive capacity and extreme adaptability. These properties of water hyacinth can create big problems in management, causing large socio-economic losses and serious damage to ecosystems.

1.1.4 Changes in the management concepts in the past and present

The direct effects of water hyacinth invasion are degradation and/or disruption of ecosystem services, including transportation, hydropower operations and fisheries. Removal (and hopefully eradication) of water hyacinth comes with substantial costs that, in most cases, are drawn from taxpayers money; furthermore, removal will not be completed in a single action due to the dynamic growth of the macrophyte. In addition, further problems occur during disposal of water hyacinth biomass, such as secondary contamination via nutrient leaching and greenhouse gas emissions, and a requirement for significant human resources (labor) with very limited financial returns.

Before 2000s, the objective of the management of this invasive species was simply to attempt to eradicate it from water bodies. The methods involving chemical, biological and/or mechanical control reflected such efforts. For example, spraying herbicides would destroy large mats of water hyacinth quickly, but resulting organic matter decomposition would degrade temporarily the water quality, and an accompanying nutrient release would affect trophic chains in the aquatic ecosystems, not to mention substantial public spending and the impacts on tourism or fisheries.

The biology of water hyacinth may imply that this species is practically impossible to eradicate, which may also imply that a practical way of management is to understand existing technologies and integrate them with further research. Even though water hyacinth presented a big problem in the past and a tough challenge at present, it may be a potential opportunity in the future. In either case, we must understand its biology and interactions with the aquatic systems in order to enable strategic planning and practical management approaches. However, one should always keep in mind that water hyacinth is a noxious invasive species.

In most natural water bodies, water hyacinth out-competes other flora and fauna. However, under intensive management of lakes and water reservoirs, the impacts of water hyacinth may shift from negative effects of a large and continuous mat of free floating macrophytes out-competing other species, to positive effects of patches of discontinuous mat co-existing with other species, thus minimizing the effects on aquatic organisms and reducing ecological risks (Zhi Wang et al. 2013).

By analyzing biology of the species and the complex interactions of the ecosystems, it can be concluded that the problem of water hyacinth invasion globally is not a simple problem of noxious weed, or global warming, or eutrophication or resources utilization. In contrast, it is a multidimensional and complex problem. The nature of this problem requires a strategic solution via systematic approach and technology integration with the knowledge of biology, chemistry, physics, ecology, management and policy. After a century of scientific research, and after recent technological advances, it is at the right time to set new directions in solving the problem by designing strategies for water hyacinth management and control based on integrated and systematic approaches.

1.2 Challenges associated with water hyacinth

The majority of management strategies so far have focused on removal of the weed by hand, mechanical equipment, biological agents, and/or chemical agents. Before the 1940s, removal of water hyacinth from small ponds and channels was mainly by hand. Heavy-duty machine removal started in 1937 by the U.S. Army Corps of Engineers (Little 1979). By the 1950s, using aquatic weed harvesters (such as Chopper and Excavator) was so popular that almost every mechanical engineer in North America knew how to build a mechanical harvester for water hyacinth (Gettys et al. 2009). However, failures in the water hyacinth management and control suggested that water hyacinth was difficult or next to impossible to eradicate (Villamagna and Murphy 2010) because of the high cost of mechanical harvesting and disposal, unsuitability of hand removal on a large scale, and incomplete effectiveness of the chemical and biological treatments (Lindsey and Hirt 2000).

Further analysis of the cost of mechanical harvesting for water hyacinth in comparison with field crops revealed that the fresh biomass of water hyacinth is about 93–95% water (Cifuentes and Bagnall 1976), whereas the fresh biomass of alfalfa only is 75–80% water (Orloff and Mueller 2008). Furthermore, water hyacinth is harvested for disposal, whereas alfalfa has value on the commercial market. Hence, two interesting ideas have been developed during the past 40 years: (1) reduce the volume (air in spongy tissues) and water content of water hyacinth during or after harvesting; (2) find uses for water hyacinth to create a commercial value to at least partially compensate for the cost of harvesting (Babourina and Rengel 2011). These two ideas are connected insofar as utilization of water hyacinth biomass may require low-cost dehydration to make it suitable for bio-energy production, silage or other valuable products (Cifuentes and Bagnall 1976).

Cifuentes and Bagnall (1976) designed and tested a press for dewatering water hyacinth and concluded that the highest pressure needed was 600 kPa (i.e., higher pressure did not remove more water from the plant tissues), with water content of the residues remaining at 84%; similar results were reported more recently (Du et al. 2010). However, dehydrated fresh biomass of water hyacinth at water content of 84% does not meet the minimum requirements for making either organic fertilizer or silage or for solid state fermentation. Further drying would inevitably increase the costs.

Over a century, many ideas on utilization of water hyacinth have been proposed, studied and progressed all over the world. The biomass of water hyacinth has been utilized as animal feed for cattle, sheep, geese, pigs, etc. as either fresh or silage. It also has been used to produce biofuel (biogas or via direct combustion), compost and proteins, and has been processed for making crafts, furniture and paper (Lindsey and Hirt 2000) in addition to being used for extracting glutathione (Bodo et al. 2004). Furthermore, water hyacinth has often been suggested as a biological agent for phytoremediation in wastewater treatment or the management of eutrophic waters. However, the literature indicated that such utilization of water hyacinth was only practiced in the laboratory or in the field on a small scale, with commercial applications being either limited (Gettys et al. 2009) or deemed economically unfeasible.

Frequently, research has focused on just one aspect of water hyacinth to elucidate its impact on ecosystems; rarely, a full-scale assessment of ecosystem changes (e.g., impacts on energy flow and nutrient dynamics at the whole-ecosystem level, covering changes in trophic dependencies among macrophytes, phytoplankton and zooplankton, invertebrates and vertebrates, benthos and birds before and after water hyacinth presence), including how water quality changes may interact with the ecosystem dynamics (Villamagna 2009). Also, there are only a few reports assessing integrated management of water hyacinth on the abiotic and biotic changes in an ecosystem (Villamagna and Murphy 2010). Also, little has been reported on using water hyacinth as a functional and biological agent in lakes and reservoirs on a large scale. Hence, there is a pressing need to not just successfully control water hyacinth but also to enhance its benefits; our capacity to do so depends on our knowledge of the water hyacinth biology and our understanding of how water hyacinth invades, inhabits, affects and alters aquatic ecosystems, and the costs and benefits of integrated management related to water purification and ecosystem recovery, with respect to social, economic as well as ecological context.

1.2.1 Challenges regarding utilization of water hyacinth as a natural resource

Water hyacinth has spongy tissues with a density of 0.3 $g \cdot cm^{-3}$, meaning that every cubic meter of fresh biomass being harvested weighs only 300 kg, and, about 285 kg of that weight is water. Mathur and Singh (2004) designed a chopper cum crusher that can reduce the volume of fresh water hyacinth by 64% (effectively increasing biomass density to 830 kg m^{-3}), processing 1.0 tonne of fresh water hyacinth per hour. However, in large lakes (such as Lake Victory in East Africa, or Lagos Lagoon in West Africa, with 1000–2000 hectares of water hyacinth) (NIFFR 2002, Wilson et al. 2007), there would be many millions of tonnes of fresh water hyacinth biomass per year, emphasizing the importance of the high-volume processing capacity. For example, a standing population of one million tonnes of water hyacinth could produce additional 70 thousand tonnes of biomass every day, which would require nine thousands choppers cum crushers (see above) to process for 8 hours a day only to remove the newly grown biomass, and not to reduce at all the standing population. Clearly, new technology and/or more efficient equipment are required. A 2010 technical report (Yan et al. 2010) described a water hyacinth harvest vessel that could reduce the volume of fresh water hyacinth by 67%, processing 44 tonnes of fresh biomass per hour on lakes. For the amount of standing population used in the example above, only 200 such vessels would be needed to prevent expansion of the standing population of one million tonnes of fresh water hyacinth when processing for 8 hours per day. Another 2011 technical report (Yan et al. 2011) described an integrated system consisting of a water hyacinth transport device using floating net, a fixed dock harvester and a dehydrating machine which is capable of harvesting and processing 75 tonnes of fresh biomass per hour from waters. For the amount of fresh biomass mentioned about, only 117-set of facilities were required at equipment cost US$90,000 per set compared to equipment cost US$224,000 per harvesting vessel (2012 price index).

The biology of water hyacinth implies that harvesting it (even with a multi-million tonne capacity) does not solve the management problem. The species has a water content of about 95% which makes the fresh biomass of water hyacinth very costly to transport after harvest and occupies a huge amount of space during its disposal. For example, removal of one million tonnes of fresh biomass (even after mechanical volume reduction) would require 75 full size (40 m^3) trucks every day to transport the biomass all year around. Also, it would require multi-million cubic meters of space at the waste dump site (about 20 standard full size football fields (about 7400 m^2) with biomass piled at a height of 2 meters if the rotting turnover rate is 100 days). However, deposition of fresh biomass at a waste dump site could have high socio-economic and ecological costs due to possible emissions of carbon dioxide, methane as well as nitrous oxide. This example highlights the scale of reductions in the harvest and dehydration costs that must be achieved to solve the water hyacinth management problems.

Fresh biomass of water hyacinth is not suitable for green feed or silage, for manufacturing organic fertilizers or for efficiently producing methane due to its high water content (about 95%). For making organic fertilizers, the water content of the fresh biomass of water hyacinth would need to be below 75% (Shi et al. 2012) to achieve good fermentation quality. For green feed or silage, the water content of the water hyacinth biomass of water hyacinth should be 70% or less (Jiang et al. 2011). Although high water content does not impede methane production specifically, the efficiency of fermentation would be reduced and the cost of biogas production increased because high water content translates into low content of total solids. However, lowering water content of the biomass may be associated with high cost of dehydration processes and technologies.

Reducing water content in water hyacinth biomass was first pioneered in the US in 1971 and was achieved by an on-board squeezer, with about 68% of water removed and good nutrient preservation (85% crude protein, 60%

potassium and 80% phosphorus) (Mitchell 1974). However, 68% of water removed from water hyacinth biomass was equivalent to decreasing water content to about 86%, which is still higher than required for most types of practical utilization (Jiang et al. 2011, Shi et al. 2012). Recent development of dehydration technology lowered water content of water hyacinth bagasse to 65% (Yan Wang et al. 2013), which could meet the requirements for production of silage or organic fertilizer, and also could increase the efficiency of methane production. However, dehydration of fresh biomass of water hyacinth created another problem that needs to be solved: a huge amount of water hyacinth extract (e.g., removing 80–90% of water from one million tonnes of fresh water hyacinth biomass would produce about 0.8 to 0.9 million tonnes of water hyacinth extract).

Recent research focused on the efficiency of the processes, i.e., quick process without reduction of the gas yield. For example, the fermentation of water hyacinth extract revealed that a Continuously-Stirred Tank Reactor (CSTR) could reduce Hydraulic Retention Time (HRT) to 2.5 days compared to normal 25 days, which increased efficiency 10-fold compared to a batch fermentation reactor (Hu et al. 2008), and could result in a gas yield of 1.1 m^3 per cubic meter of the reactor per day with a potential capability of 219 liters of gas production per kilogram of total Chemical Oxygen Demand (COD) (Ye et al. 2010). In comparison, one kilogram of Total Solids (TS) of water hyacinth bagasse has potential capacity to produce 398 L of gas with methane content of 59% (Ye et al. 2011).

1.2.2 Keys to solving problems of water hyacinth on the global scale

The above examples on water hyacinth harvest and utilization were mainly drawn from an ecological engineering project on utilization of water hyacinth for phytoremediation and natural resources in Lake Dianchi (24°45' N 102°36' E) and Lake Taihu (120°04' E, 31°27' N). These examples imply that to get the most out of phytoremediation using water hyacinth, the key is in combining the processes of phytoremediation and production of energy plus recovery of resources (water, nutrients). The challenges and the opportunities linked to this invasive species are largely related to its biology, which needs to be understood before one can establish management policies and goals for controlling water hyacinth in water bodies, which are public properties in most countries, and therefore any decisions must be transparent and explained publicly. This book will contribute to linking the biology of water hyacinth with public interests in policies and management decisions regarding water hyacinth as a phytoremediation agent and as natural resources.

Solving problems associated with water hyacinth also depends on cooperating more effectively across the globe, especially linking social and economic development and policy making with innovative research and technological development and demonstration.

References cited

Atwell, B. J., D. Eamus and G. Farquhar. 1999. Temperature: a driving variable for plant growth and development. In: Plant in Action: Adaptation in Nature, Performance in Cultivation, ed. B. J. Atwell, D. Eamus and G. Farquhar, 384–416. Melbourne, Australia: Macmillan Education Australia Pty.

Babourina, O. and Z. Rengel. 2011. Nitrogen removal from eutrophicated water by aquatic plants. In: Eutrophication: Causes, Consequences and Control, ed. A. A. Ansari, S. S. Gill, G. R. Lanza and W. Rast, 355–372. Dordrecht, Netherlands: Springer Netherlands.

Bodo, R., A. Azzouz and R. Hausler. 2004. Antioxidative activity of water hyacinth components. *Plant Science* 166(4): 893–899.

Carlson, R. E. and J. Simpson. 1996. A coordinator's guide to volunteer lake monitoring methods. Madison, WI, USA: North American Lake Management Society.

Cifuentes, J. and L. O. Bagnall. 1976. Pressing characteristics of water hyacinth. *Journal of Aquatic Plant Management* 14(0): 71–75.

Du, J., Z. Chan, H. Huang, X. Ye, Y. Ma, Y. Xu et al. 2010. Optimization of dewatering parameters for water hyacinth (*E. crassipes*). *Jiangsu Agricultual Sciences* 2(0): 267–269 (In Chinese).

Forster, P., V. Ramaswamy, P. Artaxo, T. Berntsen, R. Betts, D. W. Fahey et al. 2007. Changes in atmospheric constituents and in radiative forcing. In: Climate Change 2007: The Physical Science Basis. Contribution of Working Group I to the Fourth Assessment Report of the Intergovernmental Panel on Climate Change, ed. S. Solomon, D. Qin, M. Manning, Z. Chen, M. Marquis, K. B. Averyt et al. Cambridge, United Kingdom and New York, NY, USA: Cambridge University Press.

Gao, Y., N. Yi, Z. Zhang, H. Liu and S. Yan. 2012. Fate of $^{15}NO_3^-$ and $^{15}NH_4^+$ in the treatment of eutrophic water using the floating macrophyte, *Eichhornia crassipes*. *Journal of Environmental Quality* 41(5): 1653–60.

Gettys, L. A., W. T. Haller and M. Bellaud. 2009. Biology and control of aquatic plants: a best management practices handbook. 2nd ed. Marietta GA, USA: Aquatic Ecosystem Restoration Foundation.

Hartmann, D. L., A. M. G. K. Tank, M. Rusticucci, L. V. Alexander, S. Brönnimann, Y. Charabi et al. 2013. Observations: atmosphere and surface. In: Climate Change 2013: The Physical Science Basis. Contribution of Working Group I to the Fifth Assessment Report of the Intergovernmental Panel on Climate Change, ed. T. F. Stocker, D. Qin, G. -K. Plattner, M. Tignor, S. K. Allen, J. Boschung et al. 159–254. Cambridge, United Kingdom and New York, NY, USA: Cambridge University Press.

Hu, X., G. Zha, W. Zhang, F. Yi and R. Xu. 2008. Experimental study on mesophilic biogas fermentation with juice of *Eichhornia crassipes*. *Energy Engineering* 2: 36–38 (In Chinese with English Abstract).

Institute of Environmental Science. 1992. Survey on eutrophication of Lake Dianchi. 1st ed. Kunming, China: Kunming Science and Technology Press.

Jiang, L., Y. Bai, S. Yan, H. Zhang, J. Liu and L. Tu. 2011. Effects of additives on dehydrated water hyacinth for silage. *Jiangsu Agricultural Sciences* 39(6): 337–340 (In Chinese).

Lindsey, K. and H. M. Hirt. 2000. Use water hyacinth—a practical handbook of uses for water hyacinth from across the world. Winnenden, Germany: Anamed International.

Little, E. C. S. 1979. Handbook of utilization of aquatic plants. Rome, Italy: Food and Agriculture Organization of The United Nations.

Liu, J. Z., Y. M. Ge, Y. F. Zhou and G. M. Tian. 2010. Effects of elevated CO_2 on growth and nutrient uptake of *Eichhornia crassipes* under four different nutrient levels. *Water, Air, & Soil Pollution* 212(1-4): 387–394.

Mathur, S. and P. Singh. 2004. Development and performance evaluation of a water hyacinth chopper cum crusher. *Biosystems Engineering* 88(4): 411–418.

Meehl, G. A., W. M. Washington, W. D. Collins, J. M. Arblaster, A. Hu, L. E. Buja et al. 2005. How much more global warming and sea level rise? *Science* (*New York, N.Y.*) 307(5716): 1769–1772.

Meybeck, M. 1982. Carbon, nitrogen, and phosphorus transport by World Rivers. *American Journal of Science* 282(April): 401–450.

Mitchell, D. S. 1974. Aquatic vegetation and its use and control. Paris, France: UNESCO.

NIFFR. 2002. National surveys of infestation of water hyacinth, typha grass and other noxious weeds in water Bodies of Nigeria. occasional paper #5. New Bussa, Nigeria: National Institute for Freshwater Fisheries Research, Federal Ministry of Agriculture & Rural Development.

Orloff, S. B. and S. C. Mueller. 2008. Harvesting, curing, and preservation of alfalfa. In: Irrigated alfalfa management in Mediterranean and Desert zones, ed. C. G. Summers and D. H. Putnam, 1–18. Oakland, California, USA: University of California Agriculture and Natural Resources Publication 8300.

Porcella, D. B., S. A. Petersen and D. P. Larsen. 1979. Proposed method for evaluating the effects of restoring lakes. In: Limnological and Socioeconomic Evaluation of Lake Restoration projects: approaches and preliminary results: workshop held 28 February–2 March 1978, ed. Corvallis Environmental Research Laboratory, 265–310. Washington, DC. USA: U.S. Environmental Protection Agency.

Shi, L. L., M. X. Shen, Z. Z. Chang, H. H. Wang, C. Y. Lu, F. S. Chen et al. 2012. Effect of water content on composition of water hyacinth *Eichhornia crassipes* (Mart.) Solms residue and greenhouse gas emission. *Chinese Journal of Eco-Agriculture* 20(3): 337–342 (In Chinese with English Abstract).

Trenberth, K. E., P. D. Jones, P. Ambenje, R. Bojariu, D. Easterling, A. K. Tank et al. 2007. Observations : Surface and Atmospheric Climate Change. In: Climate Change 2007: The Physical Science Basis. Contribution of Working Group I to the Fourth Assessment Report of the Intergovernmental Panel on Climate Change, ed. S. Solomon, D. Qin, M. Manning, Z. Chen, M. Marquis, K. B. Averyt et al., 235–336. 1st ed. Cambridge, United Kingdom and New York, NY, USA: Cambridge University Press.

UNEP. 1994. *The pollution of lakes and reservoirs.* Nairobi, Kenya: UNEP Environment Library.

Villamagna, A. M. 2009. Ecological effects of water hyacinth (*Eichhornia crassipes*) on Lake Chapala, Mexico. Ph.D. Thesis, Virginia Polytechnic Institute and State University, Blacksburg, Virginia, USA.

Villamagna, A. M. and B. R. Murphy. 2010. Ecological and socio-economic impacts of invasive water hyacinth (*Eichhornia crassipes*): a review. *Freshwater Biology* 55(2): 282–298.

Wang, M., X. Liu and J. Zhang. 2002. Evaluate method and classification standard on lake eutrophication. *Environmental Monitoring in China* 18(5): 47–49 (In Chinese with English Abstract).

Wang, Y., Z. Zhang, Y. Zhang, X. Weng, X. Wang and S. Yan. 2013. A new method of dehydration of water hyacinth. *Jiangsu Agricultural Sciences* 41(10): 286–288 (In Chinese).

Wang, Z., Z. Zhang, Y. Zhang, J. Zhang, S. Yan and J. Guo. 2013. Nitrogen removal from Lake Caohai, a typical ultra-eutrophic lake in China with large scale confined growth of *Eichhornia crassipes*. *Chemosphere* 92(2): 177–183.

Wilson, J. R. U., O. Ajuonu, T. D. Center, M. P. Hill, M. H. Julien, F. F. Katagira et al. 2007. The decline of water hyacinth on Lake Victoria was due to biological control by *Neochetina* spp. *Aquatic Botany* 87(1): 90–93.

Wilson, J. R., N. Holst and M. Rees. 2005. Determinants and patterns of population growth in water hyacinth. *Aquatic Botany* 81(1): 51–67.

Yan, S., H. Liu, Z. Zhang, Y. Zhang and G. Liu. 2010. Automated water hyacinth volume reduction and harvester. China Patent: SIPO, China Patent #2010 2 0100681.6 (In Chinese).

Yan, S., Z. Zhang, Y. Zhang, X. Yang and H. Liu. 2011. Floating macrophyte on shore harvesting system. China: SIPO, China Patent #2011 2 0373563.7.

Ye, X., J. Du, Z. Chang, Y. Qian, Y. Xu and J. Zhang. 2011. Anaerobic digestion of solid residue of water hyacinth. *Jiangsu Journal of Agricultural Sciences* 27(6): 1261–1266 (In Chinese with English Abstract).

Ye, X., L. Zhou, S. Yan, Z. Chang and J. Du. 2010. Anaerobic digestion of water hyacinth juice in CSTR reactor. *Fujian Journal of Agricultural Science* 25(1): 100–103 (In Chinese with English Abstract).

Zhang, Z., Y. Gao, J. Guo and S. Yan. 2014. Practice and reflections of remediation of eutrophicated waters: a case study of haptophyte remediation of the ecology of Dianchi. *Journal of Ecology and Rural Environment* 30(1): 15–21 (In Chinese with English Abstract).

Part One
Biology of Water Hyacinth and Environment

CHAPTER 2

Biology of Water Hyacinth

Z. H. Zhang[1] and *J. Y. Guo*[2,*]

2.1 Introduction

Water hyacinth *Eichhornia crassipes* (Mart. and Zucc.) Solms belongs to Family Pontederiaceae and is an aquatic perennial free-floating plant and also a noxious, unique, useful, fast growing, and persistent invasive macrophyte. The word "noxious" refers to water hyacinth having caused extensive damages due to its biology and function in the aquatic ecosystem, with the technology of its control and management not fully understood. The macrophyte can be used as feed for animals even though it contains a large proportion of water in its cells. Duke (1983) quoted Wolverton (1976) in the context of discussion about the water hyacinth: "I fully intend to solve a major pollution problem, a major energy problem, a major food problem, and a major fertilizer problem". But if now, would the declaration have changed to "I fully intend to solve a major weed problem, a major pollution problem, a major..., etc." This will also be explained by the biology of the plant.

The main characteristics of the biology of the macrophyte can be summarized as: morphological elasticity, extremely fast growth, changing chemical composition of different parts of plant in various habitats, adaptive phenology for ecological invasion, and vegetative and sexual reproduction for persistent distribution. These basic characteristics and the interactions of this species with the environment, including temperature, light, pH, dissolved oxygen and salinity were well described (Penfound and Earle 1948). However, over the decades, new characteristics and behaviors of the plant have been elucidated to assist in-depth understanding of this macrophyte, and to help deal with environmental challenges, management and control of the species and utilization of its biomass.

[1] 50 Zhong Ling Street, Nanjing, China.
Email: zhenhuaz70@hotmail.com
[2] 5 Armagh Way, Ottawa, Canada.
* Corresponding author: guoj1210@hotmail.com

2.2 Morphology and environment

2.2.1 Description

The morphology of mature water hyacinth growing at low density is described in Fig. 2.2.1-1.

The morphological elasticity of water hyacinth is defined as the morphological adaption to the environment, underscoring a species-specific strategy for successful competition in a given habitat and a capacity to invade ecosystems. When space is available, the plant expands horizontally on the water surface to produce numerous new generations to maximize utilization of light and available nutrients. When space is not available, water hyacinth tends to grow vertically with elongated floats and large blades to accumulate nutrients and energy for further development and sexual reproduction. When growing in a high-density mat, the ball type floats change to elongated float as shown in Fig. 2.2.1-2.

The non-branched root is fibrous and about 1 mm in diameter (Hadad et al. 2009); each fibrous root has numerous lateral hairy rootlets about 2 to 3 mm in length and with functional root tips. The fibrous roots of macrophytes, dangling underwater in a dense mat, can effectively trap suspended detritus and enhance the development of microbial communities (Yi et al. 2014) to form active biofilms. Water hyacinth root length may vary from 200 mm to 2000 mm depending on the nutrient conditions (Li, Ren et al. 2011, Rodríguez

Fig. 2.2.1-1. Morphological description of a young water hyacinth (*Eichhornia crassipes*) growing in water containing adequate nutrients. l: ligule; sf: subfloat; s: stolon; ps: pseudo-lamina; i: isthmus; f: float; rh: rhizome; r: root; st: sponge tissue (Photo by Lin Shang 2015).

Fig. 2.2.1-2. Morphological description of mature water hyacinth (*Eichhornia crassipes*) growing in dense mat in water containing adequate nutrients showing elongated floats (Photo by Lin Shang 2015).

et al. 2012); the root surface area may vary from 30 to 60 m^2 per individual macrophyte (Zhou et al. 2012). Water hyacinth can grow rapidly and effectively in water with both low nutrients (NH_4^+ 0.12 mg L^{-1}, NO_3^- 0.16 mg L^{-1}, PO_4^{3-} 0.09 mg L^{-1}) (Zhang et al. 2011, Ma et al. 2013) and high nutrient concentrations [total nitrogen 160 mg L^{-1} (Li 2012), phosphate 40 mg L^{-1} (Haller and Sutton 1973)]. Although the upper limit of toxic concentration of ammonium is not clear, water hyacinth starts dying when NH_4^+ concentration in water reaches 370 mg L^{-1} (Qin et al. 2015).

This root morphology explains why water hyacinth can out-compete other aquatic plants for nutrients and can grow well in waters with low nutrient concentration. This is important because people expect the quality of reclaimed water to meet the specific surface water standards, even though the standards differ from country to country (Table 2.2.1-1).

The comparison in Table 2.2.1-1 showed that the quality of water reclaimed by water hyacinth growth (Rodríguez et al. 2012) was similar to the standard defined by US Environmental Protection Agency (EPA 2007) and Ministry of Environmental Protection of the People's Republic of China (MEP-PRC 2002).

The morphology of roots plus fast growth rate under suitable conditions enables a hectare of water surface fully covered by water hyacinth to daily

Table 2.2.1-1. Surface water standard for drinking water reservation and reclaimed water quality focused on nutrients and oxygen (EPA 2007, MEP-PRC 2002, Rodríguez et al. 2012).

Item	North America standard (mg L^{-1})	China standard (mg L^{-1})	Quality of reclaimed water (mg L^{-1})
TN (N)	≤0.90	≤ 1.0	1.0
Ammonium (N)	NR	≤ 1.0	0.3
Nitrate (N)	NR	≤10	0.1
TP (P)	≤0.076	≤0.2	0.1
Oxygen	NR	≥5	6
COD	NR	≤20	12
Chlorophyll-*a*	≤0.004	NR	NM

Note: NR: not required; NM: not measured

absorb the amount of nitrogen and phosphorus emitted by 800 people in one day (Wooten and Dodd 1976). Water hyacinth roots can also remove heavy metals, such as cadmium, lead, mercury, thallium, silver, cobalt and strontium, as well as organic pollutants, including antibiotics, that can be present in sewage (Patel 2012).

2.2.2 Shoot-root ratio

The shoot-root ratios (length/length) (range 7.1 to 0.4 in Table 2.2.2-1) vary inversely with nutrient concentrations, especially with concentrations of nitrogen and phosphorus and nitrogen-phosphorus ratio (Reddy and Tucker 1983, Li, Ren et al. 2011). The shoot-root ratios averaged from 3.2 to 2.1 at mean nitrogen concentration 2.1 mg N L^{-1} and phosphorus concentration 1.1 mg P L^{-1} (Dellarossa et al. 2001). With a continuous supply of available nitrogen ≥2 mg N L^{-1} plus available phosphorus ≥ 0.3 mg P L^{-1} at suitable temperature between 25–30°C and in high-density mat, the high shoot-root ratios can be obtained. The shoot-root ratio is an important factor in making management

Table 2.2.2-1. Comparison of plant types of *Eichhornia crassipes* cultivated from a single branch at different locations (Li, Ren et al. 2011).

Location	Plant height (cm)	Dry weight per plant (g)	Shoot height (cm)	Root length (cm)	Shoot/root ratio (length/length)
Taihu Lake	60.4 ± 5.2Bb	53.2 ± 5.3Aa	40.2 ± 2.3Aa	20.2 ± 0.5Bb	2.0 ± 0.2Bb
Nanjing	45.1 ± 3.3Cc	42.1 ± 2.9Bb	36.3 ± 1.5Aa	5.1 ± 0.2Cc	7.1 ± 0.3Aa
Dianchi Lake	78.4 ± 6.1Aa	37.5 ± 4.3Cc	22.1 ± 1.1Bb	57.3 ± 1.2Aa	0.4 ± 0.1Cc

Note: Different small and capital letters mean significant difference at $p < 0.05$ and $p < 0.01$, respectively.

decisions, in designing harvest machines as well as in biomass utilization because increasing shoot proportions are associated with high crude protein or high carbon content, whereas higher root proportions mean high cellulose content making it difficult for machine harvest, albeit resulting in good water quality.

2.2.3 Sponge tissue

The interesting characteristic of water hyacinth biology is its sponge tissue in root, stolon, rhizome, subfloat, float and pseudo-lamina. The sponge tissue creates floating force and reduces efficiency of harvesting because a vessel not only harvests biomass of water hyacinth but also harvests air in the sponge tissue. The floating capacity of the water hyacinth biomass can be expressed as the density of different plant organs. The lower the density of a plant organ, the higher the floating capacity and the lower the harvesting efficiency is. For example, harvesting 1 cubic meter of whole water hyacinth plants only collected 8.4 kg biomass on a dried basis, while a harvesting vessel was at its carrying capacity (Table 2.2.3-1).

Table 2.2.3-1. Densities of different water hyacinth parts—illustrating the scale of the reduced efficiency on harvesting due to sponge tissue.

Plant part	Density (kg m^{-3})	Estimated dried biomass on harvest (kg m^{-3})[a]	Ref.
root	782	39.1	1
rhizome	805	40.3	1
stolon	818	40.9	1
float	136	6.8	1
pseudo-lamina	741	37.1	1
whole plant	167	8.4	2

[a] Estimation based on dry matter content 50 g kg^{-1} fresh biomass; Reference 1: (Penfound and Earle 1948); Reference 2: (Bagnall 1982).

2.2.4 Water content and chemical compositions of different parts of water hyacinth

Water content and chemical composition of water hyacinth are important biological characteristics relevant to the management of the weed, especially regarding cost of the harvesting and dehydrating processes. Water hyacinth has high water content in its fresh biomass. According to the growth stage and environmental conditions, the water content of fresh water hyacinth biomass ranges from 90% (Lindsey and Hirt 2000a) to 95% (Hronich et al. 2008). Generally, mature and older plants have lower water content than the new and younger shoots. Poor nutrient conditions lower the plant water content due to slow growth and high accumulation of carbohydrates in tissue.

Another interesting biological characteristic of water hyacinth is that it can hyper-accumulate nutrients in different plant parts, leading to a large variation in its chemical composition (Table 2.2.4-1 and Table 2.2.4-2). At physiological nutrient conditions, the crude protein content of dried biomass was 213 g kg^{-1}, but at low nitrogen concentration in water it was only 151 g kg^{-1} in leaves and 74 g kg^{-1} in roots (Zhang et al. 2010), which was similar to another separate study (Rodríguez et al. 2012). The chemical composition of the macrophytes attracted both research and application interests regarding utilization of the biomass for bio-energy, feed and organic fertilizer.

The literature reported that protein content in the dry matter of water hyacinth juice from the dewatering process was as high as 350 g kg^{-1} dry weight (Du et al. 2012), which implied that in the process of dewatering, a substantial proportion of protein may move from tissues to the juice, although the screw-extracted juice may have only 14.7–21.2 g dry matter kg^{-1} fresh juice (Du et al. 2010).

Table 2.2.4-1. Chemical composition of dried water hyacinth biomass (g kg^{-1}).

Plant parts	Crude protein	Fiber	Fat	Ash	Kjeldahl nitrogen	Phosphorus	Ref.
whole plant	97–234	171–282	15.9–36.0	111–204	15.6–37.4	3.1–8.9	1
roots	33.3–115	—	—	—	—	—	2
leaves	55.7–213	—	—	—	—	—	2

Note: Refers to data not available; References 1: (Boyd 1974, Wolverton and McDonald 1978, Lindsey and Hirt 2000a, Aboud et al. 2005, Tham and Udén 2013); References 2: (Mishra and Tripathi 2009, Zhang et al. 2010).

Table 2.2.4-2. The vitamins and minerals daily allowances recommended by United States (USRDA) compared with that in dried water hyacinth leaves (DWHL) (Wolverton and McDonald 1978).

Vitamin	USRDA (per day)	DWHL (per Kg)	Mineral	USRDA (mg day^{-1})	DWHL (mg kg^{-1})
Thiamine	1.5 mg	5.91 mg	Calcium	1000	7560
Riboflavin	1.7 mg	30.7 mg	Iron	18	143
Niacin	20 mg	79.4 mg	Phosphorus	1000	9270
Vitamin E	30 I.U.	206 I.U.	Magnesium	400	8490
Pantothenic acid	10 mg	55.6 mg	Zinc	15	23
Pyroxidine HCI (Vitamin B6)	2 mg	15.2 mg	Copper	2	8
Vitamin B12	6 µg	12.6 µg	Sodium	200–4400	18300
			Potassium	3300	36000
			Sulfur	850	4500

2.3 Habitat and invasion

The biology characteristics of water hyacinth enable it to invade, establish and expand in almost every wet site, including on land with enough moisture when temperature and salinity (< 600 mg L^{-1}) allow, although it may grow very slowly in spring-fed extremely clean waters due to lack of nutrients (Penfound and Earle 1948). Water hyacinth in warm fresh waters with sufficient nutrient supply can grow as high as 1.50 meters in dense mat (Howard and Harley 1998) and double the number of new plants or biomass within about 2 weeks under favorable growing conditions in moderate temperature in lakes both at low and high elevations (Sheng et al. 2011). Water hyacinth has flexible morphology that enables it to adapt easily to a habitat with particular physicochemical and biological characteristics, particularly with changes in its root length, petiole length and shape and shoot-root ratio. In shallow water or on wet land, water hyacinth can grow roots into soil and survive at least long enough to flower and produce seeds. These are the main distribution and invasive strategies of the macrophyte and may partially answer the question on why water hyacinth is difficult, if not impossible, to eradicate.

The free-floating character can also be altered by a habitat. When water hyacinth grows in shallow water or along river embankments, it can develop roots in soil so that it is difficult to move, whereas in deep water, its fast growth rate, floating and driven by wind and water currents on the surface of rivers, lakes and water reservoirs can make it free moving and taking more habitat space. When water hyacinth is moving, the estimation of the plant population and biomass is very difficult. Nevertheless, management options are often prioritized to understand its distribution, population size and biomass in weight. However, a manual survey is either tedious or impossible because large amounts of macrophytes driven by wind are hard to track. In these cases, hyperspectral remote sensing technology with GPS correction can be used to assess the dynamics of water hyacinth population growth (Sun et al. 2011) and the amount of nitrogen being assimilated (Jing-jing Wang et al. 2011). For instance, in a large-scale survey in Lake Taihu (120°04′ E, 31°27′ N) in China, the remote sensing technology provided good estimations of the population size and the growth rate (r) using Logistic model. The rate was estimated to be 0.0124 m^2 d^{-1} with square of correlation coefficient (R^2) at 0.975. The technology was also used in a large-scale ecological engineering project in Lake Dianchi (24°45′ N 102°36′ E) in China (Sun et al. 2011).

2.3.1 Habitat pH and salinity

Water hyacinth grows best at neutral pH, but can adapt well to pH extremes from 2 to 12 (Low et al. 1995), and even grow in cyanide toxic waters (Ebel et al. 2007). Nevertheless, water hyacinth changes physiology in response to extreme pH conditions and the interactions between pH and salinity. The most important changes are: (1) reduced growth rate in waters of extreme pH values, especially when combined with salinity; (2) reduced absorption of heavy metal ions; and (3) altered competition among water hyacinth and other water macrophytes.

For example, at pH value 9, when conductivity[1] increased to 2850 μS cm^{-1}, water hyacinth growth stopped (Yan et al. 1994). Water hyacinth growing in either acidic or alkaline water can gradually alter pH towards neutral (Penfound and Earle 1948, Dai and Che 1987). At different water pHs, the macrophyte had a differential capacity to remove heavy metals from water. At pH 9, the enrichment of lead and cadmium in water hyacinth root reached the maximum, whereas at pH 4, the enrichment of chromium in water hyacinth root was maximal (Dai and Che 1987). Hence, when using water hyacinth to purify water, the target elements and growth conditions need to be clearly defined in order to achieve maximum benefits from the processes.

In most field conditions in natural fresh waters, pH near neutral may favor water hyacinth in competition with other macrophytes such as *Pistia stratiotes* that prefers pH 4 (Gopal and Goel 1993). At near neutral pH, salinity can severely impact growth, development and survival of water hyacinth (Yan et al. 1994). The growth of water hyacinth appeared to slow down and plants finally died at salt concentration 10 g L^{-1} after 21 days (Penfound and Earle 1948). When conductivity increased from 830 to 2800 μS cm^{-1}, water hyacinth growth rate dropped from 77 to 34 g kg^{-1} fresh biomass per day (Wu et al. 1990). Other responses of water hyacinth to salinity are listed in Table 2.3.1-1.

Table 2.3.1-1 showed that the extent and the speed to the damages were related positively to an increase in salinity. The lethal toxicity concentration (2 g L^{-1}) (Penfound and Earle 1948, Haller et al. 1974) was lower than that (4 g L^{-1}) reported by de Casabianca and Laugier (1995). In comparison, the nutrient levels were much lower in the studies by Penfound and Earle (1948) and Haller et al. (1974) (sea water + tap water with no nutrients added, or 0.70 mg N L^{-1} and 0.44 mg P L^{-1}) than in the report by de Casabianca and Laugier (1995) (6.59 mg N L^{-1} and 0.033 mg P L^{-1}). It is reasonable to assume that nitrogen enhanced the survival and growth of water hyacinth in waters with high salinity concentration.

The response of water hyacinth to salinity is important for making management decisions because it is not uncommon that huge amounts of fresh biomass of the macrophyte flush down to estuaries during high water or in rainy seasons. Table 2.3.1-1 indicated that large quantities of water hyacinth take time to die off, so that transportation blockage and environmental damages including those on aquatic flora and fauna need to be taken into consideration. It is also important for managers to consider using water hyacinth as a biological agent for wastewater treatment or water reclamation

[1] Conductivity (μS cm^{-1}) is directly proportional to salinity (mg L^{-1}) by applying a factor 0.5 in most freshwaters, but in sea water and wastewaters, the factor may range from 0.6–0.8 depending on the ion concentrations (Kim et al. 2013).

Table 2.3.1-1. Response of water hyacinth to different salinity levels.

Salinity ($g\ L^{-1}$)[a]	Days reaching <20% damage[b]	Days reaching 50% damage	Days reaching 100% damage	Days reaching plants died[c]	Ref.
0.17	none	none	none	none	1
0.6	28	none	none	none	2
0.83	none	none	none	none	1
1.2	14	28	none	none	2
1.66	14	none	none	none	1
2.1	none	none	none	none	3
2.5	7	14	28	28	4
2.9	7	10	none	none	3
3.33	-[d]	-[d]	14	14	1
4.1	3	8	>28	>28	3
4.16	-[d]	-[d]	14	14	1
5.0	2	10	<28	<28	4
5.9	<1	3	-[d]	-[d]	3
9.3	<1	3	-[d]	-[d]	3
10.0	<2	5	21	-[d]	2
13.7	<0.5	3	-[d]	-[d]	3
20.0	<2	2	7	-[d]	2

[a] Sea water salinity is on average 35 $g\ L^{-1}$ (~50,000 $\mu S\ cm^{-1}$) and ranges from 31–38 $g\ L^{-1}$, whereas fresh water in most lakes is less than 0.5 $g\ L^{-1}$ (~1,000 $\mu S\ cm^{-1}$) and ranges from 0.05–0.5 $g\ L^{-1}$;
[b] Damage is defined as epinasty necrosis symptom or reduced growth rate;
[c] 'Died' is defined as 100% epinasty necrosis symptom and plants not re-growing after being moved into fresh water;
[d] Data was not available because either plants died or experiment discontinued; 'none' refers to no symptom, or no further reduction in biomass productivity; Reference 1 (Haller et al. 1974); Reference 2 (Penfound and Earle 1948); Reference 3 (de Casabianca and Laugier 1995); Reference 4 (Penfound and Earle 1948, Haller et al. 1974).

via phytoremediation because wastewaters from domestic sewage treatment plants or industry discharge usually have high salinity. When water hyacinth is used in wastewater treatment, the conductivity of wastewater may need to be diluted to less than 2000 $\mu S\ cm^{-1}$ in order to have a good result, i.e., vigorous growth of water hyacinth, but the wastewaters with conductivity between 6000 and 12,000 $\mu S\ cm^{-1}$ may also be used for removal of suspended solid and chemical oxygen demand. The literature also suggested that the optimum pH and salinity for water hyacinth growth is not the same as for the optimum removal of heavy metals; hence, the management targets need to be clearly defined and the pH and salinity need to be adjusted accordingly.

2.3.2 Light and other physical factors

The invasion strategy of water hyacinth is closely related to its adaptation to light. The photosynthesis physiology of this species is well developed to enable it to thrive at both low and high incident sunlight, with general principles that young and old leaves have low photosynthesis and mature vigorously-growing leaves have high photosynthetic rate (Li et al. 2010). At favorable temperature (27–30°C), Li et al. (2011) tested the response of water hyacinth to light intensity at different levels of humidity (summarized in Table 2.3.2-1).

Photosynthesis rates are also related to environmental factors such as humidity, temperature and incident sunlight. Literature showed that high photosynthesis rates were obtained with high humidity and intensive incident sunlight (Fig. 2.3.2-1) (Li, Ren et al. 2011).

When the leaf relative humidity ranged from ~55–90%, the photosynthesis rate at Lake Taihu region ranked the highest (Fig. 2.3.2-1), which might have resulted from a favorite environment such as low evapotranspiration, stomatal limitation[2] (*Ls*) and leaf surface temperature (Fig. 2.3.2-1). These environmental factors contributed to the highest net photosynthetic rate (25 μmol CO_2 m^{-2} s^{-1}) at Lake Taihu (Fig. 2.3.2-1). Another factor contributing to high net photosynthetic rate was incident sunlight. At similar leaf relative humidity (55–60%, Fig. 2.3.2-1), averaged incident sunlight (902 μmol m^{-2} s^{-1}) at Lake Dianchi was higher than that (824 μmol m^{-2} s^{-1}) at Nanjing (not shown in the Fig. 2.3.2-1), which might have resulted in a higher net photosynthetic rate at the Lake Dianchi region (15 to 25 μmol CO_2 m^{-2} s^{-1}) compared to 4~15 μmol CO_2 m^{-2} s^{-1} in the Nanjing region (Li, Ren et al. 2011).

The analysis of the literature suggests that tested nutrient concentrations were generally within a low range, especially for phosphorus. The optimum phosphorus concentration is >1 mg P L^{-1} (Wilson et al. 2005), with nitrogen/phosphorus ratio between 5 to 7 (Reddy and Tucker 1983, Wilson et al. 2005). Also, the tested photon flux in the above experiments ranged from 2100 to 1400 μmol m^{-2} s^{-1}; with respect to Photosynthetically Active Radiation (PAR) intensity, water hyacinth is classified as heliophilous. Its fastest growth is at

Table 2.3.2-1. Response of *E. crassipes* to incident sunlight under favorable temperature and different humidity levels (Li, Ren et al. 2011).

Location	Maximum photosynthesis rates (P_{max}) [μmol CO_2 m^{-2} s^{-1}]	Light-saturation point (*LSP*) [μmol m^{-2} s^{-1}]	Light compensation point (*LCP*) [μmol m^{-2} s^{-1}]	Apparent quantum efficiency (AQE)	Dark respiration rate [μmol m^{-2} s^{-1}]
Lake Taihu	36.3±1.2Aa	2350±69Aa	36.4±3.8Bb	0.053±0.003Aa	−2.64±0.31Aa
Nanjing	25.1±1.1Bb	2250±83Aa	20.3±3.1Cc	0.035±0.002Bb	−0.84±0.22Aa
Lake Dianchi	28.5±1.3Bb	1993±27Bb	60.5±5.3Aa	0.065±0.002Aa	−4.17±0.51Bb

Note: Different small and capital letters mean significant difference at $p<0.05$ and $p<0.01$, respectively.

[2] Stomatal limitation (*Ls*) is defined as $Ls = (1 - C_i/C_a)$, where C_i and C_a are partial pressures of CO_2 inside the leaf and in the air (Farquhar and Sharkey 1982).

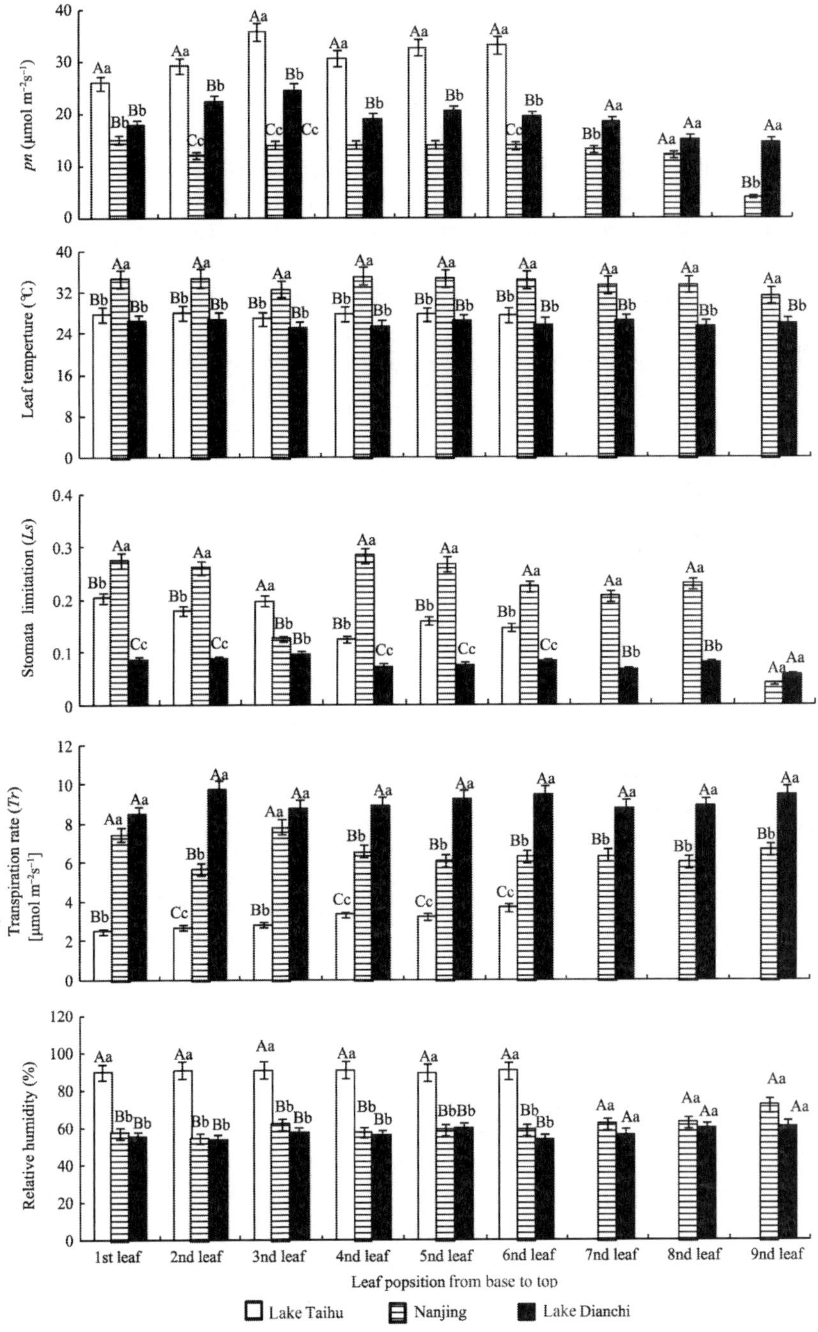

Fig. 2.3.2-1. Net photosynthetic rate (*Pn*), stomatal limitation (*Ls*), transpiration rate (*Tr*), temperature and relative humidity of leaves at different positions on *E. crassipes* at Lake Taihu, Nanjing and Lake Dianchi Lake (Li, Ren et al. 2011).

4138 μmol m^{-2} s^{-1} at full spectrum (Téllez et al. 2008) or 3310 μmol m^{-2} s^{-1} at photosynthetically-active spectrum (Barnes et al. 1993), and its minimum requirement (PAR) is 331 μmol m^{-2} s^{-1}; clearly, it can grow under a broad range of light intensities. This implies that the photosynthetic rates of water hyacinth derived from the experiments conducted by Li, Ren et al. (2011) may not be optimal or fastest rates and were possibly closer to the lower limit.

The above experiments revealed that the driving force of water hyacinth extremely fast growth rate was due to its biological characteristics related to photosynthesis, with maximum photosynthetic rate (P_m) 38 μmol CO_2 m^{-2} s^{-1}, light saturation point (LSP) 2503 μmol m^{-2} s^{-1}, light compensation point (LCP) 16 μmol $m^{-2}s^{-1}$ and the Apparent Quantum Efficiency (AQE) 0.053 in temperate climate zone in Lake Taihu region in China (Li, Cong et al. 2011). It should also be taken into account that water hyacinth has excellent adaptability to a wide range of environmental conditions such as temperature, humidity, nutrient concentrations, and incident photon flux (Li, Ren et al. 2011).

Other important environmental factors related to water hyacinth habitat are Dissolved Oxygen (DO), Chemical Oxygen Demand (COD), hydrological current and wind. These habitat factors are important for water hyacinth management, particularly making decisions on control, potential utilizations of the biomass for resources, and the use of this species in phytoremediation. Generally speaking, the presence of water hyacinth lowers concentration of dissolved oxygen to as low as 0.1–1.0 mg L^{-1} under a closed-surface mat (Penfound and Earle 1948). The reported extremely low dissolved oxygen resulted in elimination of most fish species and macroinvertebrates, but not in case with the partial surface coverage (defined as less than 25% of the total water surface covered) or patched coverage (defined as discontinued mats with wide gaps around each mat) that enables interaction between water hyacinth population and algae with improved water quality such as reducing suspended solids, turbidity, chemical oxygen demand (Masifwa et al. 2001, Toft et al. 2003, Wang et al. 2012, Wang et al. 2013, Zhang et al. 2014).

2.4 Temperature and vegetative reproduction

Water and air temperature can greatly impact water hyacinth growth and development. However, water temperature is less variable under water hyacinth mat compared to the open areas of a water body (Penfound and Earle 1948). This implies that water temperature under water hyacinth mat changes according to the seasonal pattern of a particular habitat. Generally, water temperature variation may lag behind the air temperature changes, i.e., water temperature rises in spring and drops in autumn lag behind the air temperature changes.

Water hyacinth is a perennial plant and can grow and develop all year round in the tropical and subtropical climate zones. In the temperate climate zone, it can grow well during spring, summer and autumn depending on temperature and nutrient concentration at a particular location. It can start to grow at 10°C (Gettys et al. 2009) and tolerate temperature as high as 43°C (Howard and Harley 1998).

In natural habitat, during spring, the rapidly increased air temperature can promote plant growth in terms of both plant height and leaf blade width. Following population growth, the plant density increases rapidly, which may result in changing the growth pattern of individual plants to grow vertically because interconnected stolons prevent free movement of individual plants within a mat, and the mat edge may quickly expand towards the open water surface. During summer, the plant height may reach the maximum, and growth rate then decreases in autumn with temperatures gradually falling till whole plants die in the temperate climate zones. The main difference in plant growth between tropical and temperate climate zones occur in winter. With low winter temperatures in the temperate climate zone, although the floats and pseudo-lamina may wither and become yellow, the rhizome, stolon and root may still be alive depending on water temperature and duration of the freezing period. The plants may completely die (without re-growing) after a 12-hour exposure to air temperature –7.2°C, or a 48-hour exposure to air temperature –5.0°C (Penfound and Earle 1948).

A pattern of water hyacinth vegetative reproduction is similar to that of biomass growth, except that vegetative production is not dependent just simply on temperature, but follows effective accumulated temperature (Table 2.4-1) (Wang et al. 2011a).

In an early growth period, water hyacinths need effective accumulated temperature about 37°C on average for growing one leaf (about 2 days at daily mean temperature of 28°C and daily mean effective temperature of 18°C). With an increase in leaf age, the effective accumulated temperature required for growing one new leaf had a tendency to increase from 2 to 3.2 days to develop new leaves ordered after the 3rd leaf up to 13th leaf. Even though representing the same concept of growth and development of water hyacinths being dominated by temperature, Table 2.4-1 may be more convenient for management purposes.

Another important character related to the vegetative reproduction is the individual leaf life-span that is closely linked to decision-making on harvesting date and subsequently on utilization because if the planned utilization is for feed, the biomass should be harvested when it is young. The maximum life-span of a single leaf blade was about 34 days at temperature ~28°C; a leaf would die when the accumulated effective temperature reached 544°C (Wang et al. 2011a). However, under natural conditions, most leaf blades would die and gradually fall off before reaching this accumulated temperature due to damage by biotic and abiotic factors. When this happens, using water hyacinth for phytoremediation of chemical oxygen demand becomes ineffective because died-off leaves decay and increase chemical oxygen demand in water.

The process of a leaf maturing and dying can be assessed by leaf chlorophyll content using for example a Single Photon Avalanche Diode (SPAD) chlorophyll analyzer (Pasquardini et al. 2015). Although the data

Table 2.4-1. Days needed for the emergence of each individual leaf of water hyacinth plant at ambient air temperature 28°C and effective accumulated temperatures needed for the emergence of water hyacinth leaf (Wang, Zhu et al. 2011).

Leaf order	Range of days for emergence each leaf	Average days for emergence each leaf	Effective accumulated temp for emergence each leaf °C	Average effective accumulated temp for emergence each leaf °C
3rd	2.0–3.0	2.5	29.0–45.6	37.7
4th	2.0–3.5	2.4	29.0–54.3	37.3
5th	2.0–4.5	3.5	31.3–57.1	48.2
6th	3.0–4.0	3.8	42.2–54.2	50.3
7th	2.0–4.0	2.8	34.4–60.4	44.7
8th	2.0–4.0	3.5	34.2–68.6	60.0
9th	3.0–4.0	3.3	50.8–68.5	56.3
10th	2.0–4.0	3.1	31.6–66.0	51.0
11th	3.0–4.5	3.7	55.9–80.3	66.5
12th	2.5–3.5	2.8	40.9–70.0	54.4
13th	2.5–5.0	3.6	44.6–96.5	65.9
average	3.0–3.3	3.2	48.2–53.6	52.0

Note: Leaf order was counted from base to top according to the emergence.

collected from a SPAD meter reading (an index value from 0 to 100) does not directly represent chlorophyll content the same as laboratory methods (μg L^{-1}), it is directly correlated to the result from the traditional methods; hence, many reports use SPAD values to represent chlorophyll content (Martínez et al. 2015). The SPAD values from the fourth leaf blade of water hyacinth in the whole life cycle appeared to increase first and then decrease with an increase in effective accumulated temperature. When effective accumulated temperature reached 544°C, leaf blades grew old and died off naturally, and the SPAD values of leaf blades decreased to the lowest (Fig. 2.4-1).

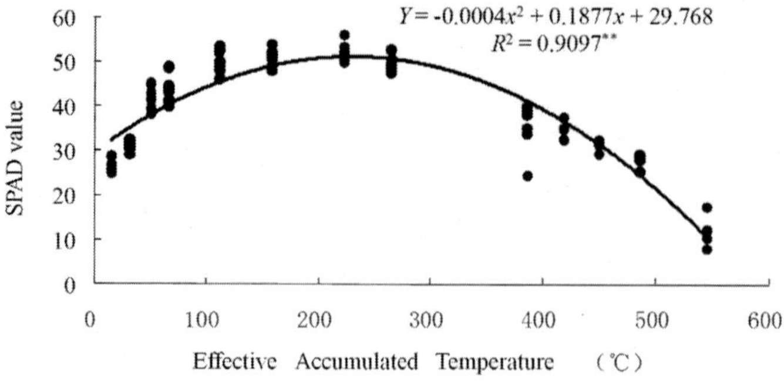

Fig. 2.4-1. SPAD values of the fourth leaf during the whole life cycle, ** refers to a very significant ($p<0.01$) curve fit.

The vegetative reproduction follows different patterns in different physical, chemical and biological conditions such as temperature, available growing space and nutrients, etc. In an uncommitted growth space, the vegetative reproduction of a single plant is dominated by temperature (Figs. 2.4-2 and 2.4-3). Figure 2.4-2 showed that the number of new plants from a single mother plant follows an exponential pattern when space allows, with the number of new plants reaching as high as 215 during 30-day growth at temperature ~28°C. Figure 2.4-3 showed that 215 new plants from a single mother plant require effective accumulated temperature of ~474°C. Although there are similar concepts represented in Figs. 2.4-2 and 2.4-3, the practical applications are totally different. Figure 2.4-2 may only serve as a rough approximation of a number of new plants at certain temperature and space, whereas Fig. 2.4-3 could be used to estimate plant population at a wide range of temperature when space is available, especially when an automated environmental monitoring equipment is available.

A relationship between the new ramet growth and the mother plant leaf development stage showed an nth leaf minus 3 leaf development pattern and a synchronous relationship between the new ramet growth and the leaf stage (Table 2.4-2).

When a single plant has four leaves (nth = 4), a new ramet will theoretically appear at the first leaf (4 minus 3); and at the 8th leaf stage of a mother plant, new ramet should appear at the 5th, 4th, 3rd, 2nd, and 1st leaves. For example, Table 2.4-2 showed six single plants that had developed ~8 leaves (7.9–8.2) and theoretically should have first generation ramet at 5th, 4th, 3rd, 2nd and

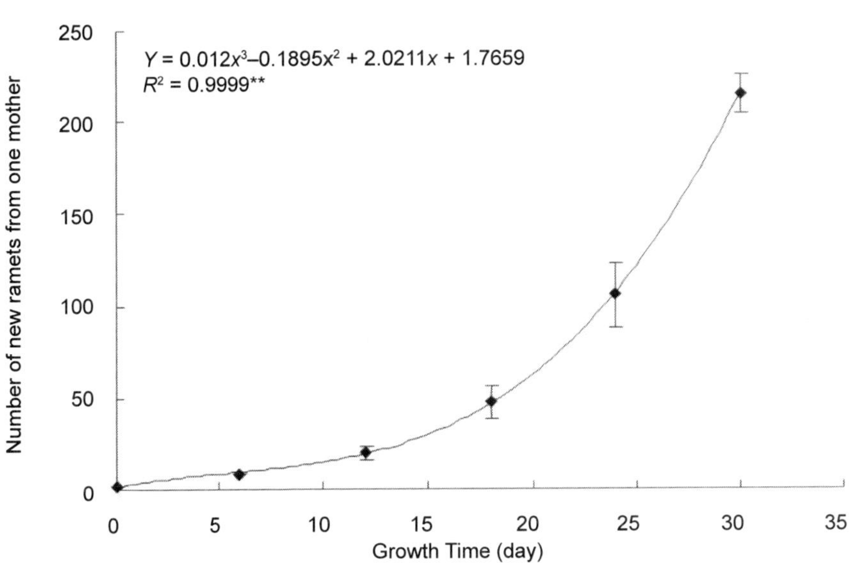

Fig. 2.4-2. Vegetative reproduction during a single mother plant growth time.

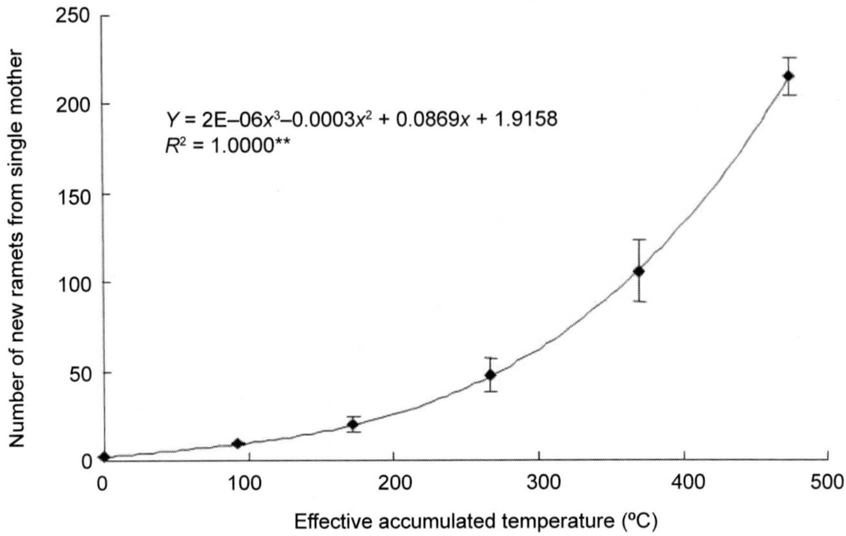

Fig. 2.4-3. Vegetative reproduction and effective accumulated temperature a single mother plant required.

Table 2.4-2. Relationship between new ramet leaf age (days) and leaf development at ~8th leaf stages of six single water hyacinth.

Plant No.	leaf stage (no.)	Generation	New ramet leaf age at the different leaf positions of mother plant (days)							
			1st	2nd	3rd	4th	5th	6th	7th	8th
1	8.0	first	4.2	4.6	4.3	3.4	0.2	0	0	0
		second	0	0	1.1, 1.0	0.9	0	0	0	0
2	8.0	first	0	3.9	4.0	3.1	1.2	0.7	0	0
		second	0	1.0	1.2, 1.0	1.0	0	0	0	0
3	8.2	first	0	5.0	4.9	4.2	2.3	0	0	0
		second	0	3.2, 1.2, 0.3	2.5, 2.7	2.2	0	0	0	0
		third	0	1.0	0	0	0	0	0	0
4	7.9	first	5.0	5.0	4.7	3.6	2.0	0	0	0
		second	0	4.0, 1.5	2.3, 1.0	0.6	0	0	0	0
5	8.1	first	4.0	5.0	4.7	4.0	1.7	0	0	0
		second	0	3.0, 1.7	2.0, 1.5	1.4, 0.2	0	0	0	0
6	8.1	first	5.9	5.3	4.7	3.9	1.0	0.2	0	0
		second	3.7, 3.0	3.3, 2.0	2.2, 1.1	1.0	0	0	0	0
		third	1.0, 0.2	1.0	0	0	0	0	0	0

Note: Digits 0 refers to no new ramet at that leaf or at that branch.

1st leaves, except plants number 2 and 3 missing ramet at the 1st leaf. All observed plants had second generations; and plants number 3 and 6 had third generations.

The relationship can be illustrated further by mother plants number 3 and 6. When the mother plant number 3 had 8.2 leaves, its 5th leaf position had the first generation new plant ramet with a leaf age of 2.3 days. For the mother plant number 6, one new ramet (with leaf age of 5.9 days) of the first generation appeared at the 1st leaf position; and the first-generation ramet had two second-generation ramets with leaf age 3.7 and 3.0 days; the second-generation ramet had the new ramet of the third generation with leaf age 1.0 and 0.2 day. The variations in Table 2.4-2 may have been caused by the micro-environment effects or nutrient conditions of a particular leaf. The observation data generally showed that in a small population, the vegetative reproduction showed exponential growth in the early period with low plant density (Fig. 2.4-4, A).

A

B

Fig. 2.4-4. Vegetative reproduction, A: after 15 days with unlimited space (one plant m^{-2}); B: after 15 days with limited space (20 plants m^{-2}), A and B both under average air temperature of 24.3°C (Photo by Lin Shang 2015).

The vegetative reproduction rate showed that a single plant produced 19 new ramets during a 15-day period at averaged air temperature of 24.3°C without space limitation. Under limited growth space or with an increase in population density of water hyacinth, however, the vegetative reproduction pattern would be totally different. The plant vegetative reproduction could slow down gradually until no new plant ramet appears. For instance, the reproduction rate slowed down gradually with an increase in population density (space limitation intensified). On average, single plant produced two new ramets in the same period and the same environmental conditions (Fig. 2.4-4, B).

Theoretically, the biology of water hyacinth showed an indeterminate growth type, i.e., under suitable environmental conditions, water hyacinth can produce leaves and grow all year round and vegetatively generate new ramets. Under natural conditions, the growth rate of leaf blades becomes slow with an increase in leaf age; an increased effective accumulated temperature is required for growing new leaf blades. The reason may be that water hyacinth growth is restrained due to internal competition for nutrients and space so that most leaf blades die before reaching its life-span (effective accumulated temperature 544°C).

The strong invasion by water hyacinth relies on its vigorous vegetative reproduction by producing stolons from leaf axils, growing a new plant at the tip of a stolon. Under suitable conditions, the quantity of plants could be doubled within 7–15 days. For example, an initial stocking of 0.06 kg m^{-2} of dry weight would increase to 0.14 kg m^{-2} after 7 days and reach 1.22 kg m^{-2} after 42 days (Zheng et al. 2011). Each plant produced several stolons and each stolon produced a new plant. Circumstances permitting, one water hyacinth plant could generate 4.7×10^7 plant divisions within a year, which could cover 47 hectare of water surface, with a fresh weight of 9000 metric tonnes (Njoka 2004).

2.5 Sexual reproduction and persistence

Water hyacinth needs a basic vegetative growth period to develop from seedling to flowering. Under natural conditions, water hyacinth would not develop inflorescence when vegetative growth does not produce 13 leaf blades (pseudo-lamina). After the plant completed 13th pseudo-lamina, an inflorescence may be developed if daily mean temperature exceeds 31°C for 5 consecutive days (Zheng et al. 2011). The inflorescence consists of a naked axis surmounted by two bracts that subtend the flower cluster. Between two events of inflorescence development, the plant needs another vegetative growth to accumulate 13 leaf blades and to meet a daily mean temperature over 31°C for 5 consecutive days. Sexual reproduction slows down vegetative growth because of competition for photosynthetic products. For example, the plants in vegetative growth phase need effective accumulated temperature of 341.0°C to grow on average 7.04 leaf blades, but after the onset of inflorescence, the same effective accumulated temperature would only support growth of 6.09 leaf blades. When effective accumulated temperature reached 532°C, the plant grew 10.3 leaf blades in vegetative growth phase, but only 8.3 leaf blades after flowering (Fig. 2.5-1) (Zi-chen Wang et al. 2011).

At the start of the inflorescence development, the bracts appear first, then a naked axis designated as subrachis, whereas the flowering portion distally is referred to as the actual rachis (Fig. 2.5-2). The rachis may commonly bear 8–15 flowers in a population (Lindsey and Hirt 2000c), but it may be from 5–23 flowers among different populations. The flower of water hyacinth is bisexual and has floral trimorphism distinguished by the style and the stamen length (Barrett 1977). The perianth has a tubular base and a limb with six colored lobes; the upper lobe is larger than the rest and bears a violet blotch with a yellow center (Barrett 1980a).

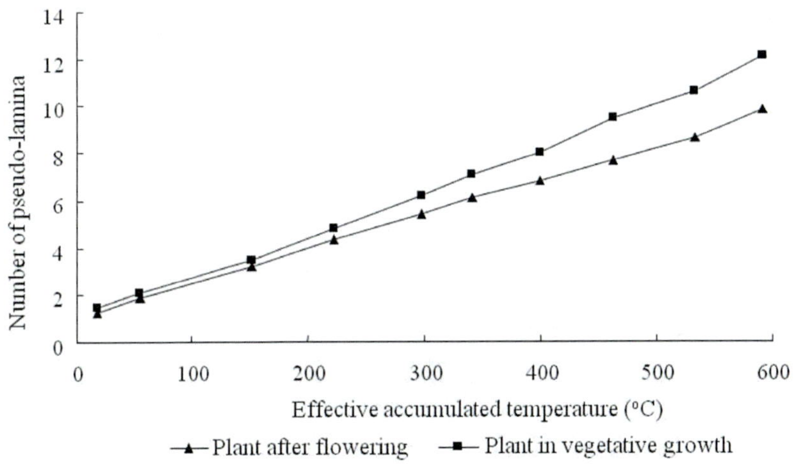

Fig. 2.5-1. Effects of flowering on the water hyacinth growth of pseudo-lamina.

Fig. 2.5-2. Water hyacinth in flower (Photo by Lin Shang 2015).

The inflorescence development takes about 14 days to complete, including the bracts, naked axis, buds and floral development. The anthokinetic cycle of water hyacinth consists of a flowering phase and a bending phase and is usually completed in 23–40 hours (Penfound and Earle 1948, Kohji et al. 1995) depending on night and day temperatures (Figs. 2.5-3 and 2.5-4). When habitat is wet land, the floats of the macrophytes may disappear, and flower size becomes small but capable of bearing normal and viable seeds to prepare the next life cycle. The pollination is completed by insects during the flowering stage (before bending), e.g., by honey bees (*Apis mellifera* L.) (Téllez et al. 2008), or by self-pollination (Barrett 1977, 1980a). In some specific habitats and natural conditions, the entomophilous pollination may have as low as ~4.0% success rate (Ren et al. 2004) in Chongqing, China, and about 5% at Lake Dianchi (Zhang et al. 2012).

Fig. 2.5-3. Water hyacinth rachis is in flowering stage before bending at average temperatures 24°C (night) to 32°C (day) (Photo by Lin Shang 2015).

Fig. 2.5-4. Water hyacinth peduncle bends gravitropically at average temperatures 24°C (night) to 32°C (day) (Photo by Lin Shang 2015).

Fig. 2.5-5. Water hyacinth peduncle bends gravitropically (final position) at average temperatures 24°C (night) to 32°C (day) (Photo by Lin Shang 2015).

The ovary of water hyacinth is three-celled with numerous ovules in each cell; the fruit is a three-celled dehiscent capsule surrounded by marcescent perianth (Singh 1962). Following successful pollination, capsules are produced with an average of 44 seeds per capsule (Barrett 1980b), but could contain as high as 450 seeds per capsule (Lindsey and Hirt 2000c). Up to 46% of flowers may produce capsules. After gravitropic bending of the peduncle to the final position, the capsule maturation continues for between 16 and 23 days until ripeness (Penfound and Earle 1948). In ripe capsules, the thin-walled pericarp dehisces and hypanthium splits to release seeds into the water. The seeds promptly sink to the bottom where they remain viable up to 28 years (Sullivan and Wood 2012). This extreme longevity of the water hyacinth seed is due to its external and internal structures (Figs. 2.5-6 to 2.5-11) (Zhang et al. 2012).

Fig. 2.5-6. Dehiscent capsules surrounded by marcescent perianth: ovate and brown, ~13 mm in length and ~4 mm in width, contain 63–153 seeds (Photo by Ying-ying Zhang 2010).

Fig. 2.5-7. Mature seeds: 1.4–1.9 mm in length and 0.7–0.9 mm in width; mature seeds are dark brown (left) and immature ones are yellow-green (right); 1000 seeds weigh 0.37–0.43 g (Photo by Ying-ying Zhang 2010).

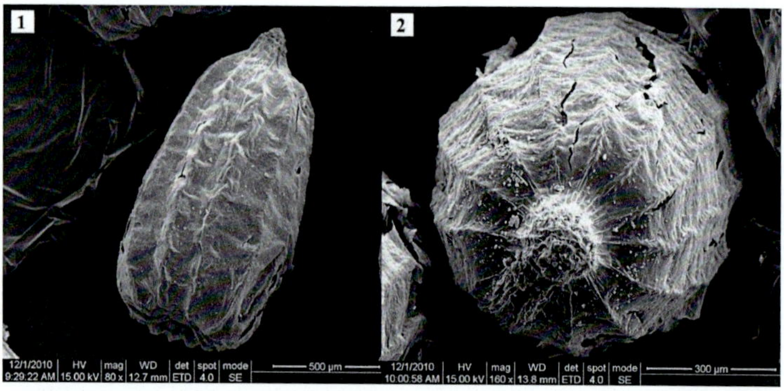

Fig. 2.5-8. Appearance of *E. crassipes* seed: (1) Oval seed shape with one round and one pointy end; (2) Pointy end, hilum, caruncle and micropyle (FEI Quanta 200 Scanning Electron Microscope photo by Ying-ying Zhang 2010).

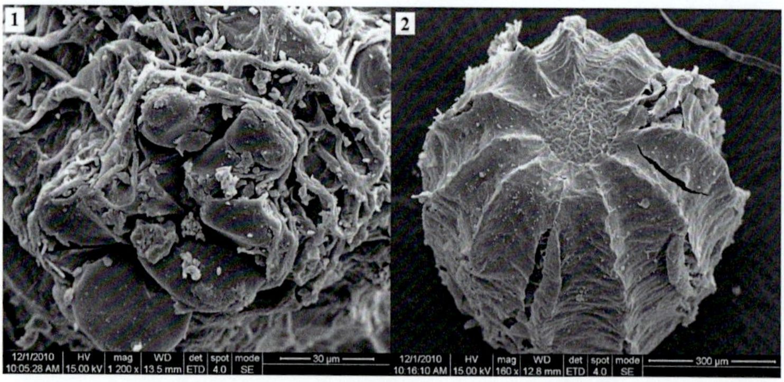

Fig. 2.5-9. Appearance of *E. crassipes* seed: (1) Enlarged pointy end: hilum, caruncle and micropyle at the tip; (2) Enlarged round end with a groove and 11 ridges uniformly distributed; scars may have been caused by damage during seed preparation (FEI Quanta 200 Scanning Electron Microscope photo by Ying-ying Zhang 2010).

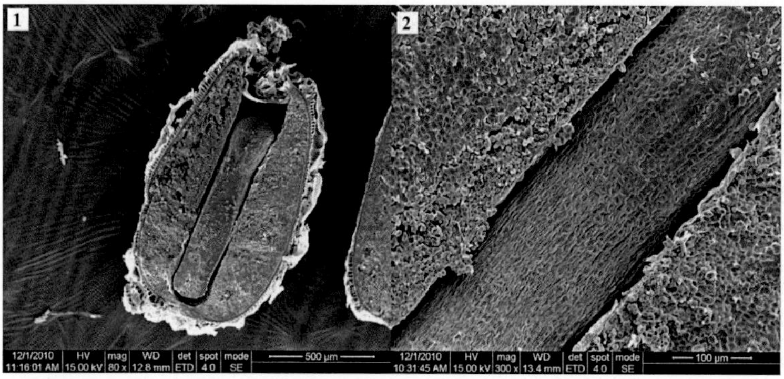

Fig. 2.5-10. Internal structure of water hyacinth seed: (1) Longitudinal section: an embryo at the center surrounded by endosperm; (2) Surface texture of embryo (FEI Quanta 200 Scanning Electron Microscope photo by Ying-ying Zhang 2010).

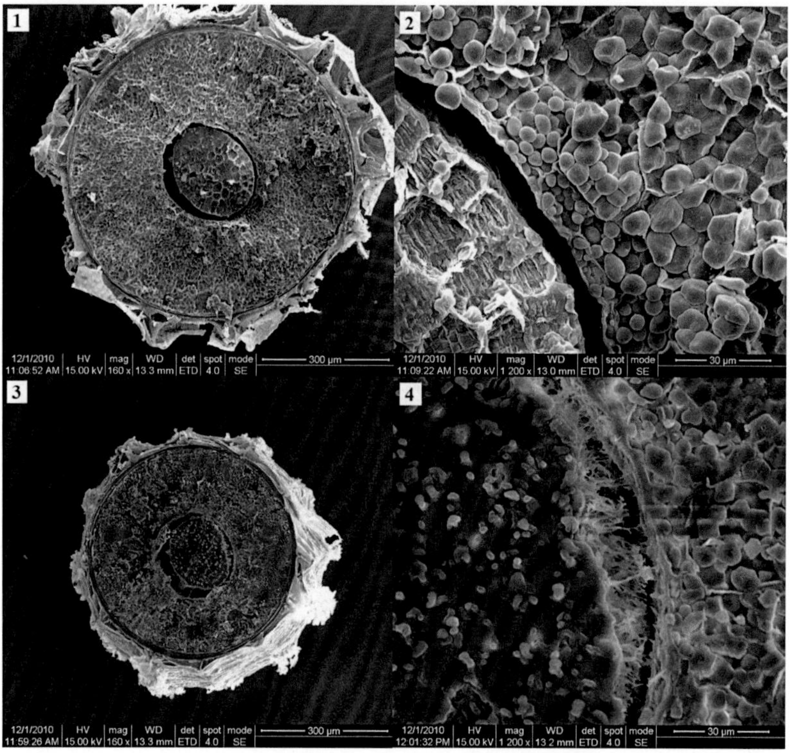

Fig. 2.5-11. Internal structure of water hyacinth seeds: (1) Cross section of mature seed, embryo at the center, endosperm around embryo, and seed coat surrounding endosperm; (2) Cross section of mature seed, embryo at the lower left corner clearly separated from endosperm at right; (3) Cross section of immature seed with embryo at the center; (4) Cross section of immature seed (embryo at left with filiform connections to endosperm) (FEI Quanta 200 Scanning Electron Microscope photo by Ying-ying Zhang 2010).

The sexual reproduction involves many steps from initializing inflorescence on a mother plant to young seedling establishment and successful float on the water surface. In the biology of this species, germination is designed as the reproduction steps from initializing embryo growth (dormancy break), then successful breaching of the seed coat to expose the primary root and leaf to external environment (Fig. 2.5-12).

The water hyacinth seeds require a complex environment for germination in natural conditions, but generally, water temperature 28–36°C, light at 41.4–44.1 μmol m^{-2} s^{-1} (Tang et al. 1987), shallow water and oxygen are necessary (Barrett 1980b, Sullivan and Wood 2012). When these conditions are available, a mature seed of this species can successfully germinate within 4 to 5 days (Pérez, Coetzee et al. 2011). In contrast, when exposed to gibberellic acid at 1000 mg L^{-1}, seeds can germinate in dark within 14 days at a germination rate of 85% (Tang et al. 1987).

Considering the environment in aquatic systems, it may be suggested that the shoreline of a water body would easily provide suitable germination requirements, which is supported by observations that river banks and lake shorelines with changing water levels and alternation of wetting and drying conditions make water hyacinth germinate easily (Barrett 1980b, Sullivan and Wood 2012). Literature also suggested that increased concentration of phosphorus and boron promoted seed germination rate (Pérez, Téllez et al. 2011), but other experiments suggested that an increase in nutrient concentrations may facilitate the survival of germinated seedlings rather than the germination rate (Sullivan and Wood 2012) because water hyacinth seedlings need an initial growth period of 40–60 days to develop float leaves and finally float on the water surface (Penfound and Earle 1948). Another characteristic related to water hyacinth seeds and their germination is that

Fig. 2.5-12. Water hyacinth seed germination (Photo by Rod Brayne and the property of G Sainty at Sainty Associates Pty Ltd. Permission obtained by pers. comm.).

seeds cannot survive at ambient temperature above 57°C (such as during biomass composting) (Montoya et al. 2013). This is also very important in the context of water hyacinth management, control and policy making to ensure limits to the water hyacinth seed disposal, especially in utilizing water hyacinth biomass for composted organic fertilizers.

To identify a proportion of viable seeds in a seed bank, topographical tetrazolium (2,3,5-triphenyl-2H-tetrazolium chloride) staining test with the embryo *in vitro* culture method (Duncan and Widholm 2004, Li et al. 2012) can be used. Tetrazolium chloride salt (TTC) is water soluble and colorless and can be dissolved in 0.1 mol L^{-1} phosphate buffer (pH 7.0) to get concentrations between 2.5 and 10 g L^{-1}. The solution should be kept in the dark before use. The assay, including soaking and staining the seeds, could be conducted at room temperatures or in an incubator from 18 to 40°C for 35–90 minutes in the dark (light affects color development). At the end of the assay, the embryo of viable seeds develops a stable and vivid red color; usually, a light microscope is sufficient for reliable observations. Zhang et al. (2012) reported viable percentages of seed collected in natural conditions at Lake Dianchi and tested in the laboratory (Table 2.5-1), and suggested that soaking and staining temperatures are critical; using soaking and staining temperature of 40°C gave the highest percentage of colored embryos.

Table 2.5-1. Topographical tetrazolium staining test with embryo *in vitro* culture (100 seeds per treatment with three replicates) at Lake Dianchi (Zhang et al. 2012).

Soaking temperature (°C)	Staining temperature (°C)	Embryo color development	Percentage color developed (%)
18	18	Pale red at embryo tip	88
18	40	Vivid red at embryo tip	87
40	18	Vivid red at embryo tip	92
40	40	Vivid red at whole embryo	95

References cited

Aboud, A. A. O., R. S. Kidunda and J. Osarya. 2005. Potential of water hyacinth (*Eicchornia crassipes*) in ruminant nutrition in Tanzania. *Livestock Research for Rural Development* 17(8): Art. #96.

Bagnall, L. O. 1982. Bulk mechanical properties of waterhyacinth. *Journal of Aquatic Plant Management* 20: 49–53.

Barnes, C., T. Tibbitts, J. Sager, G. Deitzer, D. Bubenheim, G. Koerner et al. 1993. Accuracy of quantum sensors measuring yield photon flux and photosynthetic photon flux. *HortScience* 28(12): 1197–1200.

Barrett, S. C. H. 1977. Tristyly in *Eichhornia crassipes* (Mart.) Solms (water hyacinth). *Biotropica* 9(4): 230–238.

Barrett, S. C. H. 1980a. Sexual reproduction in *Eichhornia crassipes* (water hyacinth). I. Fertility of clones from diverse regions. *Journal of Applied Ecology* 17(1): 101–112.

Barrett, S. C. H. 1980b. Sexual reproduction in *Eichhornia crassipes* (water hyacinth). II. Seed production in natural populations. *Journal of Applied Ecology* 17(1): 113–124.

Boyd, C. E. 1974. Chapter 7 Utilization of aquatic plants. In *Aquatic Vegetation and its Use and Control*, ed. D. S. Mitchell, 104–112. Paris, France: UNESCO.

de Casabianca, M. and T. Laugier. 1995. *Eichhornia crassipes* production on petroliferous waste waters: effects of salinity. *Bioresource Technology* 54: 39–43.

Dai, S. and G. Che. 1987. Removal of some heavy metals from wastewater by waterhyacinth. *Environmental Chemistry* 6(2): 43–50 (In Chinese with English Abstract).

Dellarossa, V., J. Céspedes and C. Zaror. 2001. *Eichhornia crassipes*-based tertiary treatment of Kraft pulp mill effluents in Chilean Central Region. *Hydrobiologia* 443: 187–191.

Du, J., Z. Chang, X. Ye, Y. Xu and J. Zhang. 2012. Pilot-scale study on dehydration effect of water hyacinth with different pulverization degree. *Transactions of the Chinese Society of Agricultural Engineering* 28(5): 207–212 (In Chinese with English Abstract).

Du, J., Z. Z. Chang, X. Ye and H. Huang. 2010. Losses in nitrogen, phosphorus and potassium of water hyacinth dehydrated by mechanical press. *Fujian Journal of Agricultural Science* 25(1): 104–107 (In Chinese with English Abstract).

Duncan, D. R. and J. M. Widholm. 2004. Osmotic induced stimulation of the reduction of the viability dye 2,3,5-triphenyltetrazolium chloride by maize roots and callus cultures. *Journal of plant physiology* 161(4): 397–403.

Ebel, M., M. W. H. Evangelou and A. Schaeffer. 2007. Cyanide phytoremediation by water hyacinths (*Eichhornia crassipes*). *Chemosphere* 66(5): 816–23.

EPA. 2007. Summary table for the nutrient criteria documents. *Environmental Protection Agency*. Washington DC, US: Office of Science and Technology.

Farquhar, G. D. and T. D. Sharkey. 1982. Stomatal conductance and photosynthesis. *Annual Review of Plant Physiology* 33(1): 317–345.

Gettys, L. A., W. T. Haller and M. Bellaud. 2009. *Biology and control of aquatic plants: a best management practices handbook*. 2nd ed. Marietta GA, USA: Aquatic Ecosystem Restoration Foundation.

Gopal, B. and U. Goel. 1993. Competition and allelopathy in aquatic plant communities. *The Botanical Review* 59(3): 155–210.

Hadad, H. R., M. A. Maine, M. Pinciroli and M. M. Mufarrege. 2009. Nickel and phosphorous sorption efficiencies, tissue accumulation kinetics and morphological effects on Eichhornia crassipes. *Ecotoxicology (London, England)* 18(5): 504–13.

Haller, W. T. and D. L. Sutton. 1973. Effect of pH and high phosphorus concentrations on growth of waterhyacinth. *Hyacinth Control Journal* 11: 59–61.

Haller, W. T., D. L. Sutton and W. C. Barlowe. 1974. Effects of salinity on growth of several aquatic macrophytes. *Ecology* 55(4): 891–894.

Howard, G. W. and K. L. S. Harley. 1998. How do floating aquatic weeds affect wetland conservation and development? How can these effects be minimised? *Wetlands Ecology and Management* 5(3): 215–225.

Hronich, J. E., L. Martin, J. Plawsky and H. R. Bungay. 2008. Potential of *Eichhornia crassipes* for biomass refining. *Journal of Industrial Microbiology and Biotechnology* 35(5): 393–402.

Kim, M., W. Choi, H. Lim and S. Yang. 2013. Integrated microfluidic-based sensor module for real-time measurement of temperature, conductivity, and salinity to monitor reverse osmosis. *Desalination* 317: 166–174.

Kohji, J., R. Yamamoto and Y. Masuda. 1995. Gravitropic response in *Eichhornia crassipes* (water hyacinth) I. process of gravitropic bending in the peduncle. *Journal of Plant Research* 108(3): 387–393.

Li, C. 2012. A feasibility study on blue algae pollution control by water hyacinth in Lake Dianchi. *Environmental Science Survey* 31(3): 64–68 (In Chinese with English Abstract).

Li, S., Y. Mao and Y. Jiang. 2012. Detection method of seed germination percentage of *Fraxinus chinensis*. *Journal of Northeast Forestry University* 40(3): 1–4 (In Chinese with English Abstract).

Li, X., W. Cong, C. Ren, J. Sheng, P. Zhu, J. Zheng et al. 2011. Photosynthetic productivity and the potential of carbon sink in cultivated water hyacinth (*Eichhornia crassipes*) in Taihu Lake. *Jiangsu Journal of Agricultural Sciences* 27(3): 500–504 (In Chinese with English Abstract).

Li, X., C. Ren, M. Wang, W. Cong, J. Sheng, P. Zhu et al. 2011. Comparison of photosynthesis eco-function of water hyacinth and their environmental factors in different areas. *Chinese Journal of Eco-Agriculture* 19(4): 823–830 (In Chinese with English Abstract).

Li, X., C. Ren, M. Wang, J. Sheng and J. Zheng. 2010. Response of photosynthesis of leaves to light and temperature in *Eichhornia crassipes* in Jiangsu Province. *Jiangsu Journal of Agricultural Sciences* 26(5): 943–947 (In Chinese with English Abstract).

Lindsey, K. and H.-M. Hirt. 2000a. Chapter 1: Introduction. In *Use Water Hyacinth!—A Practical Handbook of Uses for Water Hyacinth from Across the World*, ed. K. Lindsey and H. -M. Hirt, 1–4. Winnenden, Germany: Anamed International.

Lindsey, K. 2000b. Chapter 5 Utilisation: general principles. In *Use Water Hyacinth!—A Practical Handbook of Uses for Water Hyacinth from Across the World*, ed. K. Lindsey and H. -M. Hirt, 31–38. Winnenden, Germany: Anamed International.

Lindsey, K. 2000c. Chapter 2 The water hyacinth. In: *Use Water Hyacinth!—A Practical Handbook of Uses for Water Hyacinth from Across the World*, ed. K. Lindsey and H. -M. Hirt, 5–8. Winnenden, Germany: Anamed International.

Low, K. S., C. K. Lee and K. K. Tan. 1995. Biosorption of basic dyes by water hyacinth roots. *Bioresource Technology* 52(1): 79–83.

Ma, T., Z. Zhang, N. Yi, X. Liu, Y. Wang, S. Yan et al. 2013. Nitrogen removal via denitrification from eutrophic water as influenced by *Eichhornia crassipes* and sediment. *Journal of Agro-Environment Science* 32(12): 2451–2459 (In Chinese with English Abstract).

Martínez, F., P. Palencia, C. M. Weiland, D. Alonso and J. A. Oliveira. 2015. Influence of nitrification inhibitor DMPP on yield, fruit quality and SPAD values of strawberry plants. *Scientia Horticulturae* 185: 233–239.

Masifwa, W. F., T. Twongo and P. Denny. 2001. The impact of water hyacinth, *Eichhornia crassipes* (Mart) Solms on the abundance and diversity of aquatic macroinvertebrates along the shores of northern Lake Victoria, Uganda. *Hydrobiologia* 452: 79–88.

MEP-PRC. 2002. Environmental quality standards for surface water (GB3838-2002). Baijing, China: Ministry of Environmental Protection of The Peoples's Republic of China.

Mishra, V. K. and B. D. Tripathi. 2009. Accumulation of chromium and zinc from aqueous solutions using water hyacinth (*Eichhornia crassipes*). *Journal of Hazardous Materials* 164(2-3): 1059–63.

Montoya, J. E., T. M. Waliczek and M. L. Abbott. 2013. Large scale composting as a means of managing water hyacinth (*Eichhornia crassipes*). *Invasive Plant Science and Management* 6(2): 243–249.

Njoka, S. W. 2004. The biology and impact of *Neochetina* weevils on water hyacinth, *Eichhornia crassipes* in Lake Victoria Basin, Kenya. Ph.D. Thesis, Department of Environmental Studies (Biological Sciences). School of Graduate Studies at Moi University.

Pasquardini, L., L. Pancheri, C. Potrich, A. Ferri, C. Piemonte, L. Lunelli et al. 2015. SPAD aptasensor for the detection of circulating protein biomarkers. *Biosensors and Bioelectronics* 68: 500–507.

Patel, S. 2012. Threats, management and envisaged utilizations of aquatic weed *Eichhornia crassipes*: an overview. *Reviews in Environmental Science and Bio/Technology* 11(3): 249–259.

Penfound, W. T. and T. T. Earle. 1948. *The biology of the water hyacinth. Ecological Monographs.* New York City and Ann Arbor, Michigan, USA: Ecological Society of America.

Pérez, E. A., J. A. Coetzee, T. R. Téllez and M. P. Hill. 2011. A first report of water hyacinth (*Eichhornia crassipes*) soil seed banks in South Africa. *South African Journal of Botany* 77(3): 795–800.

Pérez, E. A., T. R. Téllez and J. M. S. Guzmán. 2011. Influence of physico-chemical parameters of the aquatic medium on germination of *Eichhornia crassipes* seeds. *Plant Biology* 13(4): 643–648.

Qin, H., Z. Zhang, Z. Zhang, X. Wen, H. Liu and S. Yan. 2015. Analysis of the death causes of water hyacinth planted in large-scale enclosures in Dianchi Lake. *Resources and Environment in the Yangtze Basin* 24(4): 594–602 (In Chinese with English Abstract).

Reddy, K. R. and J. C. Tucker. 1983. Productivity and nutrient uptake of water hyacinth, *Eichhornia crassipes* I. effect of nitrogen source. *Economic Botany* 37(2): 237–247.

Ren, M. X., Q. G. Zhang and D. Y. Zhang. 2004. Geographical variation in the breeding systems of an invasive plant, *Eichhornia crassipes*, within China. *Acta Phytoecologica Sinica* 28(6): 753–760 (In Chinese with English Abstract).

Rodríguez, M., J. Brisson, G. Rueda and M. S. Rodríguez. 2012. Water quality improvement of a reservoir invaded by an exotic macrophyte. *Invasive Plant Science and Management* 5(2): 290–299.

Sheng, J., J. Zheng, L. Chen, P. Zhu and W. Zhou. 2011. Study on planting and harvest conditions of *Eichhornia crassipes* for eutrophic water remediation. *Journal of Plant Resources and Environment* 20(2): 73–78 (In Chinese with English Abstract).
Singh, V. 1962. Vascular anatomy of the flower of some species of the pontederiaceae. *Proceedings of the Indian Academy of Sciences - Section B* 56(6): 339–353.
Sullivan, P. and R. Wood. 2012. Water hyacinth [*Eichhornia crassipes* (Mart.) Solms] seed longevity and the implications for management. In *Eighteenth Australasian Weeds Conference*, ed. V. Eldershaw, 37–40. Melbourne, Australia: Weed Society of Victoria Inc.
Sun, L., Z. Zhu, J. Wang and H. Liu. 2011. Dynamic growth model of water hyacinth using hyperspectral remote sensing. *Ecology and Environmental Sciences* 20(4): 623–628 (In Chinese with English Abstract).
Tang, P., J. Sun, Y. Liu and G. Huang. 1987. Studies on sexual reproduction of *Eichhornia crassipes*. *Acta Agronomica Sinica* 13(1): 53–58 (In Chinese with English Abstract).
Téllez, T. R., E. López, G. Granado, E. Pérez, R. López and J. Guzmán. 2008. The water hyacinth, *Eichhornia crassipes*: an invasive plant in the Guadiana River Basin (Spain). *Aquatic Invasions* 3(1): 42–53.
Tham, H. T. and P. Udén. 2015. Effect of water hyacinth (*Eichhornia crassipes*) on intake and digestibility in cattle fed rice straw and molasses-urea cake. *Nova Journal of Engineering and Applied Sciences* 4(1): 1–8.
Toft, J. D., C. A. Simenstad, J. R. Cordell and L. F. Grimaldo. 2003. The effects of introduced water hyacinth on habitat structure, invertebrate assemblages, and fish diets. *Estuaries* 26(3): 746–758.
Wang, J., L. Sun and H. Liu. 2011. Estimation of the nitrogen concentration of water hyacinth by hyperspectral remote sensing. In *Proceedings of the forth national conference on agricultural and environmental sciences, China*, 1020–1026 (In Chinese with English Abstract). Hohhot City, China: Journal of Agro-Environment Science.
Wang, Z., Z. Zhang, Y. Zhang and S. Yan. 2012. Effects of large-area planting water hyacinth (*Eichhornia crassipes*) on water quality in the bay of Lake Dianchi. *Chinese Journal of Environmental Engineering* 6(11): 3827–3832 (In Chinese with English Abstract).
Wang, Z., Z. Zhang, Y. Zhang, J. Zhang and S. Yan. 2013. Water quality effects of two aquatic macrophytes on eutrophic water from Lake Dianchi Caohai. *China Environmental Science* 33(2): 328–335 (In Chinese with English Abstract).
Wang, Z., P. Zhu, J. Sheng and J. Zheng. 2011a. Biological characteristics of water hyacinth. *Jiangsu Journal of Agricultural Sciences* 27(3): 531–536 (In Chinese with English Abstract).
Wilson, J. R., N. Holst and M. Rees. 2005. Determinants and patterns of population growth in water hyacinth. *Aquatic Botany* 81(1): 51–67.
Wolverton, B. C. and R. C. McDonald. 1978. Nutritional composition of water hyacinths grown on domestic sewage. *Economic Botany* 32(4): 363–370.
Wooten, J. W. and J. D. Dodd. 1976. Growth of water hyacinths in treated sewage effluent. *Economic Botany* 30(1): 29–37.
Wu, Z., C. Qiu, Y. Xia and D. Wang. 1990. Effects of the salinity in petrochemical wastewater on the growth and purification efficiency of water hyacinth. *Acta Hydrobiologica Sinica* 14(3): 239–246 (In Chinese with English Abstract).
Yan, G., N. Ren and Y. Li. 1994. Effects of environmental factors on water hyacinth growth and water purification. *Environment Science and Technology* 64(1): 2–5, 27 (In Chinese).
Yi, N., Y. Gao, X. Long, Z. Zhang, J. Guo, H. Shao et al. 2014. *Eichhornia crassipes* cleans wetlands by enhancing the nitrogen removal and modulating denitrifying bacteria community. *CLEAN - Soil, Air, Water* 42(5): 664–673.
Zhang, Y., F. Wu, Z. Zhang, H. Liu, Y. Wang, Z. Wang et al. 2012. Research on sexual reproduction, seed structure and it vigor of *Eichhornia crassipes*. *Journal of Nanjing Agricultural University* 35(1): 135–138 (In Chinese with English Abstract).
Zhang, Z., Y. Gao, J. Guo and S. Yan. 2014. Practice and reflections of remediation of eutrophicated waters: a case study of haptophyte remediation of the ecology of Dianchi. *Journal of Ecology and Rural Environment* 30(1): 15–21 (In Chinese with English Abstract).

Zhang, Z. Y., J. C. Zheng, H. Q. Liu, Z. Z. Chang, L. G. Chen and S. H. Yan. 2010. Role of *Eichhornia crassipes* uptake in the removal of nitrogen and phosphorus from eutrophic waters. *Chinese Journal of Eco-Agriculture* 18(1): 152–157 (In Chinese with English Abstract).

Zhang, Z., J. Zhang, H. Liu, L. Chen and S. Yan. 2011. Apparent removal contributions of *Eichhornia crassipes* to nitrogen and phosphorous from eutrophic water under different hydraulic loadings. *Jiangsu Journal of Agricultural Sciences* 27(2): 288–294 (In Chinese with English Abstract).

Zheng, J., J. Sheng, Z. Zhang, X. Li, Y. Bai and P. Zhu. 2011. Ecological function of hycinth and its utilization. *Jiangsu Journal of Agricultural Sciences* 27(2): 426–429 (In Chinese with English Abstract).

Zhou, Q., S. Q. Han, S. H. Yan, W. Song and J. P. Huang. 2012. The mutual effect between phytoplankton and water hyacinth planted on a large scale in the eutrophic lake. *Acta Hydrobiologica Sinica* 36(4): 873–791 (In Chinese with English Abstract).

CHAPTER 3

Direct and Strong Influence of Water Hyacinth on Aquatic Communities in Natural Waters

Z. Wang[1] and *S. H. Yan*[2,*]

3.1 Introduction

Freshwater ecosystems consist of communities of different types of aquatic organisms and their ambient environment, namely biotic and abiotic components. The abiotic component mainly refers to the water-air interface, water body and bottom including sediment where many organisms live, and energy flows and nutrient cycle occurs. Various abiotic factors (such as light, inorganic and organic substances) provide a necessary source of energy and materials for the needs of living organisms within an ecosystem. Living organisms comprise primary producers, consumers and decomposers. In fresh waters, primary producers mainly refer to aquatic plants (hydrophytes or macrophytes), phytoplankton and photosynthetic bacteria (PSBs) with chlorophyll in their living cells; decomposers consist of bacteria and fungi that decompose organic substances to derive inorganic ones; all aquatic animals in water are consumers, including zooplankton, benthos, invertebrates, macro-invertebrates, fish and other aquatic animals.

Water hyacinth, as a large and floating vascular plant, grows intensively in suitable conditions (temperature, nutrition and light), may cover a large area of water surface, reduce underwater light intensity and hinder air exchange at the water-air interface, thus negatively affecting water ecosystem (Fontanarrosa et al. 2010). On the other hand, water hyacinth could absorb pollutants from

[1] 340 Xudong dajie Road, Wuhan, China.
Email: Wazh519@hotmail.com
[2] 50 Zhong Ling Street, Nanjing, China.
* Corresponding author: shyan@jaas.ac.cn

water, and its well-developed fibrous root system could provide shelter and habitat for aquatic animals and microorganisms; hence, it may have positive effects on the structure of aquatic animal community if the mat is relatively small (Wang et al. 2012b). In addition to these direct effects, water hyacinth can also have indirect effects on the ecosystem through altering energy flows and nutrient cycles (Villamagna and Murphy 2010). Without full understanding of the effects of water hyacinth on all components within an aquatic ecosystem, it is difficult to predict the direction of the processes in the system because of the complex relationships comprising strong interactions, interdependence and competition (Villamagna and Murphy 2010).

3.2 Effects of water hyacinth on phytoplankton

Phytoplankton, as a convenient rather than scientific classification, includes the small or microscopic autotrophic organisms that drift in water. Phytoplankton refers to most types of algae (including cyanobacteria and diatoms) and to some photosynthetic species of eukaryotic microorganisms. As the most important primary producers of aquatic ecosystems, phytoplankton play a vital role in energy flows, nutrient cycling and biological information transfer in waters. The biological processes, including growth, development and decay of phytoplankton are directly affected by biotic and abiotic factors, such as temperature, light, concentration and availability of nutrients and trace elements, and ambient biological interactions such as predation and competition, which may have important effects on the abundance and biodiversity of phytoplankton communities.

General patterns and interactions of water hyacinth with phytoplankton under natural conditions were well presented by Villamagna and Murphy (2010). In eutrophic waters, the relationship may be very different, especially when blue-green algae are present. The interactions between blue-green algae and water hyacinth can greatly affect the decision-making processes on management of aquatic ecosystems. Although there was little direct evidence on how blue-green algae move from windward to lee, it was common sense that blue-green algae can accumulate on lee sites (Li et al. 2012). This implies that the nutrients assimilated by the blue-green algae at other locations in the water could be transported by wind and water current to the lee sites, where the floating macrophytes would be naturally located due to the same forces. Zhou et al. (2014a) reported the influence of water hyacinth on blue-green algae with respect to the algal growth, physiological characteristics and production as well as the nutrient release from algal cells. The report indicated that cell death of algae was quickly provoked by the stress from a two-day water hyacinth shading together with physical and chemical environmental changes that caused direct damage to phycocyanin and a change of phycocyanin/allophycocyanin ratio inside the algae, although the photosystem II-Hill reaction in algae was not affected significantly. In other literature reports, the algal population was 1.7 to 30 times higher in the root zone of a water hyacinth mat compared with an open area (Zhou et al. 2012); nutrient release from decaying algal cells was also evident (Zhou et al. 2014a), leading to an increase in concentrations of dissolved nitrogen and phosphorus below and within the root zone of the water hyacinth mat. Based on dry microalgae yields of 15–25 t ha^{-1} yr^{-1} (Lam and Lee 2012) with an average nitrogen content of 10.4% on dry biomass basis (Han et al. 2009), the amount of nitrogen and phosphorus being transported from other locations to lee sites may be huge; however, the methods for accurate quantification are missing.

The dominant mechanisms underlying the effects of water hyacinth on phytoplankton are: (1) water hyacinth grows to cover the water surface, blocking or at least reducing incident light in the water column and hampering photosynthesis of phytoplankton, thus limiting their growth and reproduction; (2) water hyacinth can effectively absorb nutrients from water and create unfavorable nutritional conditions for phytoplankton, thus inhibiting their growth and reproduction; (3) water hyacinth can produce bioactive compounds to inhibit or promote growth of algae; (4) the well-developed root system of water hyacinth can serve as adsorbent of phytoplankton.

Water hyacinth floats and grows on water surface and absorbs dissolved nitrogen and phosphorus from water, which leaves less nutrients available for phytoplankton to perform adequate photosynthesis in the process of growth and development, thus leading to depletion in numbers or even dying out of phytoplankton communities (McVea and Boyd 1975). In addition, the fibrous root system of water hyacinth could produce bioactive chemicals to inhibit or promote the growth of plants. From the early 1960s to the early 1980s, there were many reports on growth-regulating substances extracted from water hyacinth roots having growth-promoting effects on rice (*Oryza* spp.), chickpea (*Cicer arietinum*) (Sircar and Ray 1961, Sircar and Chakraverty 1962), fungi and yeasts (Sheikh et al. 1964), whereas the effects on phytoplankton were relatively unknown. Later, the authors confirmed that the substances were the gibberellins-like growth-promoting substances released by water hyacinth roots, enhancing germination and growth of rice and wheat (*Triticum* spp.) as well as other plants, such as eggplant (*Solanum melongena*), tomato (*Lycopersicum esculentum*) and blackgram (*Phaseolus mungo*) (Gopal and Goel 1993). A hypothesis is that the gibberellin-like growth-promoting substances may promote growth and enhance adaptive morphological changes of water hyacinth, which in turn would affect phytoplankton. Some of the biological compounds extracted from water hyacinth are listed in Table 3.2-1.

Sun et al. (1989) collected mixtures of algal species (mainly *Scenedesmus*) from natural waters near the city of Shanghai, China, and conducted an experiment with three treatments: 1/2 surface artificial shading without water hyacinth and without additional nutrients, 1/2 surface water hyacinth shading without additional nutrients, and 1/2 surface water hyacinth shading with additional nutrients. Hence, the experiment accounted for the effects of shading and nutrients competition by water hyacinth, but still found the significant limitation on algal growth by water hyacinth. The authors

Table 3.2-1. Bio-active substances extracted from water hyacinth (*Eichhornia crassipes*).

Allelochemicals	Molecular formula	Ref.
Gibberellins or Gibberellic Acid	$C_9H_{22}O_6$	(Sircar and Ray 1961, Sircar and Chakraverty 1962)
4α-methyl-5α-ergosta-8,14,24(28)-triene-3β,4β-diol	$C_{29}H_{46}O_2$	(Greca et al. 1991b)
4α-methyl-5α-ergosta-8,24(28)-diene-3β,4β-diol	$C_{29}H_{48}O_2$	
4α-methyl-5α-ergosta-7,24(28)-diene-3β,4β-diol	$C_{29}H_{46}O_2$	
N-phenyl-2-naphthylamine	$C_{16}H_{13}N$	(Yang et al. 1992)
Linoleic acid (omega-6 fatty acid)	$C_{18}H_{32}O_2$	
1,3-dihydroxy-2-propanyl (9Z,12Z)-9,12-octadecadienoate	$C_{21}H_{38}O_4$	
β-D dehydrated pyranose	not determined	(Jin et al. 2003)
Isocyanoethyl acetate	$C_3H_8O_2N$	
2-2-dimethylcyclopentanone	$C_7H_{12}O$	
Propanamide	C_3H_7NO	
Pelargonic acid	$C_9H_{18}O_2$	
18,19-secoyohimban-19-oic acid, 16,17,20,21-teradehydro-16-(hydroxymethyl) methyl ester	$C_{21}H_{24}N_2O_3$	(Shanab et al. 2010)
1,2-benzenedicarboxylic acid, mono(2-ethylhexyl) ester	$C_{16}H_{22}O_4$	
1,2-benzenedicarboxylic acid, diisooctyl ester	$C_{24}H_{38}O_4$	
1,2-benzenedicarboxylic acid, dioctyl ester	$C_{24}H_{38}O_4$	
Diamino-dinitro-methyl dioctyl phthalate	$C_{33}H_{50}N_4O_{10}$	
9-(2,2-dimethyl propanoilhydrazono)-2,7-bis-[2-(diethylamino)-ethoxy] fluorine	not determined	(Aboul-Enein et al. 2014)
(3-methylphenyl)-phenyl methanol		
4-(diethylamino)-alpha-[4-(diethylamino) phenyl]		
Isooctyl phthalate		

hypothesized that water hyacinth played a prominent role in inhibiting algal growth and suggested allelopathy to algae. Further experiments excluded the above two modes of competition (light and mineral nutrition) between water hyacinth and algae by preparing the solution for algal cultivation via: (1) filtering water through a 0.45-μm membrane filter from the containers in which water hyacinth was cultured to ensure almost no algae were carried over; (2) adding a moderate amount of nutrients to the filtered water to make sure there was enough nutrients to support the growth of their naturally collected algae. After a series of experiments, the authors concluded that algal growth was inhibited significantly, which indicated that the root system of water hyacinth could produce active allelochemicals (Sun et al. 1988, 1989). More evidence was reported by other researchers, with attention having been paid to isolation and identification of inhibiting agents from the water hyacinth roots. Greca, Lanzetta et al. (1991) extracted allelochemicals from the root system of water hyacinth with ethyl acetate and identified a set of 4-methylated sterols substances.

Yang et al. (1992) isolated and purified from water hyacinth roots three compounds [N-phenyl-2-naphthylamine, linoleic acid (omega-6 fatty acid) and glyceryl linoleate] with strong antialgal activity. In separate experiments, the authors also isolated N-phenyl-2-naphthylamine from the solution in which water hyacinth was cultured. Jin et al. (2003) investigated the acetone extract from water hyacinth roots and identified β-D dehydrated pyranose, isocyanoethyl acetate, 2,2-dimethylcyclopentanone, propanamide and pelargonic acid.

At present, there are two aspects of the study of algal inhibition by allelochemicals from water hyacinth. One is the study of algal inhibition via either water culture solution of water hyacinth or extracts from plants (mainly the root system). For instance, Hu et al. (2010) applied the water in which water hyacinth was cultured in comparison with methanol and acetone extracts from different parts of living or dry water hyacinth to research allelopathic inhibitory effects on *Microcystis aeruginosa*. They found that water from water hyacinth culture as well as extracts from different parts had certain allelopathic inhibitory effects on *Microcystis aeruginosa*, lowering the production of *Microcystis aeruginosa* in the stable growth phase.

The other aspect is to apply the isolated and identified main allelochemicals from water hyacinth, including N-phenyl-2-naphthylamine, linoleic acid, glyceryl linoleate and pelargonic acid, to research their inhibition effects and mechanisms on pure culture of algae. For instance, Liu et al. (2006) compared and analyzed the algal inhibition effects of different substances in acetone extract from roots of water hyacinth. The results indicated that N-phenyl-2-naphthylamine at concentration higher than 5 mg L^{-1} maintained more than 50% inhibition of *Alexandrium tamarense* after 3 days; the inhibition by linoleic acid at 70 ppm was about 40% and that of pelargonic acid at the same concentration reached 85% on the 3rd day, but the algal density rebounded subsequently. A smaller inhibitory effect on *Alexandrium tamarense* was found with glyceryl linoleate and propionamide in the concentration range 10–70 ppm. The study by Geng et al. (2009) showed that N-phenyl-2-naphthylamine had significant inhibitory effects on *Microcystis aeruginosa* with EC_{50} = 5 mg L^{-1} after 7 days. The content of chlorophyll-a decreased with increasing N-phenyl-2-naphthylamine concentration in culture solution; for example, the content of chlorophyll-a at 10 mg L^{-1} N-phenyl-2-naphthylamine decreased 67.4 and 75.9% after 8 and 24 hours, respectively, in comparison with the control group. These findings indicate that water hyacinth has allelopathic properties to inhibit growth of the tested algae.

The concentration of allelochemicals in natural waters may be far lower than the EC_{50} concentrations; similarly, the concentration of allelochemicals secreted by large aquatic plants appears unlikely to reach the concentrations used to produce algal inhibition in indoor toxicity experiments (Lu et al. 2013). However, in natural waters as well as laboratory simulations, the phenomenon of allelopathic algal inhibition by water hyacinth was reported widely under conditions that excluded confounding factors of light and nutrition competition (Sun et al. 1988, 1989, Zhou et al. 2014). Hence, we can conclude:

1) water hyacinth roots could secrete many kinds of compounds that act together to create synergistic effects on algae. For instance, Zhang (2012) used *Microcystis aeruginosa* as experimental material to prove the synergistic effects of two allelochemicals, namely linoleic acid and pelargonic acid, on algal inhibition;
2) some substances with potentially strong algal inhibition effects may not have been discovered yet due to inappropriate extraction and/or identification methods or poor stability;
3) the well-developed root system of water hyacinth provides a large surface area. Zhou et al. (2012) found that water hyacinth growing in Zhushan Bay of Lake Taihu (31°27' N 120°4' E) had an average root surface area of 30 m^2 per plant, with the largest root surface area being twice as big.

The well-developed root system of water hyacinth could adsorb suspended algae and other particulates in water (Zhou et al. 2012). It could also trap suspended alga in the rhizosphere and inhibit algal growth with high concentration of exuded allelochemicals. However, this specific mechanism is yet to be tested rigorously.

In natural waters, water hyacinth, on one hand, inhibits growth or decreases the biomass of phytoplankton and changes community structure through the light and nutrition competition and allelopathic effects; on the other hand, the trap and adsorption functions of its root system could increase the biomass of phytoplankton in surrounding waters. However, the effects of water hyacinth on phytoplankton are dependent on the properties of the aquatic environment. For instance, in a shallow lake in Portugal, water hyacinth selectively decreased the quantity and growth of green algae in natural conditions (Almeida et al. 2006). When water hyacinth was removed from the natural water, Bicudo et al. (2007) found that the total amount of phytoplankton increased significantly in a shallow lake in Brazil. Moreover, Lugo et al. (1998) and Mangas-Ramírez and Elías-Gutiérrez (2004) found a similar phenomenon in two water reservoirs in Mexico in separate investigations. In contrast, Cai (2006) indicated that the biomass propagation of floating plants (*Eichhornia crassipes* and *Pistia stratiotes*) led to a decrease in quantity and biodiversity of phytoplankton in Fujian Minjiang Shuikou Reservoir, with eutrophic-tolerant algae becoming dominant. However, Brendonck et al. (2003) found that root system of water hyacinth could trap phytoplankton and detritus in the water body of the Lake Chivero in Uganda, resulting in the density of phytoplankton in the area covered by water hyacinth being 10–30 times of that in water body without water hyacinth.

Some phytoplankton species may suppress the growth of water hyacinth. Sharma reported in 1985 that the growth of water hyacinth was inhibited when grown in water containing algae species such as *Scenedesmus bijugatus, Chlorella pyrenoidosa, Aphanothece* spp., *Euglena* spp., *Merismopedia* spp. and *Coelastrum* spp., although the test of this hypothesis was not perfect due to failed monitoring and maintenance of proper phosphorus concentration (only traces were recorded) in algae tanks. The size, dry weight, chlorophyll-a and chlorophyll-b content and the rate of vegetative propagation of the macrophyte were effectively reduced compared with the control plants grown in algae-free waters. The hyacinth plants ultimately died after 90–100 days of growth with algae. The adverse effect on growth of water hyacinth was considered to be due to allelochemicals produced by algae, although no explanation of a putative mechanism was reported (Sharma 1985).

3.3 Effects of water hyacinth on aquatic vascular plants

The effect of water hyacinth on aquatic vascular plants is similar to that on phytoplankton. On one hand, water hyacinth floating and growth affect incident light, reducing availability of underwater light energy for photosynthesis of submerged macrophytes. On the other hand, the reduced concentration of nutrients due to growth of water hyacinth could intensify nutrition competition with other aquatic plants. Moreover, the allelochemicals secreted by different macrophytes have interactive effects on growth and development of water hyacinth and other aquatic plants.

Zhao et al. (2006) compared the growth characteristics of water hyacinth, floating primrose-willow (*Ludwigia peploides*) and hydrilla (*Hydrilla verticillata*) cultured in artificial eutrophic water with three nutrition levels. The high nutrient conditions and profligate growth of water hyacinth led to advantages in competing with the other two aquatic species, significantly inhibiting growth of *Ludwigia peploides* and *Hydrilla verticillata*. Water hyacinth quickly covered a big proportion of water surface, leaving only limited space for growth of *Ludwigia peploides* and decreasing its biomass by 57 and 73% in, respectively, nutrient Level II treatment (1 mg N L^{-1}, 0.2 mg P L^{-1}) and nutrient Level III treatment (2 mg N L^{-1}, 0.4 mg P L^{-1}). Similarly, photosynthesis of *Hydrilla verticillata*, the submerged plant, was hampered by a lack of light due to water hyacinth ($p < 0.001$), the biomass of *Hydrilla verticillata* decreased between 8.5 and 81% depending on the nutrient treatments (Zhao et al. 2006). The same study also showed that water hyacinth had a stronger inhibiting effect on the other two aquatic species with an increase in the nutrient level in water, mainly because enhanced nutrition improved growth of water hyacinth more than growth of the other two aquatic species.

Wu et al. (2011) investigated the effects of water hyacinth on submerged plants *in situ*. They selected five enclosures with water hyacinth inside and three nearby open water areas without water hyacinth as control for sampling plots in Lake Dianchi (24°45' N 102°36' E), China. Samples of phytoplankton and submerged plants (*Potamogeton pectinatus*) were collected every 2 months from April to December in 2010. The content of chlorophyll in *Potamogeton pectinatus* sampled from the water hyacinth growing area was lower than that in the control area; the authors suggested that water hyacinth had certain inhibitory effects on the growth of submerged plants, although no mechanism was described.

Existing evidence suggests that water hyacinth had negative effects on the growth of aquatic (especially submerged) plants due to shading and/or nutrient competition, but literature mainly discusses the short-term effects of water hyacinth on submerged plants at a small scale in natural habitat. Hence, it is not well understood whether long-term effects of water hyacinth on submerged vegetation would cause vanishing of submerged plants and whether aquatic plants could resume growth after the harvest and removal of water hyacinth. However, in Lake Dianchi, we found that after the harvest of water hyacinth, submerged plants, such as *Potamogeton pectinatus* and *Hydrilla verticillata*, could still grow into a large habitat area; moreover, on the periphery of water hyacinth mats, due to an increase in transparency of water body, submerged plants grew and established healthy communities at places previously without submerged plants. Unfortunately, we did not conduct a detailed study on these phenomena.

3.4 Effects of water hyacinth on zooplankton

Zooplankton is a term referring to heterotrophic plankton drifting in water, usually microscopic but some large and visible by the naked eye. As consumers, zooplanktons are an important biological component in aquatic ecosystems, playing a vital role in ecological processes such as energy flows, nutrient cycles and biological information transfer. Zooplanktons have high sensitivity to the changing environment, including temperature, incident light, pH, algae and dissolved oxygen. Their abundance and community structure are usually used as indices of aquatic ecosystem health (Chen et al. 2012). A general pattern of natural interactions among water hyacinth and aquatic animals such as zooplankton, macroinvertebrates and fish is as follows: water hyacinth can alter physical and chemical conditions in aquatic environments, can provide shelter, energy and nutrients to aquatic animals, can alter food chain dependencies and abundance of prey and predators and can have the edge effects of fragmented mats (Villamagna and Murphy 2010). Due to the complexity of the above-mentioned interactions and relationships, there are many inconsistent reports because of limited and small-scale investigations in limited sets of biotic and abiotic conditions.

The distribution of zooplankton is affected by many factors, including water transparency, light, temperature, Dissolved Oxygen (DO), phytoplankton and food resources (Villamagna and Murphy 2010). The growth of water hyacinth could change the above mentioned factors to alter the abundance and diversity of zooplankton communities; the well-developed fibrous root system of water hyacinth could provide a suitable ecological niche for zooplankton, especially with the mat's "edge effect". For example, on the periphery of a water hyacinth mat, abiotic conditions provide a good survival space and suitable physical, chemical and biological conditions for aquatic animals. Due to water quality improvement in terms of dissolved oxygen (5.5–7.9 mg L^{-1}) and transparency (0.5–2.5 meters) (Zhang et al. 2014), and stable water quality in terms of pH and COD at the edges of water hyacinth mat (Wang et al. 2012a), the organisms such as dominant species of macro-zoo-benthos and cladocerans as well as copepods were maintained or were increased in population in response to the altered environment compared to the open water area (Chen et al. 2012). Furthermore, the population of rotifers increased in the gaps between patches (Wang et al. 2012b). Rotifers fall prey to copepods as well as fries of many fish species, which can lead to the potential habitat- and trophic-dependent changes among rotifers, copepods and fish species.

In the backwater area in the Delhi section of Yamuna River in India, Arora and Mehra (2003) found that the biomass and diversity of rotifers, especially epiphytic rotifers responding to the presence of water hyacinth, were significantly higher than those in the area of giant salvinia (*Salvinia molesta*). For instance, *Beauchampia crucigera dutrochet*, *Sinantherina* sp. and *Collotheca* sp. only appeared in the area of water hyacinth root system. The further analysis suggested that the main reason was that the strong root system of water hyacinth could provide more food sources as well as shelter for rotifers. However, in a shallow hypertrophic lake (Lake Rodó, 34°55' S 56°10' W) in Uruguay, Meerhoff et al. (2003) found that the density of copepods and rotifers was significantly lower in the area of water hyacinth than open water area or areas with submerged pondweed (*Potamogeton pectinatus*). In contrast, although *Alona* sp. only appeared in the water hyacinth area, no significant differences were found in abundance and diversity of cladocerans among areas of water hyacinth, open water and submerged pondweed.

In Lake Chivero (17°54' S 30°48' E) in Zimbabwe, Brendonck et al. (2003) found that the species, density and diversity indices of micro-crustacean were all lower in the area with than without water hyacinth. However, different results were reported by Wang et al. (2012). Through sampling and analysis of zooplankton inside and outside of 70-hectare water hyacinth area in Lake Dianchi, Wang et al. (2012) found that the short-term effects (5 months) of water hyacinth were relatively small on the density and diversity indices of cladocerans and copepods, but the biomass and diversity of rotifers were lowered significantly; similar phenomena had been indicated in the study in Zhushan Bay of Lake Taihu (Chen et al. 2012). These results indicated there were differences in the effects of water hyacinth on the community structure of zooplankton, which may be related to the extent of coverage by water hyacinth in natural waters. With a relatively small coverage of water hyacinth, its fibrous root system played a leading role in trapping, sheltering, protecting and providing food for organisms; however, with an extensive water hyacinth mat, its negative effects on incident light and dissolved oxygen in a water body may dominate, especially directly underneath the mat.

The hypothesis that different size of water hyacinth coverage has various effects on zooplankton was tested through a series of simulation tests that showed significant different patterns over time after the introduction of water hyacinth. In the initial phase of the introduction, a small size of water hyacinth mat (e.g., about 15% of total water surface) had positive effects on the zooplankton communities and the biodiversity indices compared to the control group. After water hyacinth grew for a longer time (coverage reaching 90% of the total water surface area), it had adverse effects on zooplankton regarding both abundance and diversity (unpubl. data). Hence, the size of water hyacinth coverage played a key role in the dynamics of zooplankton communities.

Cai (2006) focused on Shuikou Reservoir (26°25' N 118°06' E) at the trunk stream of Minjiang River in central Fujian Province, China, and investigated the community structure of zooplankton before the invasion and after the massive growth of water hyacinth (reaching 100% of the total reservoir surface). After the invasion of the floating macrophyte, (i) the proportion of protozoa (e.g., ciliates) that feed on bacteria and organic detritus increased; (ii) the proportion of cladocerans and copepods declined sharply; (iii) rotifer species *Anuraeopsis fissa* and *Keratella cochlearis* that feed on organic detritus and bacteria became dominant; (iv) rotifer species *Epiphanes brachionus* feeding on large algae, and *Asplanchna* sp. that feed on protozoa or other rotifers, vanished; and (v) total density of zooplankton decreased significantly and population structure changed, with protozoa accounting for a large proportion and large-size zooplankton decreasing significantly.

The above research showed that water hyacinth may cause variable effects on the zooplankton community structure due to different environmental conditions and different coverage of the total water surface. Moreover, zooplankton showed horizontal and vertical migration movements in the water column, with such migration being more significant in the temperate than subtropical zones. In the temperate zone, zooplankton easily moved to avoid adverse conditions (for instance, low DO and lack of food) inside and outside the water hyacinth area (Meerhoff et al. 2007, Villamagna and Murphy 2010).

3.5 Effects of water hyacinth on macroinvertebrates

Water hyacinth could provide a good habitat for macroinvertebrates, especially epiphytic ones. For example, 22 taxa of invertebrates were associated with the growth of water hyacinth in the littoral zone, including Annelida, Turbellaria, Gastropoda and Arachnida (Brendonck et al. 2003). In Alvarado Lagoon system

(18°43'–18°52' N 95°42'–95°57' W) in Mexico, the abundance and diversity of macroinvertebrates increased in the area of water hyacinth root system (Rocha-Ramirez et al. 2007); 96 macroinvertebrate taxa were identified and species with increased abundance included acari (15 taxa), decapods (14 taxa), mollusks (12 taxa), amphipods (nine taxa) and isopods (7 taxa). In another example, Marco et al. (2001) analyzed functional feeding groups of macroinvertebrates in the Pampulha Reservoir (19°55' S 43°56' W) in Brazil and found that invertebrates obviously associated with water hyacinth were dominated by detritivores, mainly Oligochaeta, Turbellaria and Gastropoda. The main reason was that the root system of water hyacinth provided a large amount of organic detritus for detritivores to use as a food source (Marco et al. 2001).

O'Hara (1967) investigated the macroinvertebrates within the water hyacinth roots in Lake Okeechobee (26°43'–27°11' N 80°58'–80°37' W) in Florida and found that macroinvertebrates in roots of water hyacinth were the typical benthonic species; the abundance of macroinvertebrates were significantly higher in the water hyacinth roots compared with other plant-root systems as well as benthic samples. Research in Lake Victoria (0°22'–0°30' N 33°10'–33°26' E) in Uganda, showed that macroinvertebrate had higher abundance and diversity at the edge of water hyacinth mats, compared with open water or areas of emergent plants (*Cyperus papyrus*) (Masifwa et al. 2001). Similarly, research in Lake Chapala (20°15' N 103°00' W) in Mexico suggested that roots of water hyacinth had greater species diversity than those of emergent plants (Villamagna 2009).

Abundance and diversity of invertebrates may also be restricted by factors such as water physicochemical properties. For instance, the density of invertebrates in the water hyacinth roots was positively correlated with dissolved oxygen (DO) in water (Marco et al. 2001). However, invertebrate abundance and diversity was not only affected by the water hyacinth roots, but also by temperature, dissolved oxygen, turbidity and salinity in the water column (Rocha-Ramirez et al. 2007).

Dissolved oxygen may be the main physicochemical factor that affects the distribution and abundance of macroinvertebrates. Water hyacinth roots had a certain capacity of radial oxygen loss (Laskov et al. 2006, Ma et al. 2014), but such capacity was not enough to make up for the reduced dissolved oxygen due to the shading function of water hyacinth. In natural waters and also in simulation experiments, the large mats of water hyacinth usually lowered dissolved oxygen in water significantly; the concentration of dissolved oxygen gradually decreased from the edge of water hyacinth mats to the inside of the mats (Bailey and Litterick 1993, Villamagna and Murphy 2010). In general, macroinvertebrate densities tend to be higher near the edge of water hyacinth mats compared to the center. For instance, in Lake Wutchung (6°50' N 31°25' E) in the River Atem, the maximum density of macroinvertebrates occurred in the area within 6 m from the edge of the water hyacinth mats (Bailey and Litterick 1993).

Benthonic macroinvertebrates are important components of diversity of aquatic ecosystems, playing an irreplaceable role in nutrient cycling and energy metabolism. They have a relatively long life cycle and limited moving and living space, making them easy to collect. The benthonic macroinvertebrates show large differences in sensitivity to different habitats among different species; hence, macroinvertebrates are often used as indicator organisms in water quality assessment and environmental monitoring (Morse et al. 2007). So, how does the existence of water hyacinth affect benthic macroinvertebrates?

Midgley et al. (2006) studied the effects of water hyacinth on the community structure of benthic macroinvertebrates in New Tear's Dam (33°17' S 26°06' E) in South Africa. They put uniform fine stones into mesh bags, suspending them about 1.5 m deep into water and collecting the benthic macroinvertebrates every 6 weeks, to study two storage reservoirs of the same river in South Africa (one with water hyacinth and the other without as a control); the long-term existence of water hyacinth significantly lowered the richness, density and diversity indices of benthic macroinvertebrates. Using a similar collection method inside and outside of water hyacinth area in Lake Nsezi (Nseleni River 28°43' S 31°58' E) in South Africa, Coetzee et al. (2014), found the invasion of water hyacinth had long-term effects on benthic macroinvertebrates. These findings showed that water hyacinth changed the community structure of benthic macroinvertebrates, with their richness, density and diversity being significantly lower inside than outside of water hyacinth mats. However, different results were reported in an investigation of Lake Dianchi, an ultra-eutrophicated lake in China. In that study, benthic macroinvertebrates were directly collected using Peterson dredge inside and outside of water hyacinth areas and also far away from water hyacinth (Wang et al. 2012b). *Radix swinhoei*, crabs, Caridina and Gammarids only appeared in the area with water hyacinth; also, the functional feeding groups of benthic macroinvertebrates showed higher biodiversity and were more complex in the area with water hyacinth than in the area near water hyacinth and the area far away (Wang et al. 2012b). In Lake Taihu in China, the biomass and density of mollusks in sediment were higher beneath the center than on the periphery of the water hyacinth area; the diversity index was higher in the area with water hyacinth than in the area without and the area far away from water hyacinth (Liu et al. 2010, Liu et al. 2014).

The inconsistent effects of water hyacinth on benthic macroinvertebrates may be caused by various reasons. Firstly, different sampling methods: in the reports on New Year's Dam and Lake Nsezi, researchers collected benthic macroinvertebrates on artificial substrates suspended under water surface, whereas in Lake Dianchi and Lake Taihu, researchers collected sediment *in situ* to identify benthic macroinvertebrates. Secondly, different cover area of water hyacinth: in New Year's Dam and Lake Nsezi, water hyacinth grew to cover almost all area, whereas in Lake Dianchi and Lake Taihu, its cover area accounted for only 0.08%~0.25% of the water surface and the water hyacinth was almost completely harvested, i.e., covering time was short.

Thirdly, different historical community backgrounds: in New Year's Dam and Lake Nsezi, the research areas included a reservoir formed by damming rivers and a river natural conservation area, with rich community of benthic macroinvertebrates including complete functional groups; in contrast, Lake Dianchi and Lake Taihu are two eutrophic lakes with heavy pollution, mainly containing pollution-tolerant species and therefore incomplete functional groups.

3.6 Effects of water hyacinth on fish and other aquatic animals

Fish provide an important protein source for humans. Also, fish play a key role in maintaining the balance and stability of aquatic ecosystems and have important functions at all trophic levels of consumers. The community structure of fish is mainly affected by the availability of food, pressure of predation and physicochemical conditions in water. With the suitable water quality, sufficient food and no threat of predation, the fish populations increase; however, a certain magnitude of increase could lead to a shortage of food and deterioration of water quality, thus inhibiting a further increase and possibly leading to a decrease until a new balance in fish population.

The relationship between water hyacinth and the dynamics of economically important aquatic species such as turtles and eels was reported from research in aquaculture. Under uncontrolled conditions, water hyacinth negatively influenced fish production (Lei et al. 2008, Wang 2011) by either reducing fish population or blocking the way to fishing grounds. However, under well-managed conditions, water hyacinth could improve fish production. Literature reported water hyacinth improved production of soft-shell turtle (*Trionyx sinensis*) by improving water quality, reducing the infectious agents, providing shelter to juveniles, and modulating surface water temperature in hot seasons (Pei 2001, Fu 2011). In aquaculture, the production of species such as carp (*Ctenopharyngodon idella*) (Xie 2000) and yellowhead catfish (*Pseudobagrus fulvidraco*) (Zhang et al. 2012) was increased by growing water hyacinth at a coverage of less than 5% of the total pond surface area. The production of Asian swamp eel (*Monopterus albus*) and snakehead (*Channa argus*) was enhanced by growing water hyacinth up to 20% coverage of the total surface area (Xu et al. 2003), although there were suggestions that 10 to 25% coverage of water hyacinth may reduce fish (*Tilapia aurea*) production by almost 50% (McVea and Boyd 1975). This inconsistency may be due to the differences in fish species, feeding practice and/or pond management. On the non fish species, juveniles of soft-shell turtle and Asian swamp eel are typical shelter seekers for their biological development and feeding, and are obligate air-breathers, having different requirements for oxygen. One general characteristic of all reported cases was that water hyacinth greatly improved water quality, especially decreasing the concentration of dissolved ammonia that was toxic to young (juvenile) aquatic animals above 3.2 mg NH_3^+ L^{-1} (Zhao et al. 1997).

After invading a water body, water hyacinth could affect various aspects of the ecosystem food-web structure, such as changing the community structure and reducing abundance of phytoplankton, zooplankton and macroinvertebrates, which may affect the population structure and abundance of fish through a "bottom-up effect". However, as mentioned above, the effects of water hyacinth on phytoplankton, zooplankton and macroinvertebrates are restrained by various conditions (such as the cover area and density of water hyacinth and the community background of a water body); therefore, it is difficult to generalize the effects of water hyacinth on fish.

The existence of water hyacinth could change the fish diets via changing the quantity and availability of prey. For instance, in Sacramento-San Joaquin Delta (37°59'–38°4' N 121°31'–121°52' W) in California the macroinvertebrate populations differed between the area covered by water hyacinth and that covered by native pennywort (*Hydrocotyle umbellata*); the modified invertebrate assemblage structure due to invasion of water hyacinth changed the fish diet (Toft et al. 2003). A similar phenomenon was found in Lake Victoria in Uruguay, with the dietary structure of Nile perch changing (Njiru et al. 2004). The change of dietary structure of fish might have caused a change in the whole food-net/web structure in the water body (Villamagna and Murphy 2010), thus affecting the whole aquatic ecosystem.

The submerged macrophytes and roots of floating macrophytes could provide habitats and shelters for small fish and juveniles, potentially increasing the diversity of fish in an aquatic ecosystem (Johnson and Stein 1979). Furthermore, abundant epiphytic invertebrates associated with water hyacinth could provide rich food sources for fish, especially omnivorous fish. For instance, the diversity of fish at littoral sites was higher with than without water hyacinth in Lake Chivero in Zimbabwe (Brendonck et al. 2003).

Compared with submerged and emergent plants, water hyacinth could provide a more complex habitat in a surface layer of water if the mat of water hyacinth is not continuous, and the resulting changes could affect fish population and density (Meerhoff et al. 2007, Villamagna and Murphy 2010). However, the maximum abundance and density of fish species appeared in habitats with moderate complexity of vegetation (Miranda and DeVries 1996, Grenouillet et al. 2002). Similarly, fish abundance in Lake Rodo in Uruguay was greater in the area of submerged plants than water hyacinth followed by the area without vegetation (Meerhoff et al. 2003). Moreover, in Atchafalaya River Basin (29°43'–29°54' N 91°10'–91°22' W) fish composition was different in the area with water hyacinth and the area with *Hydrilla verticillata*, although the total fish biomass did not differ significantly among the habitats featuring different types of aquatic plants (submerged - *Hydrilla verticillata*, emergent - *Sagittaria lancifolia* and floating—water hyacinth) (Troutman et al. 2007). By comparison, the area with submerged species had a higher diversity and density of fish. The authors considered this phenomenon was due to the better water quality (especially the level of dissolved oxygen) and moderate habitat complexity in the *Hydrilla verticillata* area.

Concentration of dissolved oxygen (DO) in water is an important limiting factor affecting the growth and survival of aquatic animals, including fish. Water hyacinth can alter aquatic dissolved oxygen and reduce primary production of phytoplankton and temperature, which may impact niches for fish populations. The existence of large and continuous mats of water hyacinth could significantly reduce dissolved oxygen in the center beneath the mat. Decreasing of dissolved oxygen could have direct effects on fish growth and even survival. However, when mats have gaps and appear in patches, the concentration of dissolved oxygen may influence distribution of fish population spatially, especially in the gaps where dissolved oxygen concentration would increase due to improved transparency, expanding phytoplankton population and enhanced oxygen exchange across the water-air interface. According to the water quality standard of United States Environment Protection Agency (US EPA), dissolved oxygen concentration in a water body below 4.8 mg L^{-1} may have adverse effects on the growth of fish, and at less than 2.3 mg L^{-1} may directly threaten the survival of fish (Chapman 1986), with most fish species preferring the concentration of dissolved oxygen above 4.8 mg L^{-1}.

The effect of water hyacinth on dissolved oxygen in water column depends on the percent coverage and the water exchange characteristics. For instance, in a simulation test, dissolved oxygen levels could fall as low as 2.3 mg L^{-1} during the growth of water hyacinth at full coverage (Wang et al. 2013a). In the open bay of Lake Dianchi, with frequent water exchange, the area covered by water hyacinth accounted for only 0.08% of the total water area, and the average dissolved oxygen concentration under the mat of water hyacinth was 5.3 mg L^{-1} (Wang et al. 2012); even with water hyacinth coverage reaching 50% of the total surface water area, dissolved oxygen concentration in the water body still remained above 4.9 mg L^{-1} (Wang et al. 2013b). In fact, the fish sensitive to dissolved oxygen may evade the area with low dissolved oxygen concentration, while the fish not sensitive to dissolved oxygen may benefit from rich foods and habitats provided by water hyacinth (Villamagna and Murphy 2010). Although there is a lack of data to interpret direct relationships between fish and water hyacinth, the presence of water hyacinth mat can add habitat complexity and heterogeneity of fish assemblages (Padial et al. 2009). Decreased dissolved oxygen may promote a shift of fish species, from those with higher oxygen demand to those that tolerate lower oxygen such as channel catfish (*Ictalurus punctatus*) or snakehead (*Channa argus*). There is also little data to support a hypothesis that herbivorous fish such as grass carp (*Ctenopharyngodon idella*) would feed on fresh roots of water hyacinth. Indeed, grass carp avoid feeding on water hyacinth (Villamagna and Murphy 2010). However, grass carp could effectively reduce the spread of water hyacinth (Gopalakrishnan et al. 2011), but a cause of reduction, i.e., feeding on water hyacinth, or by other factors, was not reported.

What size coverage of water hyacinth on water could have positive effects on fish? Or negative effects? Previous literature showed that the abundance and growth of fish had a convex relationship with the area coverage by aquatic plants, namely that a medium coverage (less than 10% of the total water surface area) was suitable for the survival of fish (Miranda and DeVries 1996, Brown and Maceina 2002, Villamagna and Murphy 2010). However, the above references researched an emergent and a submerged plant species. Focusing on the effects of water hyacinth (floating species) on fish, the data from 12 test ponds (each 0.04 ha) at Auburn Hills University showed that 5% water hyacinth coverage had no obvious effects on tilapia production, but more than 10% coverage reduced it, even though DO remained high, with the reduced fish production mainly resulting from a decrease in phytoplankton abundance due to increased water hyacinth coverage (McVea and Boyd 1975).

3.7 Effects of water hyacinth on bacteria and fungi

In natural ecosystems, bacteria and fungi are the main decomposers and drivers of nutrient cycle, representing the end point of energy flow. The complex root system of water hyacinth can release oxygen into the rhizosphere environment to compensate for oxygen loss caused by blocked air exchange and impaired photosynthesis of phytoplankton. Water hyacinth also can release organic molecules and other allelochemicals. Its complex root structure can provide a huge surface area for biofilm development. Therefore, water hyacinth could significantly affect communities of bacteria and fungi in waters.

Exudates from water hyacinth roots were shown to have variable effects on several species of bacteria. A two-day experiment revealed that the culture solution of water hyacinth had significant bacteriostasis on *Staphylococcus aureus*, stimulated *Sarcina lutea*, and had no effect on the growth of *Bacillus subtilis* and *Pseudomonas* sp. (Zheng and He 1990). In another experiment, five fractions isolated by thin layer chromatography from water hyacinth root exudates inhibited the growth of *Bacillus subtilis*, *Streptococcus faecalis*, *Escherichia coli* and *Staphylococcus aureus*, but not that of *Aspergillus flavus* and *Aspergillus niger* (Shanab et al. 2010).

Seventy-three bacterial strains were identified in water hyacinth roots and only six bacterial strains in corresponding water column. The number of bacterial colonies was $10^8 \sim 10^9$ g^{-1} Fresh Weight (FW) of roots and $10^4 \sim 10^5$ g^{-1} of water sample, indicating greater diversity and abundance of bacteria in water hyacinth roots than the corresponding water column (Zheng et al. 1987). Similarly, 14 dominant strains in 10 genera were isolated from water hyacinth roots. The main genera included *Aeromonas* sp., *Micrococcus* sp., *Pseudomonas* sp., *Agrobacterium* sp. and *Bacillus* sp. The quantity of heterotrophic bacteria in the rhizosphere, at the root surface of water hyacinth and in water body without water hyacinth were $10^8 \sim 10^{11}$, $10^5 \sim 10^7$ and $10^3 \sim 10^6$ g^{-1} FW, respectively (Zhan et al. 1993). In another experiment in the pond of treated domestic sewage, the density of epiphytic bacteria on water hyacinth roots was obviously higher than that on the root of water spinach (*Ipomoea aquatica*) (Loan et al. 2014). Also, in Lake Dianchi in China, the water of purple-root water hyacinth (same species but with more developed root system purple in color) growing area, that of wild-type water hyacinth growing area and that of control area had 54, 49 and 40 different bacterial strains, respectively, corresponding to Shannon-Wiener diversity indices of 3.2, 3.1 and 2.73. The abundances of bacteria were 1.4×10^7, 8.4×10^6 and 2.7×10^6 colony-forming units L^{-1}, respectively. There were 10 genera of bacteria in all the treatments. These results indicated that water hyacinth increased bacterial diversity in the eutrophic water of the lake and changed the bacterial community structure (Zheng et al. 2015).

Water hyacinth increased the abundance and diversity of nitrifying-denitrifying bacteria, with the quantity of nitrifying and denitrifying bacteria in eutrophic water without water hyacinth being $0.85 \sim 3.5 \times 10^6$ and $0.34 \sim 1.5 \times 10^7$ MPN L^{-1} (MPN = Most Probable Number), respectively; in the water with water hyacinth, the quantities of nitrifying and denitrifying bacteria were significant (Gao et al. 2012). They also found that the quantities of coupled nitrifying and denitrifying bacteria on the water hyacinth roots were $0.56 \sim 2.0 \times 10^9$ and $2.3 \sim 3.7 \times 10^9$ MPN L^{-1}, respectively, which were higher than those in the water column. When DGGE (Denatured Gradient Gel Electrophoresis) method was applied to determine species diversity based on nitrite reductase genes (*nirK* and *nirS*) and nitrous oxide reductase functional genes (*nosZ*), the abundance of *nirK, nirS* and *nosZ* types of denitrifying bacteria was significantly higher in the water body with than without water hyacinth; moreover, the species diversity of denitrifying bacteria was significantly higher on the water hyacinth roots than in the water (Yi et al. 2014). Similar conclusions were reported by Gao et al. (2014).

3.8 Summary of this chapter

Aquatic ecosystems are complex, with various communities dependent on and competing with each other. Changes in each community directly or indirectly affect the responses of biotic and abiotic components and potentially change community structures of others. Although the density of phytoplankton was significantly higher in the water body with water hyacinth than on the periphery of the water hyacinth area; this was mainly caused by the trapping function of water hyacinth roots. In general, water hyacinth on water surface can block light from reaching underwater, compete nutritionally and release allelochemicals to cause adverse effects on primary producers in the water, including phytoplankton and submerged plants. Changes to primary producers may have effects on consumers in the water body through a "bottom-up effect". For example, a decrease in phytoplankton abundance may have adverse effects on zooplankton and fish that rely on phytoplankton as a food source. However, with a complex structure of its root system, water hyacinth could enhance the heterogeneity of communities in water, especially the surface water, and provide good habitat for zooplankton, macroinvertebrates (especially epiphytic ones), fish, bacteria and fungi.

The growth characteristics of water hyacinth determine the type of effects on aquatic communities in natural waters. The positive effects were in roots of water hyacinth providing complex habitats and shelters for aquatic animals, whereas the negative effects depended on physicochemical environment (for instance, light and dissolved oxygen) of water body and the food sources; of course, an important consideration was whether these effects reached the threshold values that could affect aquatic animals. The water hyacinth effects were closely related to the density and coverage of water hyacinth on the water surface. At low density and coverage of water hyacinth mat, the positive effects may play a dominant role, whereas at higher density and coverage, negative effects may become important. In addition, the effects of water hyacinth on aquatic animal communities are related to background factors before the invasion of water hyacinth. In aquatic ecosystems damaged seriously before water hyacinth invasion, such an invasion may have positive effects on aquatic animals.

The allelochemicals secreted by water hyacinth roots into water may have promoting or inhibiting effects on decomposers. Experiments, simulations and surveys showed that water hyacinth could improve the community structure and diversity of bacteria and fungi in eutrophic waters. This may be the results of root functions of water hyacinth, including compensation for oxygen loss, provision of habitat space and release of low-molecular-weight organic molecules.

References cited

Aboul-Enein, A. M., S. M. Shanab, E. A. Shalaby, M. M. Zahran, D. A. Lightfoot and H. A. El-Shemy. 2014. Cytotoxic and antioxidant properties of active principals isolated from water hyacinth against four cancer cells lines. *BMC Complementary and Alternative Medicine* 14(1): 397.

Almeida, A. S., A. M. M. Goncalves, J. L. Pereira and F. Goncalves. 2006. The impact of *Eichhornia crassipes* on green algae and cladocerans. *Fresenius Environmental Bulletin* 15(12 A): 1531–1538.

Arora, J. and N. K. Mehra. 2003. Species diversity of planktonic and epiphytic rotifers in the backwaters of the Delhi Segment of the Yamuna River, with remarks on new records from India. *Zoological Studies* 42(2): 239–247.

Bailey, R. G. and M. R. Litterick. 1993. The macroinvertebrate fauna of water hyacinth fringes in the Sudd swamps (River Nile, southern Sudan). *Hydrobiologia* 250(2): 97–103.

Bicudo, D. D. C., B. M. Fonseca, L. M. Bini, L. O. Crossetti, C. E. Bicudo and T. Araújo-Jesus. 2007. Undesirable side-effects of water hyacinth control in a shallow tropical reservoir. *Freshwater Biology* 52(6): 1120–1133.

Brendonck, L., J. Maes, W. Rommens, N. Dekeza, T. Nhiwatiwa, M. Barson et al. 2003. The impact of water hyacinth (*Eichhornia crassipes*) in a eutrophic subtropical impoundment (Lake Chivero, Zimbabwe). II. species diversity. *Archiv für Hydrobiologie* 158(3): 389–405.

Brown, S. J. and M. J. Maceina. 2002. The influence of disparate levels of submersed aquatic vegetation on largemouth bass population characteristics in a Georgia reservoir. *Journal of Aquatic Plant Management* 40: 28–35.

Cai, L. 2006. Impact of floating vegetation in Shuikou Impoundment, Minjiang River, Fujian Province. *Journal of Lake Sciences* 18(3): 250–254 (In Chinese with English Abstract).

Chapman, G. A. 1986. *Ambient water quality criteria for dissolved oxygen (freshwater aquatic life)*. Washington DC, USA: US Environmental Protection Agency.

Chen, H. G., F. Peng, Z. Y. Zhang, G. F. Liu, W. Da Xue, S. H. Yan et al. 2012. Effects of engineered use of water hyacinths (*Eichhornia crassipes*) on the zooplankton community in Lake Taihu, China. *Ecological Engineering* 38(1): 125–129.

Coetzee, J. A., R. W. Jones and M. P. Hill. 2014. Water hyacinth, *Eichhornia crassipes* (Pontederiaceae), reduces benthic macroinvertebrate diversity in a protected subtropical lake in South Africa. *Biodiversity and conservation* 23(5): 1319–1330.

Fontanarrosa, M. S., G. Chaparro, P. Tezanos Pinto, P. Rodriguez and I. O'Farrell. 2010. Zooplankton response to shading effects of free-floating plants in shallow warm temperate lakes: a field mesocosm experiment. *Hydrobiologia* 646(1): 231–242.

Fu, M. 2011. Effects on improving water quality in the culturing *Trionyx sinensis* ponds by planting *Eichhornia crassipes* solms and raising a small amount of *Aristichthys nobilis*. *Journal of Fujian Fisheries* 33(1): 14–16 (In Chinese with English Abstract).

Gao, Y., N. Yi, Y. Wang, T. Ma, Q. Zhou, Z. Zhang et al. 2014. Effect of *Eichhornia crassipes* on production of N_2 by denitrification in eutrophic water. *Ecological Engineering* 68: 14–24.

Gao, Y., N. Yi, Z. Zhang, H. Liu, L. Zou, H. Zhu et al. 2012. Effect of water hyacinth on N_2O emission through nitrification and denitrification reactions in eutrophic water. *Acta Scientiae Circumstantiae* 32(2): 349–359 (In Chinese with English Abstract).

Geng, X., Y. Fan, X. Wang, H. Fu and L. Lan. 2009. Effect of allelochemical n-phenyl-2-naphthylamine from water hyacinth on growth of *Microcystis aeruginosa*. *Journal of Sichuan University: Nature Science Edition* (5): 1493–1496 (In Chinese with English Abstract).

Gopal, B. and U. Goel. 1993. Competition and allelopathy in aquatic plant communities. *The Botanical Review* 59(3): 155–210.

Gopalakrishnan, A., M. Rajkumar, J. Sun, A. Parida and M. B. A. Venmathi. 2011. Integrated biological control of water hyacinths, *Eichhornia crassipes* by a novel combination of grass carp *Ctenopharyngodon idella* (Valenciennes, 1844), and the weevil, *Neochetina* spp. *Chinese Journal of Oceanology and Limnology* 29(1): 162–166.

Greca, M. Della, R. Lanzetta, L. Mangoni, P. Monaco and L. Previtera. 1991. A bioactive benzoindenone from *Eichhornia crassipes* solms. *Bioorganic & Medicinal Chemistry Letters* 1(11): 599–600.

Greca, M. Della, P. Monaca and L. Previtera. 1991. New oxygenated sterols from the weed *Eichhornia crassipes* SOLMS. *Tetrahedron* 47(34): 7129–7134.

Grenouillet, G., D. Pont and K. L. Seip. 2002. Abundance and species richness as a function of food resources and vegetation structure: juvenile fish assemblages in rivers. *Ecography* 25(6): 641–650.

Han, S., S. Yan, Z. Wang, W. Song, H. Liu, J. Zhang et al. 2009. Harmless disposal and resources utilizations of Taihu Lake blue algae. *Journal of Natural Resources* 24(3): 431–438 (In Chinese with English Abstract).

Hu, T., Y. Wang, F. Chen, L. Yan, Q. Lian and S. Liu. 2010. Study on allelopathic effects of *Eichhornia crassipes* to *Microcystis aeruginosa*. *Journal of Hydroecology* 3(6): 47–51 (In Chinese with English Abstract).

Jin, Z. H., Y. Y. Zhuang, S. G. Dai and T. L. Li. 2003. Isolation and identification of extracts of *Eichhornia crassipes* and their allelopathic effects on algae. *Bulletin of environmental contamination and toxicology* 71(5): 1048–1052.

Johnson, D. L. and R. A. Stein. 1979. Response of fish to habitat structure in standing water. In *proceedings of a symposium entitled Interrelationships between fish and cover in standing water held during the 40th Annual Midwest Fish and Wildlife Conference, December, 1978*, ed. D. L. Johnson and R. A. Stein, 77. Columbus, Ohio, USA: North Central Division, American Fisheries Society.

Lam, M. K. and K. T. Lee. 2012. Microalgae biofuels: a critical review of issues, problems and the way forward. *Biotechnology advances* 30(3): 673–90.

Laskov, C., O. Horn and M. Hupfer. 2006. Environmental factors regulating the radial oxygen loss from roots of *Myriophyllum spicatum* and *Potamogeton crispus*. *Aquatic Botany* 84(4): 333–340.

Lei, Y., K. Xiao and Y. He. 2008. Water hyacinth utilization and its damage to fisheries and waters. *Inland Fisheries* 33(1): 23–26 (In Chinese).

Li, Y., B. Rao, Z. Wang, H. Qin, L. Zhang and D. Li. 2012. Spatial-temporal distribution of phytoplankton in bloom-accimulation area in Lake Chaohu. *Resources and Environment in the Yangtze Basin* 21(Z2): 25–31 (In Chinese with English Abstract).

Liu, G., S. Han, J. He, S. Yan and Q. Zhou. 2014. Effects of ecological purification engineering of planting water hyacinth on macro-benthos community structure. *Ecology and Environmental Sciences* 23(8): 1311–1319 (In Chinese with English Abstract).

Liu, G., H. Liu, Z. Zhang, Y. Zhang, S. Yan, J. Zhong et al. 2010. Effects of large-area planting water hyacinth on macro-benthos community structure and biomass. *Environmental Science* 31(12): 2925–2931 (In Chinese with English Abstract).

Liu, J., Z. Chen, W. Yang and W. Chen. 2006. Inhibitory mechanism of acetone extract from *Eichhornia crassipes* root on *Alexandrium tamarense*. *Acta Phytophysiologica Sinica* 26(5): 815–820 (In Chinese with English Abstract).

Loan, N. T., N. M. Phuong and N. T. N. Anh. 2014. The role of aquatic plants and microorganisms in domestic wastewater treatment. *Environmental Engineering and Management Journal* 13(8): 2031–2038.

Lu, Z., Y. Gao, B. Liu, X. Sun, Y. Zhang and Z. Wu. 2013. Advances in research on mechanism of inhibitory effects on phytoplankton mediated by aquatic allelochemicals. *Environmental Science & Technology* 36(7): 64–69 (In Chinese with English Abstract).

Lugo, A., L. A. Bravo-Inclán, J. Alcocer, M. L. Gaytán, M. G. Oliva, M. del R. Sánchez et al. 1998. Effect on the planktonic community of the chemical program used to control water hyacinth (*Eichhornia crassipes*) in Guadalupe Dam, Mexico. *Aquatic Ecosystem Health & Management* 1(3-4): 333–343.

Ma, T., N. Yi, Z. Zhang, Y. Wang, Y. Gao and S. Yan. 2014. Oxygen and organic carbon releases from roots of *Eichhornia crassipes* and their influence on transformation of nitrogen in water. *Journal of Agro-Environment Science* 33(10): 2003–2013 (In Chinese with English Abstract).

Mangas-Ramírez, E. and M. Elías-Gutiérrez. 2004. Effect of mechanical removal of water hyacinth (*Eichhornia crassipes*) on the water quality and biological communities in a Mexican reservoir. *Aquatic Ecosystem Health & Management* 7(1): 161–168.

Marco, P., M. A. Reis Araújo, M. K. Barcelos and M. B. L. Santos. 2001. Aquatic invertebrates associated with the water-hyacinth (*Eichhornia crassipes*) in an eutrophic reservoir in tropical Brazil. *Studies on Neotropical Fauna and Environment* 36(1): 73–80.

Masifwa, W. F., T. Twongo and P. Denny. 2001. The impact of water hyacinth, *Eichhornia crassipes* (Mart) Solms on the abundance and diversity of aquatic macroinvertebrates along the shores of northern Lake Victoria, Uganda. *Hydrobiologia* 452: 79–88.

McVea, C. and C. E. Boyd. 1975. Effects of waterhyacinth cover on water chemistry, phytoplankton, and fish in ponds. *Journal of Environmental Quality* 4(3): 375–378.

Meerhoff, M., C. Iglesias, F. T. De Mello, J. M. Clemente, E. Jensen, T. L. Lauridsen et al. 2007. Effects of habitat complexity on community structure and predator avoidance behaviour of littoral zooplankton in temperate versus subtropical shallow lakes. *Freshwater Biology* 52(6): 1009–1021.

Meerhoff, M., N. Mazzeo, B. Moss and L. Rodríguez-Gallego. 2003. The structuring role of free-floating versus submerged plants in a subtropical shallow lake. *Aquatic Ecology* 37(4): 377–391.

Midgley, J. M., M. P. Hill and M. H. Villet. 2006. The effect of water hyacinth, *Eichhornia crassipes* (Martius) Solms-Laubach (Pontederiaceae), on benthic biodiversity in two impoundments on the New Year's River, South Africa. *African Journal of Aquatic Science* 31(1): 25–30.

Miranda, L. E. and D. R. DeVries. 1996. Multidimensional approaches to reservoir fisheries management. In *Proceedings of the Third National Reservoir Fisheries Symposium*, ed. M. Bethesda, 462 pp. Chattanooga, Tennessee, USA: American Fisheries Society.

Morse, J. C., Y. J. Bae, G. Munkhjargal, N. Sangpradub, K. Tanida, T. S. Vshivkova et al. 2007. Freshwater biomonitoring with macroinvertebrates in East Asia. *Frontiers in Ecology and the Environment* 5(1): 33–42.

Njiru, M., J. B. Okeyo-Owuor, M. Muchiri and I. G. Cowx. 2004. Shifts in the food of Nile tilapia, *Oreochromis niloticus* (L.) in Lake Victoria, Kenya. *African Journal of Ecology* 42(3): 163–170.

O'Hara, J. 1967. Invertebrates found in water hyacinth mats. *Quarterly Journal of the Florida Academy of Science* 30(1): 73–80.

Padial, A. A., S. M. Thomaz and A. A. Agostinho. 2009. Effects of structural heterogeneity provided by the floating macrophyte *Eichhornia azurea* on the predation efficiency and habitat use of the small Neotropical fish *Moenkhausia sanctaefilomenae*. *Hydrobiologia* 624(1): 161–170.

Pei, J. 2001. Water hyacinth reduced white abdominal shell disease in soft shell turtle production. *Fishery Guide to be Rich* (18): 47–47 (In Chinese with English Abstract).

Rocha-Ramirez, A., A. Ramirez-Rojas, R. Chaivez-Loipez and J. Alcocer. 2007. Invertebrate assemblages associated with root masses of *Eichhornia crassipes* (Mart.) Solms-Laubach 1883 in the Alvarado Lagoonal System, Veracruz, Mexico. *Aquatic Ecology* 41: 319–333.

Shanab, S. M. M., E. A. Shalaby, D. A. Lightfoot and H. A. El-Shemy. 2010. Allelopathic effects of water hyacinth (*Eichhornia crassipes*). *PloS one* 5(10): e13200.

Sharma, K. P. 1985. Allelopathic influence of algae on the growth of *Eichhornia crassipes* (Mart.) Solms. *Aquatic Botany* 22(1): 71–78.

Sheikh, N. M., S. A. Ahmed and S. Hedayetullah. 1964. The effect of root extraction of water hyacinth (*Eichhornia crassipes*) on the growth of microorganisms and mash kalai (*Phaseolus muno var. Roxburghii*) and on alcoholic fermentation. *Pakistan Journal of Scientific and Industrial Research* 7: 96–102.

Sircar, S. M. and R. Chakraverty. 1962. The effect of gibberellic acid and growth substances from the root extract of water hyacinth, *Eichhornia crassipes* on rice and gram. *Indian Journal of Plant Physiology* 5: 1–2.

Sircar, S. M. and A. Ray. 1961. Growth substances separated from the root of water hyacinth by paper chromatography. *Nature* 190(4782): 1213–1214.

Sun, W., Z. Yu and S. Yu. 1988. Inhibitory effect of *Eichhornia crassipes* (Mart.) Solms on algae. *Acta Phytophysiologica Sinica* 14(3): 294–300 (In Chinese with English Abstract).

Sun, W. 1989. The harness of an eutrophic water body by water hyacinth. *Acta Scientiae Circumstantiae* 9(2): 188–195 (In Chinese with English Abstract).

Toft, J. D., C. A. Simenstad, J. R. Cordell and L. F. Grimaldo. 2003. The effects of introduced water hyacinth on habitat structure, invertebrate assemblages, and fish diets. *Estuaries* 26(3): 746–758.

Troutman, J. P., D. A. Rutherford and W. E. Kelso. 2007. Patterns of habitat use among vegetation-dwelling littoral fishes in the Atchafalaya River Basin, Louisiana. *Transactions of the American Fisheries Society* 136(4): 1063–1075.

Villamagna, A. M. 2009. Ecological effects of water hyacinth (*Eichhornia crassipes*) on Lake Chapala, Mexico. Ph.D. Thesis, Virginia Polytechnic Institute and State University, Blacksburg, Virginia, USA.

Villamagna, A. M. and B. R. Murphy. 2010. Ecological and socio-economic impacts of invasive water hyacinth (*Eichhornia crassipes*): a review. *Freshwater Biology* 55(2): 282–298.

Wang, C. 2011. Methods to prevent damage by water hyacinth to cage culture. *Fishery Guide to be Rich* (17): 62–63 (In Chinese).

Wang, Z., Z. Zhang, Y. Zhang and S. Yan. 2012a. Effects of large-area planting water hyacinth (*Eichhornia crassipes*) on water quality in the bay of Lake Dianchi. *Chinese Journal of Environmental Engineering* 6(11): 3827–3832 (In Chinese with English Abstract).

Wang, Z. 2012b. Large-scale utilization of water hyacinth for nutrient removal in Lake Dianchi in China: the effects on the water quality, macrozoobenthos and zooplankton. *Chemosphere* 89(10): 1255–61.

Wang, Z., Z. Zhang, Y. Zhang, J. Zhang and S. Yan. 2013. Water quality effects of two aquatic macrophytes on eutrophic water from Lake Dianchi Caohai. *China Environmental Science* 33(2): 328–335 (In Chinese with English Abstract).

Wang, Z., Z. Zhang, Y. Zhang, J. Zhang, S. Yan and J. Guo. 2013. Nitrogen removal from Lake Caohai, a typical ultra-eutrophic lake in China with large scale confined growth of *Eichhornia crassipes*. *Chemosphere* 92(2): 177–183.

Wu, F., T. Liu, Z. Wang, Y. Wang and S. He. 2011. Effects of *Eichhornia crassipes* growth on aquatic plant in Dianchi Lake. *Journal of Anhui Agricultural Science* 39(15): 9167–9168 (In Chinese with English Abstract).

Xie, W. 2000. The benefit in aquaculture with water hyacinth. *Freshwater fisheries* 30(9): 25–27 (In Chinese).

Xu, Z., J. Lu, X. Chen and H. Liu. 2003. Application of water hyacinth in aquaculture. *Scientific Fish Farming* (8): 50–50 (In Chinese).

Yang, S. Y., Z. W. Yu, W. H. Sun, B. W. Zhao, S. W. Yu, H. M. Wu et al. 1992. Isolation and identification of antialgal compounds from root system of water hyacinth. *Acta Phytoecologica Sinica* 18(4): 399–402 (In Chinese with English Abstract).

Yi, N., Y. Gao, X. Long, Z. Zhang, J. Guo, H. Shao et al. 2014. *Eichhornia crassipes* cleans wetlands by enhancing the nitrogen removal and modulating denitrifying bacteria community. *CLEAN–Soil, Air, Water* 42(5): 664–673.

Zhan, F., J. Deng, Y. Xia and Z. Wu. 1993. Studies on community characteristics and heterotrophic activity of heterotrophic bacteria from root-zone of water hyacinth. *Acta Hydrobiologica Sinica* 17(2): 150–156 (In Chinese with English Abstract).

Zhang, D. 2012. The inhibition of compound allelochemicals on algae and the influence on microcystin biosynthesis. MSc. Thesis, An Hui Normal University, Benbu, An Hui, China.

Zhang, D., Q. Ling, W. Liu, W. Li, G. Chen, Y. Xue et al. 2012. Research of *Eichhornia crassipes* and *Ceratophyllum demersum* L. in the remediation of degraded aquacultural water of *Pseudobagrus fulvidraco*. *Journal of Yangzhou University (Agricultural and Life Science Edition)* 33(4): 66–71 (In Chinese with English Abstract).

Zhang, Z., Y. Gao, J. Guo and S. Yan. 2014. Practice and reflections of remediation of eutrophicated waters: a case study of haptophyte remediation of the ecology of Dianchi. *Journal of Ecology and Rural Environment* 30(1): 15–21 (In Chinese with English Abstract).

Zhao, J., T. J. Lam and Y. Guo. 1997. Acute toxicity of ammonia to the early stage-larvae and juveniles of *Eriocheir sinensis* H. Milne-Edwards, 1853 (Decapoda: Grapsidae) reared in the laboratory. *Aquaculture Research* 28(7): 517–525.

Zhao, Y., J. Lu, L. Zhu and Z. Fu. 2006. Effects of nutrient levels on growth characteristics and competitive ability of water hyacinth (*Eichhornia crassipes*), an aquatic invasive plant. *Biodiversity Science* 14(2): 159–164 (In Chinese with English Abstract).

Zheng, S. and M. He. 1990. Effect of root secretion of water-hyacinth on several bacteria. *Journal of Ecology* 9(5): 56–57 (In Chinese with English Abstract).

Zheng, S., J. Huang and M. He. 1987. Comparative study of bacteria from rhizosphere of *Eichhornia crassipes* and water. *Journal of Ecology* 6(4): 30–32 (In Chinese with English Abstract).

Zheng, Y. K., K. Liu, Z. J. Xiong, C. P. Miao, Y. W. Chen, L. H. Xu et al. 2015. Effect of large-scale planting water hyacinth on cultivable bacterial community structure in the eutrophic lake. *Microbiology China* 42(1): 42–53 (In Chinese with English Abstract).

Zhou, Q., S. Q. Han, S. H. Yan, W. Song and J. P. Huang. 2012. The mutual effect between phytoplankton and water hyacinth planted on a large scale in the eutrophic lake. *Acta Hydrobiologica Sinica* 36(4): 873–791 (In Chinese with English Abstract).

Zhou, Q., S. Han, S. Yan, W. Song and G. Liu. 2014a. Impacts of *Eichhornia crassipes* (Mart.) Solms stress on the growth characteristics, microcystins and nutrients release of *Microcystis aeruginosa*. *Environmental Science* 35(2): 597–604 (In Chinese with English Abstract).

Zhou, Q., S. Han, S. Yan, W. Song and G. Liu. 2014b. Impacts of *Eichhornia crassipes* (Mart.) Solms stress on the physiological characteristics, microcystin production and release of *Microcystis aeruginosa*. *Biochemical Systematics and Ecology* 55: 148–155.

CHAPTER 4

Impacts of Water Hyacinth on Ecosystem Services

J. Y. Guo

4.1 Introduction

Rivers, lakes and water reservoirs are essential and critical natural resources for human life and global ecosystem services, including the function and services for supplying water resources. Over the long history of civilization, people have tried to maintain the function and ecosystem services provided by rivers, lakes and reservoirs through a great effort to minimize flood damages and water shortages as well as to increase their production and transportation capacity, hydropower generation, and suitability for recreation and tourism.

Water hyacinth can exert negative effects on ecosystem services including transportation, recreation, sports, hydropower, fisheries, water resources and utilization, but also can exert positive effects by phytoremediation. The dichotomy of these effects is mainly due to the high growth rate, ecological adaptive morphological changes and great adaptability on nutrient conditions. Considering the physical aspects, water hyacinth can form a large floating mat covering water surface and causing obstruction of waterways, water inlets for hydropower plants, water abstraction units, etc. Regarding the chemical aspects, water hyacinth can (1) release nutrients and organic matter to water due to biomass decay, or (2) reduce nutrients and other pollutants in water through rapid absorption and adsorption functions, resulting in phytoremediation of eutrophic or polluted water (Masifwa et al. 2001).

Developing countries rely on water resources for their agriculture and fisheries to support people's livelihood and their water security. The effects of water hyacinth restrain the existing scarcity of global water resources

5 Armagh Way, Ottawa, Canada.
Email: guoj1210@hotmail.com

through evapotranspiration and water quality degradation during decay of large quantities of biomass. This can be clearly illustrated by the report from United Nation Environmental Protection program (UNEP). Most waters on Earth are of little direct use to humans because about 97.5% of them are salty (sea water or water in salt lakes); only the remain 2.5% of water is fresh water, of which about 69.6% is in the forms of ice and snow, about 30.1% is stored underground and replenishes very slowly (UNEP 1994), leaving only about 0.26% of the total fresh water available for direct use by humans.

Conflicts over water resources are not uncommon (Simonovic 2002) because water is a finite, mobile, precious and multi-functional resource used for domestics and industries as well as for recreational purposes, agriculture and fishery production, hydropower generation and transportation. The most available water for human utilization is found in rivers, freshwater lakes and freshwater reservoirs, which contain about 0.26% of all the water on the planet (Ahmed and Eldaw 2002). Humans mainly rely on rivers, freshwater lakes and freshwater reservoirs, giving them large economic values. For example, wetland economic values for Canada amounted to US$6.73 billion in 1994 (UNEP 1994). However, these "water containers" are severely impacted by the invasion of aquatic weeds, especially water hyacinth in tropical, subtropical and some temperate climate zones (UNEP and GEAS 2013), combined with exacerbated eutrophication due to deforestation, urbanization and agriculture development to support the rapid growth of human population.

Water hyacinth is listed by the International Union for Conservation of Nature (IUCN) as one of the 100 most aggressive Invasive Alien Species (IAS) (Téllez et al. 2008) and the top 10 of the world's worst weeds (Patel 2012) due to its rapid growth rates, extensive dispersal capacity, large and rapid reproductive output and broad environmental tolerance. Indeed, the invasions of water hyacinth have been impacting ecosystem services in many ways, causing substantial economic losses (UNEP and GEAS 2013).

Water hyacinth can form mats for a few square kilometers (Villamagna et al. 2012) on water surface that may degrade the flora and fauna beneath the mat; it can grow so thick that a powered motor vehicle or a boat may find it too difficult to move and water entry to a hydropower generator may be blocked (Government of South Australia 2013). However, despite so many negative effects, water hyacinth can also clean water pollutants through absorption of dissolved nutrients or heavy metals, and adsorption of detritus or filtration of particles by its fibrous roots dangling beneath the mat so that the water leaves the mat clearer, with good transparency and lower in plant nutrients, thus potentially suitable for irrigation or to be processed for drinking. The ultimate outcomes of this invasive species depend on the scale and cost of damage, existence of pollutants and the scale and benefits of utilization as well as the management strategies and financial resources.

The generalized effects of water hyacinth on global ecosystem services can be summarized in three aspects: (1) financial losses due to water hyacinth invasion, such as losses of fishing grounds or fish production, or of water bodies and shores for sporting services, or of sustainable water body utilization, etc.; (2) costs of removing nuisance weeds that may have only limited financial returns and become an economic burden on society (3) benefits of phytoremediation or utilization of harvested biomass as biological resources for energy, feed, organic fertilizers or other useful applications. These aspects are interrelated and interdependent, so that the challenges or the solutions will be multidimensional rather than a simple approach for control, or eradication, or utilization or phytoremediation, etc.

In Europe, the cost due to water hyacinth invasion is estimated to be at least US$9 million per year (Kettunen et al. 2009). In the United State, invasion of aquatic weeds causes economic losses estimated to total US$110 million annually (Pimentel et al. 2005). In South Africa, the cost of biological control of Invasive Alien Species (IAS) was estimated above US$760 million (South Africa Rand 6.5 billion) per annum or 0.3% of South Africa's GDP in 2011, and could rise to over 5% of GDP if invasive plants are allowed to reach their full potential (van Wilgen and Lange 2011). In Australia, water hyacinth has infested Queensland, Victoria, western Australia, northern territory and certain areas of New South Wales. Furthermore, the Australian Weeds Committee also cautioned that the actual costs, including overheads, were much higher than estimated (Wright 1996). In China, water hyacinth invaded 19 southern provinces (Duan et al. 2003). The invasion of all aquatic alien species in rivers, lakes and reservoirs caused damages estimated at US$7 billion annually, out of which water hyacinth was responsible for a major proportion, but an accurate estimate was not available (Chu et al. 2006). For example, the government of Kun Ming city spent about US$61 million over 10 years to manage water hyacinth in Lake Dianchi (Duan et al. 2003). In Fujian province, the cost of managing water hyacinth averaged US$416 per hectare per year (Hong 2004); it was about US$0.70 per tonne fresh weight of water hyacinth biomass in Shanghai district in 2002 (Chu et al. 2006). The above examples indicate the scale of global water hyacinth infestation as well as associated management costs.

4.2 Impact of water hyacinth on tourism

Rivers, lakes and water reservoirs are important resources for recreation and tourism that contribute significantly to local, regional and national economies in both developed and developing countries. Water-based international tourism has become a popular global leisure activity and is a significant source of income for the both source and host countries and is of vital importance in some developing countries. International tourism receipts grew to US$1.38 trillion in 2013 (The World Bank 2015). In The Great Lakes region of North America, the water-based tourism receipts reached US$4 billion annually (Manninen 2015).

The United Nation Environment Program (UNEP) and International Lake Environment Committee (ILEC) investigated 215 world lakes and water reservoirs in 1987 to 1990 and reported that about 62% of lakes in Africa and 92% of lakes in North America could potentially be developed for tourism, whereas 80% of lakes in the Asian-Pacific region, 91% in Europe and 83% in South America were found to have potential value for tourism. For example, Loch Ness is an important tourist attraction site in the United Kingdom and the same holds for Everglades in the United States, Lake Kariba in Zimbabwe and Zambia, Lake Toba in Indonesia, Lake Nakuru in Kenya and Dal Lake in India (UNEP 1994).

The value of the tourist attraction and its potential may be greatly increased if a water body has rare or threatened plant and animal species, unusual ecosystem, landscape and natural processes, a high diversity of habitats and significant altitude changes across the site (UNEP 2006). However, these values and potentials are greatly diminished by notorious and invasive water hyacinth that impacts aquatic ecosystems on a global scale, affects aquatic fauna and flora and threaten rare native aquatic species.

Another example is Lake Victoria that is now unattractive for sunbathing, swimming and water sports such as wind-surfing and sailing due to the invasion of water hyacinth (Lindsey and Hirt 2000, Nkandu 2013). The Lake Victoria Environmental Management Project Phase II (LVEMP II) have tried to make the lake a sustainable living resource and a popular area for holiday makers, especially given the striking beauty of 62 islands of the Sese Archipelago in the northwest, and Mwanza in the southwest of the lake (Lindsey and Hirt 2000, FETA 2014).

The conversion of native biotic communities to invasive-weed-dominated communities also directly affects tourism and has aesthetic and cultural impacts. Water hyacinth directly threatens the habitat of aquatic species that are keys to the tourism industry; species loss reduces the attraction and ecological value for tourism. For example, Christmas Bush (*Chromolaena odorata*) affects the nesting sites of crocodiles, directly placing these populations at risk, which may devalue the site for tourism (Munyaradzi Chenje and Mohamed-Katerere 2006).

Water hyacinth, by clogging waterways, affects water-based recreational activities and thrives well in waters that are important source for livelihood and tourism such as Lake Victoria in East Africa (PHS 2014). Although there is a lack of reliable data on tourism income over the Lake Victoria region, the value of tourism development is undoubted due to ecological importance of biodiversity, unique flora and fauna, history, humanity and natural beauty. Moreover, Lake Victoria is the second large freshwater lake in the world and is used as a source of food, energy, drinking and irrigation water, transport, and as a repository for human, agricultural and industrial waste and is of great economic worth to the sub-region and of great scientific and cultural significance to the global community. The invasion of water hyacinth in the region could be traced back to early 1988 and reached a peak spread during 1996 to 1998 (Kateregga and Sterner 2007), and was then successfully restricted to only small patches (2000 hectares) in areas rich in nutrients by 2000

(Williams et al. 2005). However, the lake is severely impacted by pollution from land-based activities (heavy pollution by domestic and industrial waste as well as agricultural run-off high in nitrogen and phosphorus), introduction of invasive alien species (both fish and plants), and excessive exploitation of living resources. As a consequence, the lake is facing potentially irreversible environmental damage, causing hardships among the poor, and serious health concerns. Therefore, there is a danger that the continued organic and nutrient pollution of the region may reverse the success achieved in controlling water hyacinth (UNEP 2006).

Although it seems the above examples are all concerned with negative impacts, water hyacinth might have positive effects on tourism if it could be successfully controlled to achieve removal of excessive nutrients, especially nitrogen, phosphorus and organic pollutants as well as detritus, sand and mud from the runoffs. With respect to Lake Victoria as a relatively shallow lake (average depth 40 meters) (Lindsey and Hirt 2000), total nitrogen concentration of 0.54–0.98 mg N L^{-1} plus total phosphorus concentration of 0.21–0.58 mg P L^{-1} is inevitably going to lead to algal blooms (Gichuki et al. 2012). Algal blooms are likely to have a more severe impact on tourism than the boring scenery of water hyacinth mat. But, why would we not use water hyacinth to suppress algae and thus minimize the effects on tourism development?

4.3 Impact of water hyacinth on rivers and transportation

Mature water hyacinth has a fibrous root system with length from 0.09 meters in nutrient-abundant (Penfound and Earle 1948) to 2 meters in low-nutrient waters (Rodríguez et al. 2012), and a spongy stolon, petioles and pseudo-lamina 0.22 (Petrell et al. 1991) to 1.5 meters above the water surface (Howard and Harley 1998, Solms et al. 2009); fresh biomass may be 40–60 kg m^{-2} in dense mats (Patil et al. 2014, Rezania et al. 2015). Individual plants are connected via stolons and rhizomes to form a tight mat. The connectivity[1] of a dense mat can be as high as 182 Newton m^{-2} (Petrell et al. 1991). The continuous mats of water hyacinth can grow big enough to block commercial harbors and waterways to fishing grounds, causing disruptions, delays and rising costs of inland waterway transport, thus greatly affecting the daily life of people who rely on the waterways for communication and transportation as well as a source of livelihood.

For instance, the communities along the River Niger are benefiting from the river transportation that plays an important role in socio-economic activities and gives access to remote areas, thereby promoting local development. About 6,000 km of the River Niger are estimated to be navigable including about 3,800 km of well-developed waterway from Kouroussa (Guinea) to the

[1] Defined as the difference in pressure (Newton per square meter) required to submerge connected and disconnected mat (Petrell et al. 1991).

river mouth in Nigeria via Koulikoro (Mali), Niamey (Niger) and Garoua (Cameron). However, water hyacinth affects more than 550 km of the Niger River, causing major problems in West Africa. In Benin, water hyacinth was first observed in 1977 (Neuenschwander et al. 1996) and gradually expanded into large areas along rivers, affecting the lower reaches of the Oueme River and the So River within a 1000-km^2 flood-plain and the connected lagoons after 1988. Each dry season, the salinity in the lagoon rises above the tolerance limit of water hyacinth and the plants die. Water hyacinth also occurs in the far north of Benin along the Niger River; water hyacinth mats increased in size up to a maximum in October, gradually blocking the narrow parts of the river each year. The largest single blockage was 6.2 km long in 1992. When waters started receding in November 1992, many water hyacinth plants were stranded and eventually died. From November 1992 to April 1993, winds ('harmattan') blew from the north, and mats of water hyacinth increased in size and density until salinity crept up and plants again started to die. During 1994 and 1995, essentially the same cycles were observed, yet water hyacinth densities were reduced (Neuenschwander et al. 1996). Another instance was observed in Nigeria where the weed was first reported in 1982 (Akinyemiju 1987) and was spreading fast along Badagry Creek in 1984 (Kusemiju et al. 2002). Based on the national survey conducted in 1995, more than 20 states (out of 36) and Federal Capital Territory of Nigeria were infested by water hyacinth (NIFFR 2002). The weed has spread along the entire coastline causing serious problems especially in transportation (Olomoda 2002, Uka et al. 2007).

The economic losses of the waterway transportation and livelihood of the communities due to the impacts of weed infestation were estimated at US$20–50 million per year in the six West African countries (Niger, Mali, Republic of Ivory Coast, Benin, Nigeria and Senegal) along the Niger River (UNEP 2006). The costs may be as much as US$100 million annually across the whole of Africa (UNEP 2006). This situation encouraged the relevant regional governments to develop co-operative projects to enhance the co-ordination, monitoring and control of the basin resources at the regional scale, including the development of a regional program for eradication and control of water hyacinth in the basin (Olomoda 2002).

The same story was replicated in Northeast Africa where 6853 kilometers of the world's longest river (River Nile) serve the important trade routes on Lake Kyoga, Lake Victoria and River Nile Basin in Kenya, Uganda, Tanzania, Burundi, Rwanda, Democratic Republic of the Congo, Ethiopia, Eritrea, South Sudan, Sudan and Egypt for railway-wagon ferries across rivers and channels and for inland canoe transport. However, those activities have been widely affected by heavy proliferation of water hyacinth (Orach-Meza 1996). The activities of the Kenya Railways have been closed since 1997 at all the piers in Asembo, Homa Bay, Kendu Bay, Kowor, Mbita and Mfangano, except for Kisumu port that was operational only for larger vessels, though with some difficulties. Such vessels (weighing over 700 tonnes) have their propellers deep, avoiding entanglement with water hyacinth. The outgoing cargo transportation at Kisumu port was reduced by 60% in the period 1996–1998 when water hyacinth infestation was at the peak (Mailu 2001, Pietersen and Beekman 2006).

Water hyacinth mats blocked access to fishing grounds, delayed access to markets and increased fishing costs (effort and materials) due to impedance of transportation (Kateregga and Sterner 2009). In Uganda, maintaining a clear passage for ships to dock at the Pot Bell cost US$3–5 million per annum after 1995 (Mailu 2001). In Kenya, there was a 70% decline in activities at the port of Kisumu as a result of water hyacinth choking the port and fish landing grounds. Overall, the economic losses due to water hyacinth infestation covering 12,000 ha in Kenya, Tanzania and Uganda in the Lake Victoria region were roughly estimated to be of the order of billions of US dollars in all sectors of economic activities (Mailu 2001).

In Asia and South East Asia, impacts of water hyacinth infestation on transportation have also been a critical concern. In North-east India, it affected navigation on the Brahmaputra River by entangling motor boats and blocking many irrigation channels that obstructed flow of water to crop fields and caused siltation and floods. The climate in North-east India allowed the weed to prevail in many important canal and irrigation systems, causing staggering losses in agriculture in West Bengal (Patel 2012).

In China, the situation is similar. Lu et al. (2007) summarized Chinese publications and characterized the route and history of water hyacinth growth in China: from Japan to Taiwan in 1901, from Taiwan to the Mainland in the 1930s, and a booming spread all over the country after the 1950s; by 2006, water hyacinth has infested the water bodies of 19 provinces in China, including heavily impacted Hainan Island, Fujian, Zhejiang, Shanghai, Yunnan, Sichuan, Hunan, Hubei, and Jiangsu (Tan et al. 2005). Although the early days of water hyacinth in China were not well documented, the outbreaks and serious damages caused by infestation of the macrophyte after 1980s were well reported and substantiated. Water hyacinth mat frequently blocked transportation on Huangpu River in Shanghai after 1995. In 2002, a cost of US$2.9 million was incurred in cleaning water hyacinth, and a total grew to US$12.3 million when the cost of post-treatments in Huangpu River region were included (Gao and Li 2004). Water hyacinth biomass that needed to be removed increased from 0.5 t d^{-1} in 1975, to 50 t d^{-1} in 1995 and 255 t d^{-1} in 2001 (Yang et al. 2002). In the rainy season or flooding days, it reached up to 400 t d^{-1} in the region (Li et al. 2004). In other places in China, water hyacinth mats caused similar damages in Bailianhe Reservoir in Hubei Province (Xu et al. 2003), Shuikou Reservoir in Fujian Province and 25 rivers in Haining region in Zhejiang Province (Lu et al. 2007), the Yaojiang, Fenghua and Yongjiang rivers in Ningbo region in Zhejiang Province and Hangjiang River in Hubai Province (Duan et al. 2003), Minjiang River in Fujian Province (Huang 2009) and to many other rivers countrywide.

In North America, water hyacinth was first introduced into the United States as early as 1884 (Ellis 2011) and was widespread along east coast by 1900 (Penfound and Earle 1948); it was widely distributed in the Pacific

Northwest US more recently (Ellis 2011). In Florida, water hyacinth, along with other aquatic weeds, has regularly choked waterways (Fig. 4.2.2-1) so that the long-term and active-control measures became necessary to remove exotic aquatic weed. For that purpose, the Florida government spends about US$14.5 million annually (Pimentel et al. 2005).

In Mexico, López et al. (1996) estimated in 1993 that more than one hundred reservoirs and lakes (approximately 62,000 ha) were infested by aquatic weeds, and the most prevalent species was water hyacinth, present on 40,000 ha of the infested surface. Also, in the irrigation districts aquatic weeds were a problem on 12,000 km (27%) of irrigation canals and 19,000 km (63%) of drainage canals.

In Europe, Téllez et al. (2008) reported infestation of water hyacinth in River Guadiana in the SW Iberian Peninsula, Spain. The greatest damage due to water hyacinth's fast expansion was in the middle reaches of 75 km of the river. The expansion was noticed in the autumn of 2004 and continued via a strong regeneration of the fragments that had been left on the banks in April 2005. By October and November 2005, it occupied an area of approximately 200 ha and produced a biomass of 175,000 tonnes in that period. Due to rapid expansion of water hyacinth, mechanical extraction was carried out by the Confederación Hidrográfica del Guadiana (CHG) of Spain's Ministry of the Environment since the affected zone is an important area for irrigation farming, and this weed provoked acute social alarm.

Australia may be among the countries that were affected by water hyacinth infestation early and severely, but achieved good management targets over the decades. The weed was first reported as early as the 1890's in New South Wales

Fig. 4.2.2-1. Water hyacinth infestation in southern Florida (Photo by T.D. Center, USDA, ARS. Permission obtained by pers. comm.).

and Queensland (Australian Weeds Committee 2012) and, by 2013, was spread in northern territory, western Australia, South Australia and Victoria. Among all the Australia states, only Tasmania (as an island) avoided the infestation (Australia Weed Committee 2013). Water hyacinth has caused severe damage and economic losses by covering entire water surface and waterways in freshwater rivers, wetlands, dams, lakes, irrigation and drainage channels such as Wappa Dam, Gingham Watercourse, Mary River and others (Osmond and Petroeschevsky 2013). The weed was declared National Weed of Significance and put on the list of highly alarming and nationwide monitored Invasive Alien Species in 2012 (Australian Weeds Committee 2012, Petroeschevsky 2012).

4.4 Impact of water hyacinth on fisheries and hydropower

Impacts of water hyacinth on fisheries

Freshwater fisheries in rivers, lakes and freshwater reservoirs support livelihoods of hundreds of millions worldwide. Fish provide 25% of the protein intake of many people in developing countries (UNEP 1994). However, water hyacinth has had big effects on fisheries in rivers, lakes and reservoirs where it exists. The most commonly documented effects of water hyacinth on fisheries were related to (1) accessing fishing grounds and markets, (2) lowering phytoplankton productivity so that feed resources for fish were reduced and in consequence, fish yield was reduced, (3) lowering dissolved oxygen concentration beneath the mat so that it may impose a limitation on fish surviving (Villamagna and Murphy 2010).

For example, in India water hyacinth hampered fish culture and caused many fishery tanks and ponds to become unsuitable due to overgrowth of the weed (Patel 2012). In Malawi (Southeast Africa), dense mats of floating water hyacinth formed an almost impenetrable barrier to boats and covered fishing areas near the shore, necessitating long journeys to reach open water for fishing; instead of canoe journeys of half an hour or less, fishermen were forced to spend two or three hours to reach their destinations at Ndinde Marsh (17°03' S 35°16' E) in 1996 (Terry 1996). At Lower Shire Valley (Malawi, 15°55'–16°33' S 34°45'–35°08' E), fisheries were threatened by infestation of water hyacinth (Lindsey and Hirt 2000). At Bangula Lagoon (Malawi, 16°31' S 35°07' E), fishing was abandoned at one time in 1991 because of water hyacinth. Fishermen had abandoned Alumenda (Malawi, 16°13' S 34°56' E) for the same reason and moved to the eastern side of Elephant Marsh (Malawi, 16°28' S 35°08' E). They claimed that more time could be spent fishing if it was easier to reach the fishing grounds.

Another example was reported in Benin, West Africa. About 1000 square kilometers of Queme flood plain (6°35' N 2°26' E) was the main dry-season agricultural production area of the country. In addition, 23,747 tonnes of fish and crustaceans were caught by 24,360 full-time fishermen; its rivers and swamps produced 65% of the country's fisheries catch (80% of the inland production) in 1990. This production was in jeopardy by 1996 because waterways were blocked by water hyacinth; village transportation to markets by boat was often impeded. In addition, water hyacinth affected fishing directly by covering the open water surface or clogging fish traps (Neuenschwander et al. 1996). It was also reported that various types of nets were difficult to place where water hyacinth was present. There was insufficient clear water for the use of cast nets and seine nets. Clear patches in the weed had to be cut for placement of gill nets. Long lines and nets left in the water were swept away by the weed, reducing catches and causing losses of fishing gear (Terry 1996).

The situation was similar in East Africa. The concerns over the threat of water hyacinth to the thriving fisheries in Uganda have intensified during the late 1980s to late 1990s. During that time, the lakes and rivers in Uganda supplied on average 220,000 tonnes of fish per year. The subsector contributed some US$88 million to GDP and US$39 million to export earnings at the time. Furthermore, the subsector contributed to food security, income generation and employment opportunities to the members of the communities. Water hyacinth mats impeded fishing activities and deteriorated swamp fish breeding areas, as reflected in decreased production, a reduction in species diversity, poor quality of fish, rising costs of operation, lower incomes to the fishermen and higher prices to consumers (Orach-Meza 1996).

Kateregga and Sterner (2007) summarized the impacts of water hyacinth infestation in Lake Victoria using relatively reliable data source from 1992 to 2001 and described that water hyacinth started to grow rapidly in 1993 and reached a maximum cover in 1997–1998 (covering more than 174 square kilometers of the lake's total 68,000 square kilometers of water surface (Albright et al. 2004, World Agroforestry Centre 2006). Although this coverage was a very small proportion (about 0.25% of total lake surface area), it did cause big problems in Uganda, Tanzania, Kenya and others sharing the world second largest freshwater lake. Uganda, Tanzania and Kenya have a large portion of economic activities associated with agriculture and fisheries, which support 25 million people living in the basin with an estimated gross economic product of US3–4 billion annually. In 2007, fish catch from the lake was estimated at 400–500 thousand tonnes annually; this economic activity directly employed ~100,000 people plus additional 2 million people involved in associated activities.

Although the water hyacinth coverage of 174 square kilometers consisted of numerous patches ranging from 2.5 hectares to 5000 hectares (Mailu et al. 1998, Ochiel et al. 1998), most mat sizes were big enough to block commercial harbors, choke the fish nursery grounds and block the transport to fishing grounds or increase the cost of fish transport (Albright et al. 2004, Kateregga and Sterner 2007). Over the water hyacinth infestation period (1991–1997), the catches of three important fish species of tilapia (*Oreochromis* spp.), catfish (*Clarias* spp.) and bream (*Mormyrus* spp.) declined by 14, 37 and 59% compared to the period 1986–1991, though the total fish yield was not affected (Mailu 2001).

The above examples illustrated the negative impacts of water hyacinth infestation on fisheries worldwide. However, these impacts are complex. In natural ecosystems, the fish abundance and population structure are impacted by many biotic and abiotic factors and their interactions. For example, where water hyacinth was present, water quality was improved (positive impact), especially regarding ammonium and oxygen concentration in open water area into which water flew from water hyacinth mats (Rodríguez et al. 2012). This result was similar with the observations in Lake Chivero (17°54' S 30°48' E) in Zimbabwe in 2003 (Rommens et al. 2003). The same team assessed fish abundance and population structure and concluded that the presence of water hyacinth supported a sheltering or nursery function for small-size classes of fish and at the same time did not have any clear negative impacts on fish populations regarding abundance and diversity (Brendonck et al. 2003). Similar findings were also reported in Lake Dianchi (24°45' N 102°36' E) in China (Wang et al. 2012) and Lake Rodó (34°55' S 56°10' W) in Uruguay (Meerhoff et al. 2003). Further analysis of these cases clarified that the water hyacinth mats were relatively small in size: about 10–30 meters in diameter (<3.2% coverage of the total area) in Lake Chivero, 500 meters in diameter (<1% coverage of the total area) in Lake Dianchi, and three patches of 45 square meters each (about 1% cover) in Lake Rodó. When the mat size was small, the edge effects on aquatic animal population were beneficial (Villamagna 2009). The edge effects were clearly illustrated by an inverse relationship between dissolved oxygen concentration in open water and the distance from the water hyacinth mats in Lake Victoria, Uganda (Masifwa et al. 2001). Kateregga and Sterner (2009) noted that decreased catching rate of certain overfished species led to increased fishery stocks for large-sized species of tilapia (*Oreochromis* spp.), catfish (*Clarias* spp.) and bream (*Mormyrus* spp.), potentially benefiting fisheries and human society (Villamagna 2009).

An impact of water hyacinth on fisheries varied according to different biotic, abiotic, location and management strategies. The usual result was reduced fish yield, increased cost of operation and transportation and increased time to market if the site was not well managed. When a continuous mat of water hyacinth covered more than 25% of water surface (McVea and Boyd 1975), it would decrease dissolved oxygen concentration beneath mats by preventing the transfer of oxygen from the air to the water surface (Hunt and Christiansen 2000) and by blocking light used for photosynthesis by phytoplankton and submerged vegetation. Dissolved oxygen is critical for all aquatic fauna, influencing the abundance and population structure of aquatic animals. Compared with phytoplankton and submerged vegetation, water hyacinth releases less oxygen into the water column via its rot surface (maximum 68 mg O_2 h^{-1} kg^{-1} fresh root) (Ma et al. 2014). In natural waters, submerged vegetation could release 220 mg O_2 m^{-2} h^{-1}, whereas water hyacinth could release only 90 mg O_2 m^{-2} h^{-1} (Zhang et al. 2014). If we assume that dissolved oxygen (DO) concentration in non-plant area was mainly controlled by phytoplankton, the submerged vegetation may increase dissolved oxygen concentration by 7% over phytoplankton and 20% over water hyacinth (Meerhoff et al. 2003). Similarly, areas covered by water hyacinth (~ average 31 square meters) had the lowest dissolved oxygen concentration when compared to areas covered (~ average 31 square meters) by submerged aquatic macrophytes in the Sacramento-San Joaquin Delta, California, US; water hyacinth was the only plant species associated with average dissolved oxygen concentrations dropping below 5 mg L^{-1} (Toft et al. 2003).

Even though the above experiments were conducted in relatively small coverage (30–132 square meters), so that the water exchange and diffusion may minimize the impact, the general trends were clear: water hyacinth could decrease oxygen supply in waters and thus reduce fish yield. However, the impact of aquatic vegetation on dissolved oxygen in whole ecosystem of a lake and freshwater reservoir is very complex and depends on hydraulics, limnological characteristics, levels of eutrophication and nutrient loading in a catchment, and biotic and abiotic factors, including weather. Hence, the management strategies are extremely important to balance the overall outcomes of water quality, fishery and invasive weed control.

Impacts of water hyacinth on hydropower

Hydropower is another important service provided by the ecosystems of rivers, lakes and water reservoirs. Where temperature and nutrient concentrations are suitable, water hyacinth impacts hydropower generation by blocking water intake facilities (McVea and Boyd 1975). Literature reported that the macrophyte affected the production of hydro-electricity throughout Africa (e.g., hydropower station at the Kafue Gorge Dam in Zambia and the hydropower dam on the Shire River in Malawi). The Kafue Gorge power station is responsible for supplying 900 MW power to Zambia. Water hyacinth at the location forced at least one of five turbines to be shut down for one day per week for cleaning (EPPO 2008). A financial loss due to the effect was estimated at several million USD per annum (Wise et al. 2007). Another example was reported at Owen Fall Hydroelectric station at the northern part of Lake Victoria (Uganda) where water coolers and generators were often damaged by the presence of patches of water hyacinth mats. When this happened, a generator had to be switched off for maintenance and this meant the loss of 15 MW of electricity for a while, i.e., a blackout in one of the urban areas of Uganda (Labrada 1996, Orach-Meza 1996).

In a separate investigation, correlations were calculated between areas of water hyacinth infestation and the hydropower outage using reliable data (Table 4.4-1).

Table 4.4-1. Water hyacinth-induced outages and estimated mat area, re-arranged data and statistics sourced and data re-arranged from (Kateregga and Sterner 2007).

Year	**Outages (h)**	**Mat area (ha) in Uganda**
1990	54	0
Year	Outages (h)	Mat area (ha) in Uganda
1991	68	0
1992	76	0
1993	63	800
1994	226	2130
1995	293	3080
1996	367	3670
1997	336	3880
1998	563	3720
1999	148	3190
2000	3	2280

Correlation coefficients between power outages and mat areas in Uganda were 0.76 (data till 2000, $p < 0.05$) and 0.92 (data till 1998, $p < 0.05$).

The River Nile brought down masses of water hyacinth to the dam. While weed-harvesting efforts freed the dam from water hyacinth biomass, the debris that it brought down with it continued to flow down into the protective screens of turbines and the cooling system so that the turbines were often shut down in order to clean the screens and filters (Orach-Meza 1996). Table 4.4-1 indicated that the longest outages occurred in 1996–1998 (Kateregga and Sterner 2007). According to the Uganda Electric Board (UEB), power outages and manual removal of water hyacinth were costing about US$1 million per annum (Mailu 2001, Nang'alelwa 2008).

Other cases have also been reported. In China, blockage of hydroelectric power generation was reported in Shuikou Reservoir in Fujian Province (Lu et al. 2007), Bailianhe Reservoir (60% of which was covered by water hyacinth) in Hubei Province (Xu et al. 2003), on River Nanliujiang in Guang Xi Province (Su 2008), in Brokopondo Reservoir in Suriname in South America (Pringle et al. 2000), in the hydroelectric systems in Spain (Téllez et al. 2008), in Kerala in India (Jayan and Sathyanathan 2012), etc. The impact differed according to various locations and management strategies, but reduced hydropower operation time (less electricity output) and increased operation costs were common.

4.5 Impact of water hyacinth on human health

Water hyacinth is blamed for effects on human health such as poisonous snake bites and infections of malaria, schistosomiasis, amoebic dysentery and typhoid (Orach-Meza 1996), although there was a lack of data from actual case studies to support these assertions. Information largely relied on local newspaper reports such as from The East African Network and The East Africa Environment Network (Navarro and Phiri 2000). It was believed that water hyacinth created an environment suitable as a breeding ground for pathogens or habitat for poisonous snakes. The indirect effects may have been caused by blocking transportation or communication channels for riparian communities. There was a lack of scientific research on the subject, with no report on human pathogens hosted, e.g., by mosquitoes or snails living on water.

Another negative impact of water hyacinth on human health may be linked with deteriorating water quality caused by rotting of massive biomass after the death of water hyacinth. This can happen in shallow non-managed waters or after spreading herbicides. When it happens, the quality of water drinking or other domestic uses is compromised, with especially severe effects in developing countries.

4.6 Impact of water hyacinth on water resources

Fresh water is an essential resource for humans and a source of conflict and tension between tribes, regions and nations. Due to climate change and variability, ecological strain, socio-economic development and human population growth, the surface fresh water from rivers, lakes and reservoirs as well as ground fresh water in shallow aquifers are showing a limitation trend, with about 1.2 billion people living in areas where water is scarce (UN-Water/FAO 2007). The extreme examples were the conflicts between India and Pakistan in the 1950s and between Lebanon and Israel during 1964–1967 (Ahmed and Eldaw 2002). About 20% of world's freshwater aquifers are currently over-exploited (WWAP-United Nations World Water Assessment Programme 2015).

The problems of freshwater supply are not only reflected in quantity, but also in quality. Globally, UNESCO estimated that at least 748 million people do not have access to a source of drinking water in 2015; in addition, about a 400% increase in demand for freshwater supply is expected in the manufacturing sector by 2050 (WWAP-United Nations World Water Assessment Programme 2015). United Nation Environment Protection Agency estimated that every day 25,000 people died as a result of poor water quality in 1994; and during the same period, about one-third of the world's populations were suffering from unsafe drinking waters, which may, in serious situations, kill children or even adults because of diarrhea (UNEP 1994).

Rivers, lakes and water reservoirs are key components of the Earth hydrological cycle, water storage and supply. Additionally, they provide important ecosystem services while supporting significant aquatic biodiversity and recreational activities (Ballatore and Muhandiki 2002).

The impacts of water hyacinth on water quantity and quality were reflected in three aspects:

1) water hyacinth is a superior and approved macrophyte to reclaim clear water from polluted sources such as effluent from Wastewater Treatment Plants (WTP) from domestic sources or eutrophic waters (Sooknah 2000, Chunkao et al. 2012, Achi and Sridhar 2014);
2) water hyacinth may threaten water quality via massive biomass decaying due to natural dying off, or as a results of spreading herbicides for control purposes (Gorchev and Ozolins 1984, Orach-Meza 1996);
3) water hyacinth would increase evapotranspiration by 1.6 (Ndimele et al. 2011) to 6.6 times (Penfound and Earle 1948) over natural evaporation from freshwater surface.

Table 4.6-1 comprises the data on evaporation and evapotranspiration as influenced by the ambient air temperature and relative humidity.

The majority of people and in particular water resources managers may only be concerned with water quantity and quality under the influence of water hyacinth, although the interactions of water resources and water hyacinth are very complex; and the final outcomes depend on many factors. However, the general principles for management decisions are clear: (i) in natural rivers, lakes and water reservoirs, water hyacinth must be harvested, dehydrated and removed away from the water bodies; (ii) control by herbicides must be limited and applied only in emergency. Under normal temperature and nutrient supply, one square meter of water hyacinth mat with about 20 kg of standing biomass can remove 2 g of nitrogen plus other pollutants per day, with a potential loss of 20 L of water (about 2% of the total reclaimed water volume at the conditions of natural evaporation of 3.1 mm d^{-1} and evapotranspiration/evaporation ratio at 6.6 according to Table 4.6-1.

Table 4.6-1. Comparison of evaporation and evapotranspiration influenced by water hyacinth.

Temp (°C)	Relative humidity (%)	Evaporation (EP) (mm d^{-1})	Precipitation (mm y^{-1})	Evapotranspiration (SP) (mm d^{-1})	SP/EP Ratio	Ref.
24.15	79.00	1.4	1331	4.7	3.4	1
15.00	67.00	2.1	150	1.7	0.8	2
20.00	68.00	2.1	150	4.0	1.9	2
28.00	74.33	3.1	150	4.6	1.5	2

Reference 1: (Reddy and Tucker 1983); Reference 2: (Rashed 2014).

References cited

Achi, C. G. and M. K. C. Sridhar. 2014. Performance evaluation of a water hyacinth based institutional wastewater treatment plant to mitigate aquatic macrophyte growths at Ibadan, Nigeria. *International Journal of Applied Science and Technology* 4(3): 117–124.

Ahmed, A. A. and A. K. Eldaw. 2002. Challenges and future opportunities in the Nile Basin. In: International Conference "From Conflict to Co-operation in International Water Resources Management: Challenges and Opportunities," ed. S. Castelein, 49–58. Delft, Netherlands: UNESCO-IHE Institute for Water Education.

Akinyemiju, O. A. 1987. Invasion of Nigerian waters by water hyacinth. *Journal of Aquatic Plant Management* 25: 24–26.

Albright, T. P., T. G. Moorhouse and T. J. Mcnabb. 2004. The rise and fall of water hyacinth in Lake Victoria and the Kagera River Basin, 1989–2001. *Journal of Aquatic Plant Management* 42: 73–84.

Australia Weed Committee. 2013. Water hyacinth - map. *Australian Weeds Committee*. Canberra, Australia: Australian Weeds Committee.

Australian Weeds Committee. 2012. Weed of national significance, water hyacinth (*Eichhornia crassipes*) strategic plan 2012–2017. Canberra, Australia: Australian Weeds Committee.

Ballatore, T. J. and V. S. Muhandiki. 2002. The case for a world lake vision. *Hydrological Processes* 16(11): 2079–2089.

Brendonck, L., J. Maes, W. Rommens, N. Dekeza, T. Nhiwatiwa, M. Barson et al. 2003. The impact of water hyacinth (*Eichhornia crassipes*) in a eutrophic subtropical impoundment (Lake Chivero, Zimbabwe). II. species diversity. *Archiv für Hydrobiologie* 158(3): 389–405.

Chu, J., Y. Ding and Q. Zhuang. 2006. Invasion and control of water hyacinth (*Eichhornia crassipes*) in China. *Journal of Zhejiang University. Science. B* 7(8): 623–6.

Chunkao, K., C. Nimpee and K. Duangmal. 2012. The King's initiatives using water hyacinth to remove heavy metals and plant nutrients from wastewater through Bueng Makkasan in Bangkok, Thailand. *Ecological Engineering* 39: 40–52.

Duan, H., S. Qiang, H. Wu and J. Lin. 2003. Water hyacinth [*Eichhornia crassipes* (Mart.) Solms.]. *Weed Sciences* 2: 39–40 (In Chinese).

Ellis, A. T. 2011. Invasive species profile water hyacinth, *Eichhornia crassipes*. *University of Washington*. Seattle, WA, USA. http://depts.washington.edu/oldenlab/wordpress/wp-content/uploads/2013/03/Eichhoria-crassipes_Ellis.pdf.

EPPO. 2008. Data sheets on quarantine pests: *Eichhornia crassipes*. *EPPO Bulletin* 38(3): 441–449.

FETA. 2014. *Executive summary: proposed implementation of fish hatchery project under FETA and LVEMP II*. Ed. FETA. Dar Es Salaam, Tanzania: United Republic of Tanzania Ministry of Livestock Fisheries Education and Training Agency.

Gao, L. and B. Li. 2004. The study of a specious invasive plant, water hyacinth (*Eichhornia crassipes*): achievements and challenges. *Acta Phytoecologica Sinica* 28(6): 735–752 (In Chinese with English Abstract).

Gichuki, J., R. Omondi, P. Boera, T. Okorut, A. S. Matano, T. Jembe et al. 2012. Water hyacinth *Eichhornia crassipes* (Mart.) Solms-Laubach dynamics and succession in the Nyanza Gulf of Lake Victoria (East Africa): implications for water quality and biodiversity conservation. *The Scientific World Journal* 2012: 1–10.

Government of South Australia. 2013. Weed sheet - water hyacinth (*Eichhornia crassipes*). Adelaide, Australia: South Australian Murray-Darling Basin Natural Resources Management Board.

Hong, S. H. 2004. Invasion of water hyacinth and its control. *Fujian Science & Technology of Tropical Crops* 29(2): 38–40, 16 (In Chinese).

Howard, G. W. and K. L. S. Harley. 1998. How do floating aquatic weeds affect wetland conservation and development? How can these effects be minimised? *Wetlands Ecology and Management* 5(3): 215–225.

Huang, Q. 2009. Inquiring on way of administer the water hyacinth pollution in Nanping. *Chemical Engineering & Equipment* 3: 119–121 (In Chinese with English Abstract).

Hunt, R. J. and I. H. Christiansen. 2000. *Understanding dissolved oxygen in streams. In Information Kit*. 1st ed. Townsville Qld, Australia: CRC Sugar Technical Publication (CRC Sustainable Sugar Production).

Jayan, P. R. and N. Sathyanathan. 2012. Aquatic weed classification, environmental effects and the management technologies for its effective control in Kerala, India. *International Journal of Agricultural and Biological Engineering* 5(1): 76–92.

Kateregga, E. and T. Sterner. 2007. Indicators for an invasive species: water hyacinths in Lake Victoria. *Ecological Indicators* 7(2): 362–370.

Kateregga, E. and T. Sterner. 2009. Lake Victoria fish stocks and the effects of water hyacinth. *The Journal of Environment & Development* 18(1): 62–78.

Kettunen, M., P. Genovesi, S. Gollasch, S. Pagad, U. Starfinger, P. ten Brink et al. 2009. *Technical support to EU strategy on invasive species (IAS)—Assessment of the impacts of IAS in Europe and the EU (final module report for the European Commission)*. Ed. M. Kettunen, P. Genovesi, S. Gollasch, S. Pagad, U. Starfinger, P. ten Brink et al. Brussels, Belgium: Institute for European Environmental Policy (IEEP).

Kusemiju, K., T. A. Farri, F. D. Chizea and E. Ekere. 2002. Water hyacinth infestation of river Niger and Kainji Lake, Nigeria. In: Proceedings of the International Conference on Water Hyacinth, 99–104. New Bussa, Nigeria: National Institute for Freshwater Fisheries Research, Nigeria.

Labrada, R. 1996. Status of water hyacinth in developing coutries. In: Strategies for Water hyacinth Control - Report of a Panel of Experts Meeting, ed. R. Charudattan, R. Labrada, T. D. Center, and C. Kelly-Begazo, 7–15. Rome, Italy: Food and Agricultural Organization of the United Nations.

Li, B., C. Liao, L. Gao, Y. Luo and Z. Ma. 2004. Strategic management of water hyacinth (*Eichhornia crassipes*), an invasive alien plant. *Journal of Fudan University (Natural Science)* 43(2): 267–274 (In Chinese with English Abstract).

Lindsey, K. and H.-M. Hirt. 2000. The water hyacinth in East Africa. In *Use Water Hyacinth! - A Practical Handbook of Uses for Water Hyacinth from Across the World*, ed. K. Lindsey and H. -M. Hirt, 9–20. Winnenden, Germany: Anamed International.

López, E. G., R. H. Delgadillo and M. M. Jiménez. 1996. Water hyacinth problems in Mexico and practised methods for control. In *Strategies for Water hyacinth Control—Report of a Panel of Experts Meeting*, ed. R. Charudattan, R. Labrada, T. D. Center, and C. Kelly-Begazo, 125–135 pp. Rome, Italy: Food and Agricultural Organization of the United Nations.

Lu, J., J. Wu, Z. Fu and L. Zhu. 2007. Water hyacinth in China: a sustainability science-based management framework. *Environmental Management* 40(6): 823–830.

Ma, T., N. Yi, Z. Zhang, Y. Wang, Y. Gao and S. Yan. 2014. Oxygen and organic carbon releases from roots of *Eichhornia crassipes* and their influence on transformation of nitrogen in water. *Journal of Agro-Environment Science* 33(10): 2003–2013 (In Chinese with English Abstract).

Mailu, A. M. 2001. Preliminary assessment of the social, economic and environmental impacts of water hyacinth in the Lake Victoria Basin and the status of control. In: Proceedings of the Second Meeting of the Global Working Group for the Biological and Integrated Control of Water Hyacinth, ed. M. H. Julien, M. P. Hill, T. D. Center, and J. Ding, October 20: 130–139. Beijing, China: Australian Centre for International Agricultural Research (ACIAR).

Mailu, A. M., G. R. S. Ochiel, W. Gitonga and S. W. Njoka. 1998. Water hyacinth: an environmental disaster in the Winam Gulf of Lake Victoria and its control. In: Proceedings of the First IOBC Global Working Group Meeting for the Biological and Integrated control of water Hyacinth: 16–19 November 1998, ed. M. P. Hill, M. H. Julien, and T. D. Center, 101–105. Harare, Zimbabwe: International Organization for Biological Control.

Manninen, C. 2015. Tourism in the Great Lakes Region. *Great Lakes Information Network*. http://www.great-lakes.net/tourism.

Masifwa, W. F., T. Twongo and P. Denny. 2001. The impact of water hyacinth, *Eichhornia crassipes* (Mart) Solms on the abundance and diversity of aquatic macroinvertebrates along the shores of northern Lake Victoria, Uganda. *Hydrobiologia* 452: 79–88.

McVea, C. and C. E. Boyd. 1975. Effects of waterhyacinth cover on water chemistry, phytoplankton, and fish in ponds. *Journal of Environmental Quality* 4(3): 375–378.

Meerhoff, M., N. Mazzeo, B. Moss and L. Rodríguez-Gallego. 2003. The structuring role of free-floating versus submerged plants in a subtropical shallow lake. *Aquatic Ecology* 37(4): 377–391.

Munyaradzi Chenje and J. Mohamed-Katerere. 2006. Invasive alien species. In *Africa Environment Outlook 2—Our Environment, Our Wealth*, ed. UNEP, 331–349. Nairobi 00100, Kenya: United Nations Environment Programme.

Nang'alelwa, M. 2008. Environmental and socio-economic impacts of *Eichhornia crassipes* in the Victoria Falls/Musi-oa-Tunya Pilot Site. In *EPPO/CoE Workshop - How to manage invasive alien plants?*, 18 pp. Mérida, Spain: European and Mediterranean Plant Protection Organization and Council of Europe.

Navarro, L. and G. Phiri. 2000. Water hyacinth in Africa and the Middle East: a survey of problems and solutions. 1st ed. Ottawa, Canada: International Development Research Centre.

Ndimele, P. E., C. A. Kumolu-Johnson and M. A. Anetekhai. 2011. The invasive aquatic macrophyte, water hyacinth (*Eichhornia crassipes*) (Mart.) Solm-Laubach: Pontedericeae: problems and prospects. *Research Journal of Environmental Sciences* 5(6): 509–520.

Neuenschwander, P., O. Ajuonu and V. Schade. 1996. Biological control of water hyacinth in Benin, West Africa. In: Strategies for Water Hyacinth Control - Report of a Panel of Experts

Meeting, ed. R. Charudattan, R. Labrada, T. D. Center and C. Kelly-Begazo, 15–26. Rome, Italy: Food and Agricultural Organization of the United Nations.

NIFFR. 2002. National surveys of infestation of water hyacinth, typha grass and other noxious weeds in water Bodies of Nigeria. occasional paper #5. New Bussa, Nigeria: National Institute for Freshwater Fisheries Research, Federal Ministry of Agriculture & Rural Development.

Nkandu, P. 2013. Zambia: water hyacinth in Zambian Rivers—a threat to tourism. (*News Paper) The Times of Zambia, October 10, 2013.*

Ochiel, G. R. S., A. M. Mailu, W. Gitonga and S. W. Njoka. 1998. Biological control of water hyacinth on Lake Victoria, Kenya. In: Proceedings of the First IOBC Global Working Group Meeting for the Biological and Integrated control of water Hyacinth: 16–19 November 1998, ed. M. P. Hill, M. H. Julien and T. D. Center, 115–118. Harare, Zimbabwe: International Organization for Biological Control.

Olomoda, I. A. 2002. Integrated water resources management: Niger authority's experience. In: International Conference "From Conflict to Co-operation in International Water Resources Management: Challenges and Opportunities," ed. S. Castelein, 13–26. Delft, Netherlands: UNESCO-IHE Institute for Water Education.

Orach-Meza, F. L. 1996. Water hyacinth: its problems and the means of control in Uganda. In: Strategies for Water Hyacinth Control—Report of a Panel of Experts Meeting, ed. R. Charudattan, R. Labrada, T. D. Center and C. Kelly-Begazo, 95–102. Rome, Italy: Food and Agricultural Organization of the United Nations.

Osmond, R. and A. Petroeschevsky. 2013. *Water hyacinth—control modules.* Final V2. ORANGE NSW 2800, Australia: NSW Department of Primary Industries.

Patel, S. 2012. Threats, management and envisaged utilizations of aquatic weed *Eichhornia crassipes*: an overview. *Reviews in Environmental Science and Bio/Technology* 11(3): 249–259.

Patil, J. H., M. AntonyRaj, B. B. Shankar, M. K. Shetty and B. P. P. Kumar. 2014. Anaerobic co-digestion of water hyacinth and sheep waste. *Energy Procedia* 52: 572–578.

Penfound, W. T. and T. T. Earle. 1948. *The biology of the water hyacinth.* New York City and Ann Arbor, Michigan, USA: Ecological Society of America.

Petrell, R., L. Bagnall and G. Smerage. 1991. Physical description of water hyacinth mats to improve harvester design. *J. Aquat. Plant Manage.* 29: 45–50.

Petroeschevsky, A. 2012. Weed management guide, weed of national significance, water hyacinth (*Eichhornia crassipes*). Orange NSW 2800, Australia: NSW Department of Primary Industries.

PHS. 2014. *Proposed control of water hyacinth in Lake Victoria Basin, Tanzania.* Ed. PHS. 1. Mwanza, Tanzania: Plant Health Service Unit, Ministry of Agriculture, Food Security and Cooperatives.

Pietersen, K. and H. Beekman. 2006. Freshwater. In *Africa Environment Outlook 2 - Our Environment, Our Wealth*, ed. UNEP, 121–155. Nairobi 00100, Kenya: United Nations Environment Programme.

Pimentel, D., R. Zuniga and D. Morrison. 2005. Update on the environmental and economic costs associated with alien-invasive species in the United States. *Ecological Economics* 52(3): 273–288.

Pringle, C. M., M. C. Freeman and B. J. Freeman. 2000. Regional effects of hydrologic alterations on riverine macrobiota in the new world: tropical–temperate comparisons. *BioScience* 50(9): 807–823.

Rashed, A. A. 2014. Assessment of aquatic plants evapotranspiration for secondary agriculture drains (case study: Edfina drain, Egypt). *The Egyptian Journal of Aquatic Research* 40(2): 117–124.

Reddy, K. R. and J. C. Tucker. 1983. Productivity and nutrient uptake of water hyacinth, *Eichhornia crassipes* I. effect of nitrogen source. *Economic Botany* 37(2): 237–247.

Rezania, S., M. Ponraj, M. F. M. Din, A. R. Songip, F. M. Sairan and S. Chelliapan. 2015. The diverse applications of water hyacinth with main focus on sustainable energy and production for new era: an overview. *Renewable and Sustainable Energy Reviews* 41: 943–954.

Rodríguez, M., J. Brisson, G. Rueda and M. S. Rodríguez. 2012. Water quality improvement of a reservoir invaded by an exotic macrophyte. *Invasive Plant Science and Management* 5(2): 290–299.

Rommens, W., J. Maes, N. Dekeza, P. Inghelbrecht, T. Nhiwatiwa, E. Holsters et al. 2003. The impact of water hyacinth (*Eichhornia crassipes*) in a eutrophic subtropical impoundment (Lake Chivero, Zimbabwe). I. water quality. *Archiv für Hydrobiologie* 158(3): 373–388.

Simonovic, S. P. 2002. Water resources conflicts and climatic change. In: International Conference "From Conflict to Co-operation in International Water Resources Management: Challenges and Opportunities," ed. S. Castelein, 153–162. Delft, Netherlands: UNESCO-IHE Institute for Water Education.

Solms, M., J. A. Coetzee, M. P. Hill, M. H. Julien and H. A. Cordo. 2009. *Eichhornia crassipes* (Mart.) Solms-Laub. (Pontederiaceae). In: Biological Control of Tropical Weeds using Arthropods, ed. R. Muniappan, G. V. P. Reddy, and A. Raman, 183–210. Cambridge, UK: Cambridge University Press.

Sooknah, R. 2000. A review of the mechanism of pollutant removal in water hyacinth systems. *Science and Technology* 6: 49–57.

Su, S. 2008. A pilot study on the reasons' harms and preventative measures of the hyacinth in Nanliujiang River. *Enterprise Science and Technology & Development* 22(244): 200–202 (In Chinese with English Abstract).

Tan, C., Q. Dong, Y. Wang, Y. Fan, Z. Fan and B. Zhao. 2005. The harmfulness, exploitation and treating measurement of water hyacinth. *Progress in Veterinary Medicine* 26(3): 55–58 (In Chinese with English Abstract).

Téllez, T. R., E. López, G. Granado, E. Pérez, R. López and J. Guzmán. 2008. The water hyacinth, *Eichhornia crassipes*: an invasive plant in the Guadiana River Basin (Spain). *Aquatic Invasions* 3(1): 42–53.

Terry, P. J. 1996. The water hyacinth problem in Malawi and foreseen methods of control. In: Strategies for Water Hyacinth Control—Report of a Panel of Experts Meeting, ed. R. Charudattan, R. Labrada, T. D. Center, and C. Kelly-Begazo, 57–72. Rome, Italy: Food and Agricultural Organization of the United Nations.

The World Bank. 2015. International tourism, receipts (current US$). *The World Bank*. http://data.worldbank.org/indicator/ST.INT.RCPT.CD.

Toft, J. D., C. A. Simenstad, J. R. Cordell and L. F. Grimaldo. 2003. The effects of introduced water hyacinth on habitat structure, invertebrate assemblages, and fish diets. *Estuaries* 26(3): 746–758.

Uka, U. N., K. S. Chukwuka and F. Daddy. 2007. Water hyacinth infestation and management in Nigeria inland waters: a review. *Journal of Plant Sciences* 2(5): 480–488.

UNEP. 1994. *The pollution of lakes and reservoirs*. Nairobi, Kenya: UNEP Environment Library.

UNEP. 2006. *Africa Environment Outlook 2—Our Environment, Our Wealth*. Ed. Division of Early Warning and Assessment (DEWA). *Division of Early Warning and Assessment (DEWA)*. Nairobi, Kenya: United Nations Environment Programme.

UNEP and GEAS. 2013. Water hyacinth—can its aggressive invasion be controlled? *Environmental Development* 7: 139–154.

UN-Water/FAO. 2007. *2007 World Water Day: Coping with Water Scarcity: Challenge of the Twenty-First Century*. New York, USA: UN-Water/FAO.

Villamagna, A. M. 2009. Ecological effects of water hyacinth (*Eichhornia crassipes*) on Lake Chapala, Mexico. Ph.D. Thesis, Virginia Polytechnic Institute and State University, Blacksburg, Virginia, USA.

Villamagna, A. M. and B. R. Murphy. 2010. Ecological and socio-economic impacts of invasive water hyacinth (*Eichhornia crassipes*): a review. *Freshwater Biology* 55(2): 282–298.

Villamagna, A. M., B. R. Murphy and S. M. Karpanty. 2012. Community-level waterbird responses to water hyacinth (*Eichhornia crassipes*). *Invasive Plant Science and Management* 5(3): 353–362.

Wang, Z., Z. Zhang, Y. Zhang and S. Yan. 2012. Large-scale utilization of water hyacinth for nutrient removal in Lake Dianchi in China: the effects on the water quality, macrozoobenthos and zooplankton. *Chemosphere* 89(10): 1255–61.

van Wilgen, B. and Wjd. Lange. 2011. The costs and benefits of biological control of invasive alien plants in South Africa. *African Entomology* 19(2): 504–514.

Williams, A. E., H. C. Duthie and R. E. Hecky. 2005. Water hyacinth in Lake Victoria: Why did it vanish so quickly and will it return? *Aquatic Botany* 81(4): 300–314.

Wise, R., B. VanWilgen, M. Hill, F. Schulthess, D. Tweddle, A. Chabi-Olay et al. 2007. *The economic impact and appropriate management of selected invasive alien species on the African Continent*. CSIR Report Number: CSIR/NRE/RBSD/ER/2007/0044/C: Global Invasive Species Programme.

World Agroforestry Centre. 2006. *Improved land management in the Lake Victoria Basin: Final report on the TransVic project*. Ed. ICRAF. Nairobi, Kenya: World Agroforestry Centre.
World Health Organization (WHO). 2011. Guidelines for drinking-water quality - 4th Edition. Geneva, Switzerland: World Health Organization.
Wright, A. D. 1996. An outline of water hyacinth control in Australia. In: Strategies for Water Hyacinth Control - Report of a Panel of Experts Meeting, ed. R. Charudattan, R. Labrada, T. D. Center, and C. Kelly-Begazo, 105–112. Rome, Italy: Food and Agricultural Organization of the United Nations.
WWAP-United Nations World Water Assessment Programme. 2015. *The United Nations World Water Development Report 2015: Water for a Sustainable World*. Paris, France: UNESCO.
Xu, H., P. Zhu and Y. Zhong. 2003. Excessive propagation of *Eichhornia crassipes* in Bailianhe Reservoir and its preventive countermeasure. *Journal of Huanggang Normal University* 23(6): 71–73 (In Chinese with English Abstract).
Yang, F., T. Ma, J. Chen and B. Li. 2002. *Eichhornia crassipes* disaster in Huangpu River in Shanghai: causes, consequences and control strategies. *Journal of Fudan University (Natural Science)* 41(6): 599–603 (In Chinese with English Abstract).
Zhang, L., Z. Zhang, Y. Gao and S. Yan. 2014. Effect of aquatic plants on emission of gases from eutrophic water. *Journal of Ecology and Rural Environment* 30(6): 736–743 (In Chinese with English Abstract).

Part Two

Mechanism and Implications of Pollutant Removal by Water Hyacinth

CHAPTER 5

Mechanisms and Implications of Nitrogen Removal

Y. Gao, N. Yi and *S. H. Yan**

5.1 Introduction

5.1.1 Nitrogen and eutrophication

Nitrogen (N) dynamics in geochemical processes maintain the living systems in the biosphere on our planet. Application of nitrogen fertilizers in agriculture resulted in significant yield increases during the last century. However, the side effect of nitrogen fertilization is an increased potential for nitrogen leaching, which may cause contamination of groundwater and surface waters. On one side, increased yields from agriculture improve human life and social development. On the other side, world population growth has brought out challenges regarding the treatment of municipal wastewater and effluent discharge, within which nitrogen is frequently the main pollutant.

In the past, management of eutrophic waters was focused on decreasing phosphorus because the literature suggested that low phosphorus was the main limiting factor in the aquatic ecosystems. The strategy achieved great success in enhancing water quality in North America and Europe. However, there have been examples of failure in the implementation of this strategy, such as a lack of improved water quality in Lake Apopka, Lake George and Lake Okeechobee in the United States, Lake Donghu in China and Lake Kasumigaura in Japan using the policy of controlling phosphorus only (Conley et al. 2002). More recently, the attention shifted to nitrogen as the active biogenic element to elucidate its roles in eutrophication and water quality control. Studies have shown that in many aquatic ecosystems, the consumption of

50 Zhong Ling Street, Nanjing, China.
Email: jaas.gaoyan@hotmail.com, yineng5288ys@163.com
* Corresponding author: shyan@jaas.ac.cn

inorganic nitrogen is often accompanied by a bloom of nitrogen-fixing blue-green algae (Levine and Schindler 1999, Temponeras and Kristiansen 2000). In addition, a field survey at Meiliang Bay in Lake Taihu in China showed that cyanobacteria blooms in summer and fall were related to increases in nitrogen and phosphorus concentration, with nitrogen being the first limiting factor and low phosphorus limiting phytoplankton growth when nitrogen was abundant (Xu et al. 2010, Wu et al. 2014).

Nitrogen is one of the key elements causing eutrophication, but also an essential nutrient for biological processes in aquatic ecosystems. There were examples of effective phosphorus management reducing algal blooms, but also increasing nitrogen pollution (Finlay et al. 2013). The reason is simple: because algae require 10 to 40 times as much nitrogen as phosphorus, and phosphorus is typically in shorter supply than nitrogen, controlling phosphorus supply may result in excess unused nitrogen in lake waters and impaired denitrification due to declined growth of algae induced by phosphorus shortage (Bernhardt 2013). Therefore, understanding the characteristics of nitrogen pollution and eutrophication of water bodies is important.

The root cause of eutrophication is excessive nutrients causing algal blooms and destroying original ecological balance because of nutrient loading being larger than the self-purification capacity of water bodies. There are many water properties to evaluate and assess the eutrophic status, among which nitrogen concentration above 0.3 mg N L^{-1} and total phosphorus concentrations above 0.02 mg P L^{-1} are usually used to define eutrophication and surface water quality (Qi et al. 2005), although different countries and regions may have different standards. For example, in 1984 China defined eutrophic standard for surface water on total nitrogen above 1.0 mg N L^{-1} and total phosphorus above 0.2 mg P L^{-1} (MEP-PRC 2002).

The sources of nitrogen overloading can be categorized as exogenous (point and non-point) and endogenous. Exogenous inputs are mainly from agricultural land (surface runoff and leaching), and domestic and industrial wastewater discharges. A portion of nitrogen in water may go through physical, chemical and biological processes and be deposited in sediment to become an endogenous source. When exogenous nitrogen pollution is effectively controlled, the release of endogenous nitrogen may contribute to eutrophication (Qin and Fan 2002).

Nitrogen in water may be in various forms such as molecular nitrogen (N_2), ammonium (NH_4^+-N), nitrate (NO_3^--N), nitrite (NO_2^--N), organic nitrogen compounds (Org-N) and biological nitrogen (in the bodies of algae, plants and animals). These different nitrogen forms are recycling and transforming in complex processes in aquatic ecosystems, including the interaction with adjacent terrestrial ecosystems. In the aquatic environment, nitrogen transformation is mainly driven by microorganisms in autotrophic and heterotrophic processes such as nitrogen fixation, organic nitrogen mineralization, nitrification, denitrification, anaerobic ammonium oxidation as well as reduction of nitric acid to ammonium (Trimmer et al. 2012).

Nitrogen is removed from an aquatic ecosystem usually by three pathways: (1) uptake by algae, aquatic plants and animals and removal of their biomass by harvesting; (2) nitrogen in water and sediment being transformed via mineralization, nitrification, denitrification and other biological processes to various gases (e.g., N_2O, N_2, etc.) that escape from the aquatic ecosystems; and (3) deposition and trapping in the sediments to be temporarily blocked from biogeochemical cycling.

5.1.2 Phytoremediation of eutrophic waters using water hyacinth

Aquatic plants play an important role in nitrogen cycling and can therefore influence nitrogen concentration in water. Macrophyte also can change the rooting environment to impact microbial communities and regulate nitrogen biotransformation processes. Different types of aquatic plants have differential capacities as well as temporal and spatial patterns regarding nitrogen transformation. Emergent and submerged plants generally absorb nitrogen from sediment, but have relatively little effect on nitrogen concentration in the overlying water. Also, these plant types are usually found in shallow waters or near the shore and cannot be established in deep water in some lakes. Among them, the submerged aquatic plants are very sensitive to environmental conditions for survival and reproduction and often require relatively low nitrogen and phosphorus concentrations in water to avoid heavy algal growth that may block the incident light under water for normal photosynthesis. To some extent these factors limit suitability of submerged aquatic plants for remediating eutrophic waters.

Floating aquatic plants such as water hyacinth (*Eichhornia crassipes*) are good candidates for remediation of eutrophic waters due to: (1) growing on water surface (including in deep water areas) for ease of harvesting; (2) assimilating nitrogen and phosphorus directly from water to effectively decrease nutrient concentration in water; (3) accumulating biomass quickly and in large quantities; and (4) effectively accumulating nitrogen and phosphorus. As early as the 1950s to 1970s, the biology of water hyacinth and its socio-economic and ecological impact were investigated in detail. Initially, many studies focused on the relationship between water hyacinth biology and environment, and on how to utilize huge biomass of water hyacinth (Wolverton and McDonald 1979). Some researchers proposed using water hyacinth for treating domestic sewage and wastewater discharged from animal production (Cornwell et al. 1977, Wolverton and McDonald 1978) because water hyacinth has a strong reproductive capacity and can absorb large amounts of nutrients from sewage and wastewater. The macrophyte grows and reproduces very rapidly; under appropriate conditions, one plant can produce 3000 new ones in 50 days or cover 600 m^2 in one year (Madsen 1993, López et al. 1996). To support this rapid growth and reproduction, nitrogen and phosphorus are critical nutrients for water hyacinth in the water (Center and Spencer 1981, Fitzsimons and Vallejos 1986, Moorhead et al. 1988a) influencing the quantity and quality of water hyacinth biomass (i.e., yield, nutrient removal efficiency, dry biomass percentage and crude protein content) (Reddy et al. 1989).

Water hyacinth has strong capacity to tolerate pollutants, including nitrogen and phosphorus, making it an ideal species for purification of wastewater containing N, P and other pollutants at high concentrations. In addition, water hyacinth is also adapted to low nutrient concentrations to produce very clean water (Reddy et al. 1989, Rodríguez et al. 2012). These biological properties make water hyacinth a good biological agent for phytoremediation in eutrophic rivers, lakes and reservoirs. Water hyacinth can form a thick mat to inhibit the growth of algae and intercept detritus by fibrous roots, creating a stable physico-chemical and biological environment.

The well developed and strong root system of water hyacinth has a high degree of morphological flexibility and large root surface area to provide good adhesion for detritus and propagation medium for microbes. The water hyacinth roots can stimulate microbes in the processes of nitrification and denitrification to promote biological nitrogen removal from water bodies (Gao et al. 2014). However, to meet the target, a careful and skillful management plan must include periodical harvests as well as post-harvest procedures for the biomass to completely remove nutrients from water and prevent the pollutants to be returned to water (e.g., in case of on-shore decay of the harvested materials).

Since the 1980s, researchers have focused on water hyacinth biology, utilization and weed management and facilitated application of the macrophyte in purification of eutrophic water contaminated with nutrient runoff or leaching from farm land and also in treatment of sewage and aquaculture wastewater, achieving nitrogen removal rate of up to 80% (Reddy and Tucker 1983, Li et al. 2003, Yi et al. 2009). Water hyacinth has been increasingly recognized as a main macrophyte to remediate eutrophic lake in the ecological restoration/ rebuild projects. For example, Yuan et al. (1983) reported preliminary studies on using water hyacinth (*Eichhornia crassipes*), water lettuce (*Pistia stratiotes*) and alligator weed (*Alternanthera philoxeroides*) to control nutrient overload and utilization of the plant biomass as organic fertilizer in agriculture. The report concluded that in the Lake Taihu region water hyacinth achieved fresh biomass yield of 621 tonnes ha^{-1} yr^{-1} (47 tonnes dry matter ha^{-1} yr^{-1} and 740 kg N ha^{-1} yr^{-1}), which was higher than in the other two plant species. Later, some universities and research institutes in China used water hyacinth in pioneer projects to remediate severely polluted lakes such as Lake Taihu and Lake Dianchi and integrated technologies from weed control, biomass harvesting and post-harvest processing and biomass utilization on a large scale as part of the multidimensional management system. Promising results were achieved regarding water quality control, ecological stability, large-scale (million tonnes) biomass harvest, biomass dehydration and utilization of biomass as bio-energy and organic fertilizer resources (Chen et al. 2012).

Another pioneer project was launched in hyper-eutrophic Lake Dianchi in Kunming, Yunnan Province, China, in 2010. The project was large scale (760 hectares) and employed a multidimensional management strategy to control water hyacinth population, purify effluent from domestic sewage plants and non-point pollution sources, dehydrate harvested water hyacinth biomass and process the biomass for bio-energy and organic fertilizer. The project also launched a third party monitoring system to assess the outcomes of the project. The results from the third party (National Monitoring Program) indicated that during 2006 to 2009, the average concentration of total nitrogen in the water averaged 13 mg L^{-1}. After water hyacinth population was increased to 200 hectares in 2010, water quality improved significantly. When water hyacinth population was further increased to 533 hectares in 2011, nitrogen concentration rapidly dropped from 12 mg L^{-1} at water intake point to 5 mg N L^{-1} on average and to 3 mg N L^{-1} at water outflow site (Wang et al. 2013a, Zhang et al. 2014). Considering the water volume (71 million m^3 yr^{-1}), intake from the catchment and discharge at the exit point, good targets were achieved: removal of 761 tonnes of nitrogen and harvest of 211,000 tonnes of fresh biomass that was 100% dehydrated and utilized for production of bio-energy and organic fertilizer (Wang et al. 2012, Wang et al. 2013a).

5.1.3 Integrated technology for multidimensional approach

There is a large volume of literature on water hyacinth biology, the relationship between water hyacinth and water quality (positive or negative), weed control, biomass utilization, and use of water hyacinth in phytoremediation. However, to successfully control this noxious weed, effectively harvesting and utilizing its huge biomass, and thus safely using this species for phytoremediation and water resource reclamation, it is important to adopt integrated technology and multidimensional strategies. This chapter summarizes water hyacinth biology related to nitrogen, including: (1) the biological characteristics of water hyacinth related to nitrogen assimilation and distribution; (2) the relationship between water hyacinth and aquatic environment regarding nitrogen forms, concentration, accumulation and assimilating speed; (3) the efficiency and effectiveness of nitrogen removal from waters; and (4) role of water hyacinth in influencing microbiological processes involved in nitrogen cycling. The summary may pave the way for further studies or application of water hyacinth management planning and practice in the global scenarios.

5.2 Water hyacinth and nitrogen dynamics

5.2.1 Nitrogen as an essential macro-element for water hyacinth

Nitrogen is an essential macro-element for almost every plant component such as proteins, amino acids, nucleic acids, enzymes, chlorophyll, alkaloids, certain vitamins and hormones. In aquatic environments, nitrogen is often a limiting factor for plant growth and reproduction especially in case of water hyacinth because fast growth rate of the macrophyte requires relatively large quantity of nitrogen. The availability of nitrogen not only governs photosynthetic and respiration processes, but also most other biochemical reactions to determine the yield and quality of water hyacinth biomass when other nutrients are available especially phosphorus. At very high nitrogen concentration (>5.5 mg N L^{-1}), water hyacinth exhibits hyperaccumulation, i.e., absorbing nitrogen more than its physiological need and storing it in tissues (Reddy et al. 1989). Nitrogen deficiency can slow or completely halt plant growth, so that the water hyacinth population control and water quality control may be achieved at the same time.

5.2.2 Nitrogen content in different water hyacinth parts

Different parts of water hyacinth plant have different nitrogen content as influenced by biological developmental stages and the external environment properties, especially nitrogen concentration in growth media. Nitrogen assimilated by water hyacinth is usually incorporated in amino acids and protein very quickly. Due to different biological functions of different plant parts, the protein content of water hyacinth parts is significantly different. Dried water hyacinth roots have the lowest nitrogen content (5.1–23 mg N g^{-1} dry matter) (DeBusk and Dierberg 1984, Mishra and Tripathi 2009) and high fiber content (650 mg g^{-1} dry matter), whereas the leaf (aerial) tissues has high nitrogen content (24–34 mg N g^{-1} dry matter) (Zhang et al. 2010) and low fiber content (490 mg g^{-1} dry matter %).

Usually, the aerial parts of water hyacinth function as the main photosynthetic organs, with higher nitrogen content than the parts under water. Under eutrophic conditions, for example at US Coral Spring sewage lagoons with total nitrogen concentration at 9.8 mg N L^{-1}, nitrogen content in water hyacinth leaf tissues was up to 28 mg N g^{-1} dry matter, but at lower nutrient conditions, for instance at US Washington Lake with total nitrogen concentration at 1.4 mg N L^{-1}, nitrogen content in water hyacinth leaf tissues was only 13 mg N g^{-1} dry matter (DeBusk and Dierberg 1984). Wolverton and McDonald (1978) also found that, when water hyacinth grew at different nitrogen levels in sewage oxidation ponds, the plant nitrogen contents were significantly different: the US Mississippi Lucedale sewage pond with total nitrogen loading (based on an annual average concentration) 9.8 kg ha^{-1} d^{-1}, the crude protein content was 223 mg g^{-1} dry matter (nitrogen content 35.6 mg N g^{-1} dry matter); US Mississippi Orange Grove sewage pond with total nitrogen loading (based on an annual average concentration) 14.9 kg ha^{-1} d^{-1}, the crude protein content was 234 mg g^{-1} dry matter (nitrogen content 37.4 mg N g^{-1} dry matter); US Mississippi National Space Technology Laboratory (NSTL) #1 sewage pond with total nitrogen loading (based on an annual average concentration) 2.5 kg ha^{-1} d^{-1}, the crude protein content was 171 mg g^{-1} dry matter (nitrogen content 27.3 mg N g^{-1} dry matter); US Mississippi NSTL #2 sewage pond with total nitrogen loading (based on an annual average concentration) of 1.1 kg ha^{-1} d^{-1}, the crude protein content was only 97 mg g^{-1} dry matter (nitrogen content 15.6 mg N g^{-1} dry matter).

Another example is from Lake Dianchi in Kunming City, China. At different locations with different nitrogen concentration levels, water hyacinth exhibited different nitrogen content. At one area with nitrogen concentration at 6.38 mg N L^{-1} in water, nitrogen content of the whole plant was 32.9 g N kg^{-1} dried biomass. When nitrogen concentration was 2.35 mg L^{-1}, nitrogen content of the whole plant was 15.0 g N kg^{-1} dried biomass (Yingying Zhang et al. 2011). Further evidence was revealed by 15-N tracer technology for water hyacinth root, stem and leaves. Gao et al. (2012) applied stable isotope $^{15}NH_4^+$ and $^{15}NO_3^-$ separately in experiments to purify eutrophic water and found that (i) the nitrogen content in roots was much less than that in the plant tissues above water surface, and (ii) water hyacinth assimilated NH_4^+ more than NO_3^-. The ^{15}N atomic % (at%) excess in water hyacinth roots (1.1–1.5%) was significantly lower than that in leaf tissues (2.0–2.9%). The ^{15}N recovery rate was also lower in the root (19–21%) than leaf tissue (45–65%). The results also revealed that water hyacinth assimilated more nitrogen (at% excess 1.5–2.9%) in the treatment with $^{15}NH_4^+$ than $^{15}NO_3^-$ (at% excess 1.1–2.9%).

5.3 Relationship between water hyacinth and nitrogen in water

5.3.1 Nitrogen forms and water hyacinth growth

Nitrogen in water has different forms including molecular nitrogen, ammonium (NH_4^+), nitrate and nitrite (NO_3^- and NO_2^-) and organic nitrogen (Org-N). Water hyacinth can only assimilate inorganic nitrogen, although organic nitrogen may be assimilated with the cooperation of microorganisms, e.g., in co-existence with blue-green algae that can fix molecular nitrogen and release inorganic nitrogen into water upon their death and decomposition (Zhou et al. 2014).

Early studies have shown that water hyacinth can simultaneously, quickly and efficiently absorb ammonium and nitrate from water, but has a preference for ammonium (Reddy and Tucker 1983). Moorhead et al. (1988b) applied $^{15}NH_4^+$ and $^{15}NO_3^-$ separately in water and found that water hyacinth absorbed $^{15}NH_4^+$ significantly more than $^{15}NO_3^-$. In the practical application of sewage treatment process, water hyacinth also showed strong capacity to remove ammonium (Zhu and Zhu 1998, Snow and Ghaly 2008). Water hyacinth showed a two-stage mode of NH_4^+ absorption: at low concentrations it showed a saturable absorption curve, but at higher concentrations it showed linear unsaturated absorption (Fang 2006). The two-stage model implies at least two types of transporters: a High Affinity Transport System (HATS) and a Low-Affinity Transport System (LATS). In low concentration range (<500 uM or <1.0 mM NH_4^+), water hyacinth roots absorb NH_4^+ mainly by high affinity system (HATS), showing saturated absorption characteristics and assimilating enough nitrogen for growth. However, at high concentrations (>2.0 mM NH_4^+), water hyacinth roots absorb NH_4^+ by low affinity system (LATS) and show nitrogen hyperaccumulation (Fang 2006). The Michaelis-Menten kinetics of NH_4^+ by water hyacinth are 20–40 uM for half saturation constant (K_m) and 0.3–0.35 umol m^{-2} s^{-1} for maximum absorption capacity (V_{max}) in HATS. When the growth medium NH_4^+ concentrations are in an LATS range, the kinetics are higher at 205–354 uM for K_m and 0.48–1.43 umol m^{-2} s^{-1} for V_{max} (Fang 2006). The value of K_m (NH_4^+) corresponds to the water hyacinth affinity for NH_4^+; the lower the K_m value, the greater the affinity. The Michaelis-Menten kinetics of NH_4^+ suggest that water hyacinth can adapt to or tolerate a wide range of ammonium concentrations (Gopal 1987).

In eutrophic rivers, lakes and reservoirs, ammonium is usually and rapidly converted to nitrate in aerobic conditions by nitrifying bacteria, so that nitrate concentrations in waters are usually higher than ammonium concentrations. Water hyacinth absorbs nitrate across the plasma membrane via co-transport of $2H^+/1NO_3^-$ or by a primary ATP-driven pump ($2NO_3^-/1ATP$), which require energy of electrochemical potential gradient (Ritchie 2006). Depending on the concentration of NO_3^- in water, the absorption processes can be divided into high-affinity nitrate transport system (HATS) and low-affinity nitrate transport system (LATS). When in low NO_3^- concentration growth medium, absorption of NO_3^- by water hyacinth roots mainly depends on the HATS, showing a saturable uptake curve. The NO_3^- uptake by water hyacinth complies with Michaelis-Menten kinetics, averaging K_m 3.2–4.0 umol L^{-1} and V_{max} 4.6–6.1 umol g^{-1} (fresh weight) h^{-1} in 0.6 to 60 mmol L^{-1} NO_3^- concentration range (Wang et al. 2008). After nitrate is absorbed by roots, it must first be reduced to ammonium and then can be effectively included in the nitrogen metabolism in plants. This reduction can occur in roots and leaves.

5.3.2 Nitrogen concentration and water hyacinth growth

Water hyacinth can adapt to a wide range of nitrogen concentration in waters, and can grow and reproduce at low ammonium concentration of 0.05 mg N L^{-1} or nitrate concentration in a similar range (Shiralipour et al. 1981, Tucker 1981). For N concentrations from 0.5 to 5.5 mg N L^{-1}, the growth rate was significantly and positively correlated with nitrogen levels (Reddy et al. 1989). Fox et al. (2008) applied a series of nitrogen concentrations (0, 40, 80, 100, 150, 200, and 300 mg N L^{-1}) and found that in the concentration range from 0 to 80 mg N L^{-1}, the water hyacinth biomass yield increased linearly with an increase in nitrogen levels, but the dry matter yield and the amount of nitrogen taken up changed little at nitrogen concentrations above 80 mg N L^{-1}.

At very high total nitrogen concentration of 420 mg N L^{-1} as nitrate, water hyacinth still grew well (Alves et al. 2003, Li 2012), suggesting that nitrate was almost non-toxic to water hyacinth. Although the upper limit of lethal concentration of ammonium is uncertain, water hyacinth starts to die when NH_4^+ concentration in water reaches 370 mg N L^{-1} (Qin et al. 2015). The results reported may suggest that the normal growth of water hyacinth requires nitrogen concentration at least 0.05 mg N L^{-1} for survival and as high as 80 mg N L^{-1} for maximum yield (Reddy et al. 1989, Fox et al. 2008), but can tolerate very high total nitrogen concentration at 420 mg N L^{-1}. Such biological characteristics make water hyacinth a good candidate for treatment of sewage (high concentrations) and phytoremediation of eutrophic water in rivers, lakes and reservoirs (low concentrations).

Niu (2012) investigated interaction between nitrogen and phosphorus in indoor simulation experiments with the combinations of nitrogen concentrations at 1.0, 1.5, 2.0, 2.5, 3.0, 3.5, 4.0, 4.5, and 5.0 mg N L^{-1} and phosphorus concentrations at 0.2, 0.3, 0.4, 0.5, 0.6, 0.7, 0.8, and 1.0 mg P L^{-1}. All combinations of concentrations promoted rapid growth of water hyacinth, with a continuous increase in growth in the first 15 days to reach the highest growth rate, followed by a gradual decline. The best growth rate was observed at 4.5 mg N L^{-1} and 0.9 mg P L^{-1}, which was in compliance with other literature reported (Polomski et al. 2009) and indicated that interactions of nitrogen and phosphorus are affecting the growth of water hyacinth. Hence, in phytoremediation of eutrophic waters, nitrogen/phosphorus ratio can be important for water hyacinth. Niu (2012) further reported that nitrogen:phosphorus ratio (N:P) of 7:1 may result in effective growth of water hyacinth, suggesting it as an optimal ratio. This ratio was also confirmed by observations in natural waters: a better growth on one site with the ratio of 7:1 (2.8 mg N L^{-1} and 0.39 mg P L^{-1}) compared with the other site and ratio of 5:1 (2.28 mg N L^{-1} and 0.50 mg P L^{-1}) in the same ecological environment. The optimum N:P ratio of 7:1 in the Niu's report is higher than 1.5 to 5 in other literature reports (Reddy and Tucker 1983, Petrucio and Esteves 2000). The differences may come from the N and P initial concentrations in the experiments and the growing stages of the macrophyte. Basically, low nutrient concentration and more mature growing stage may result in a higher optimal N:P ratio, whereas higher nutrient concentration and young growing stage may show a lower N:P ratio. In real applications in rivers, lakes and reservoirs, a higher N:P ratio may be desirable.

5.3.3 Nitrogen assimilation and storage by water hyacinth

The absorbed NO_3^- must be reduced to NH_4^+ in the plant tissues, whereas absorbed NH_4^+ is immediately incorporated into metabolism to avoid NH_4^+ toxicity at high concentrations in tissues. The free ammonia (NH_3) in plant tissue may inhibit respiration in electron transfer systems, especially nicotianamide adenine dinucleotide (NADH) processes. Water hyacinth has the capacity to absorb large amounts of ammonium and has an efficient ammonium assimilation system to reduce the potential toxic effects. After nitrate enters a cell, it is stored in vacuole or processed in the cytoplasm by Nitrate Reductase (NR) to convert it to nitrite that then enters plastids (usually chloroplasts), within which further transformation to ammonium by nitrite reductase (NiR) in aerobic conditions takes place; NH_4^+ so produced then enters metabolic pathways the same way as NH_4^+ absorbed in that from the rooting environment.

The ammonium nitrogen in water hyacinth cells is involved in three major reactions: (1) carbamoyl-phosphate synthetase I reaction; (2) glutamate dehydrogenase (GDH) reaction producing glutamate via reductive amination of alpha-ketoglutarate; and (3) glutamine synthetase (GS) reaction using ATP-dependent amidation of gamma-carboxyl of glutamate to form glutamine. There are two major pathways for ammonium assimilation in water hyacinth, GDH/GS route for ammonium-rich condition and GS/GOGAT[1] for ammonium-limited situations. The final products are amino acids, from which proteins are synthesized. Thus, the nitrogen extracted from water is finally stored as protein in water hyacinth and can be further utilized via post-harvest processing to achieve water purification, resource recycling or bio-energy production.

Nitrogen hyper-accumulation is defined as water hyacinth absorbing and storing nitrogen in plant tissues in amounts greater than what is needed to support its biomass yield and reproduction. The hyper-accumulation of nitrogen resulted in increased protein content of water hyacinth tissues (Alves et al. 2003, Reddy and Tucker 1983, Reddy et al. 1989). However, concentration of available nitrogen (mainly inorganic nitrogen) in most eutrophic rivers, lakes and reservoirs is usually low (2.0–5.0 mg N L^{-1}) so that water hyacinth mainly absorbs nitrogen in a linear unsaturated mode (high-affinity transporters), with a powerful nitrogen removal and storage capacity. These properties of water hyacinth regarding nutrient uptake and storage are preferred in practical applications of water management strategies. For example, in eutrophic Lake Dianchi (Kunming, China) where an ecological phytoremediation project was executed, water hyacinth accumulated various nitrogen contents in the biomass at different sites with different nitrogen concentrations in water. The nitrogen content of the plant biomass was positively correlated with nitrogen concentration in water (Table 5.3.3-1) (Yingying Zhang et al. 2011). Another example is the phytoremediation practice in Lake Taihu (Jiangsu Province, China) where water hyacinth yielded 797 t (fresh weight) ha^{-1} yr^{-1} and absorbed 1.2 t N ha^{-1} yr^{-1} (Zheng et al. 2008).

Table 5.3.3-1. The relationship between nitrogen concentration in water and nitrogen content in water hyacinth biomass.

Location	Nitrogen Content (g N kg^{-1} dry weight)	Nitrogen concentration in water (mg N L^{-1})
Cao Hai	32.9	6.38
Lao Gang Yu Tang	16.5	3.05
Long Meng Cun	25.8	4.28
Hai Kou Zheng	14.1	1.37
Bai Shan Wan	15.0	2.35

[1] GOGAT stands for glutamine: 2-oxoglutarate amidotransferase.

5.4 Impact of water hyacinth on nitrogen removal

5.4.1 Efficiency of nitrogen removal by water hyacinth

To assess the outcomes of phytoremediation in practice, nitrogen removal efficiency is defined by two concepts: (1) nitrogen removed per unit area and time, i.e., g N m^{-2} yr^{-1}; and (2) nitrogen removed per unit fresh weight biomass and time, i.e., g N kg^{-1} (fresh weight biomass) yr^{-1}. These two concepts are often used to illustrate the efficiency of water hyacinth in water purification. The two concepts have some advantages and disadvantages: the efficiency per unit area can intuitively reflect the relationship between the size of water hyacinth mat and the nutrient removal, but the density of water hyacinth (standing crop) has a great impact on purification outcomes. Therefore, when using the expression of efficiency based on unit area, water hyacinth growth should be assessed via both biomass yield and coverage densities. Efficiency assessed by unit fresh weight biomass takes the relationship between water hyacinth biomass and nitrogen uptake into consideration, but different growing stages of water hyacinth at different environmental conditions cause differences in nitrogen uptake and utilization. Hence, plant growing stage and environmental conditions must be analyzed. Furthermore, nitrogen assimilation by water hyacinth is often associated with available phosphorus, so in assessing nitrogen removal efficiency, available phosphorus must be taken into consideration.

Efficiency of water hyacinth to absorb nitrogen in the practice of water purification attracted a lot of research interest to assess the capacity of the macrophyte. In early years, most experiments were conducted under static nitrogen concentration, with a lack of consideration for maintaining a continuous supply of nitrogen and phosphorus. In addition, different experiments used different forms and concentrations of nitrogen, different growing stages of water hyacinth and experimental conditions such as temperature, pH and incident light, leading to a wide range of nitrogen removal efficiencies. For example, one square meter of water hyacinth mat can remove from 416 mg N m^{-2} d^{-1} up to as high as 2316 mg N m^{-2} d^{-1} (Reddy and Tucker 1983, Petrucio and Esteves 2000). Despite a wide range, the literature still indicated that water hyacinth has a great advantage in phytoremediation of eutrophic waters or sewage.

Many reports have the efficiency expressed on unit fresh weight (fw). Zhang et al. (2011) reported efficiency of 0.95 g N kg^{-1} fw d^{-1} in the dynamic simulation test with TN 4.85 mg N L^{-1}, TP 0.50 mg P L^{-1} and a continuous flow rate of 0.14–1.00 m^3 d^{-1} (at hydraulic loadings of 0.14, 0.20, 0.33, or 1.0 m^3 m^{-2} d^{-1}). Rommens et al. (2003) tested nitrogen (NO_3^- and NH_4^+) and phosphate uptake capacity of water hyacinth in a laboratory-based experiment designed to mimic nutrient conditions of Lake Chivero, Uganda, and reported that the average water hyacinth plant absorbed 2.4 mg of ammonium, 1.1 mg of nitrate, and 0.39 mg of phosphate per kilogram of fresh weight per hour.

Water quality improvement

In the practice of phytoremediation, the final target inevitably involves the water quality improvement. To achieve this target, the final water quality standard, effluent characteristics (especially nitrogen and phosphorus loading), water hyacinth coverage and growth rate, and harvest strategies including post-harvest procedures need to be carefully investigated and planned before the implementation of a eutrophic water purification project. A usual principle is: the higher the nitrogen loading (high concentration and large quantity of effluents at a given time), or the higher the quality standard (lower final nitrogen concentration in outflow), the longer time is required, or the larger the water hyacinth coverage is needed. The longer treatment time or larger water hyacinth coverage implies an increase in the management cost, making acceptable quality standard and high efficiency procedures the main concerns in a practical project.

An increase in the hydraulic retention time (i.e., longer processing time) can mean a lower coverage of water hyacinth mat, and vice versa. For this reason, hydraulic loading is the key character of a treatment system design in order to achieve the final water quality standard for total nitrogen, nitrate, ammonium, total phosphorus, chemical oxygen demand, biological oxygen demand and detritus removal. Due to the complexities of environmental factors such as temperature, pH, dissolved oxygen content and contaminant concentrations in effluent, there is no universal model for a particular system design. Literature suggested using historical data (local and global) for a pilot design, then fine tuning it with practice.

Wooten and Dodd (1976) studied the effect of water hyacinth on the purification of effluent discharged from the secondary treatment at Ames sewage treatment plant in Iowa, the United States. During the experiment, effluent filled into a consecutive series of five ponds (each 465 square meters in area and 82 cm in depth) at the flow rate 0.48 $m^3\ min^{-1}$ at pond #1. The effluent had concentrations of 2.4 mg $NH_4^+\ L^{-1}$, 2.9 mg $NO_3^-\ L^{-1}$ and phosphate 18.8 mg L^{-1} at pH 7.5. After water hyacinth covered the surface areas of all five ponds, NH_4^+ was reduced to 0.5 mg L^{-1} and NO_3^- to 0 mg L^{-1} (below detection limit).

Another example was reported by Zhang et al. (2010) in the laboratory in Nanjing, China. With hydraulic loadings of 0.14, 0.20, 0.33 and 1.00 $m^3\ m^{-2}\ d^{-1}$ and initial nitrogen and phosphorus concentration at 4.85 (TN), 1.33 (NH_4^+), 2.92 (NO_3^-) and 0.50 (TP) mg L^{-1}, water hyacinth removed 85% NT and 81% NH_4^+ at hydraulic loadings of 0.14 and 0.20 $m^3\ m^{-2}\ d^{-1}$, but 74% of TN at hydraulic loading 0.33 $m^3\ m^{-2}\ d^{-1}$ and 73% of TN at hydraulic loading 1.00 $m^3\ m^{-2}\ d^{-1}$. The data showed a higher hydraulic loading correlated with lower water quality obtained. However, when looking at nutrient removal efficiency, TN removal efficiency was at 0.58 and 0.72 g $m^{-2}\ d^{-1}$ at hydraulic loadings 0.14 and 0.20 $m^3\ m^{-2}\ d^{-1}$, respectively, but 0.83 and 1.47 g $m^{-2}\ d^{-1}$ at hydraulic loadings 0.33 and 1.00 $m^3\ m^{-2}\ d^{-1}$, respectively. The data elucidated that a higher hydraulic loading correlated with higher efficiency of TN removal.

From 2014 to 2015, the Jiangsu Academy of Agricultural Sciences initiated *in situ* experiments with flowing water flume (Fig. 5.4.1-1) at sewage treatment pond located in Nanjing, Jiangsu Province, China.

The experiment used 24-hour continuous water flow to study the efficiency and water quality control by water hyacinth. The experiment was designed with three hydraulic loadings (0.2, 0.4 and 0.8 m^3 m^{-2} d^{-1}). Preliminary results showed that TN can be reduced from initial 14.8 mg L^{-1} to outflow 5.0 mg L^{-1} (20-day average) at hydraulic loadings 0.2 and 0.4 m^3 m^{-2} d^{-1} (and outflow 9.0 mg L^{-1} during the same period at the hydraulic loading 0.8 m^3 m^{-2} d^{-1}). Even though water quality was better in the low hydraulic loading treatment, the TN removal efficiency showed an opposite tend. Averaged daily TN removal amounts were 2.2, 3.9 and 4.7 g m^{-2} d^{-1} at hydraulic loading 0.2, 0.4 and 0.8 m^3 m^{-2} d^{-1}, respectively (unpubl. data).

In the absence of sustained supply of nitrogen and phosphorus in static culture media, water hyacinth can remove nitrogen and other pollutants to improve water quality standard. Zimmels et al. (2007) studied four macrophyte species: water hyacinth (*Eichhornia crassipes*), water lettuce (*Pistia stratiotes*), salvinia (*Salvinia rotundifolia*) and water primrose (*Ludvigia palustris*) regarding sewage treatment and found that water hyacinth was the most effective macrophyte to purify sewage to low concentrations of 0.2 mg L^{-1} (NH_4^+), 1.3 mg L^{-1} (BOD), 11.3 mg L^{-1} (COD) and 0.5 mg L^{-1} (total suspended particles, TSS) in 11 days on 40 liters of sewage with initial concentrations of 10 mg L^{-1} (NH_4^+), 10 mg L^{-1} (BOD), 30 mg L^{-1} (COD) and 16 mg L^{-1} (TSS) (Zimmels et al. 2007). Zhang et al. (2010) applied four levels of TN at 2.1, 6.2, 15.1 and 20.1 mg L^{-1} and repeated the experiment for 15 times at intervals of every 21 days to study the effects of water hyacinth on purification of eutrophic water. They found that TN was decreased to 0.28, 1.6, 5.9 and 8.9 mg L^{-1} within 21 days using 1000 liters of eutrophic waters.

Fig. 5.4.1-1. Experimental layout with a number of stainless steel flow sinks (10 m length by 1 m width by 0.5 m depth), with the depth of each flow sink being adjusted by foam floats on both sides of the frame. Wastewater was continuously flowing into each sink through a steel pipe (inner diameter 50 mm) placed at one end 10 cm below the water surface. Large size detritus was filtered by a 1.3-mm nylon mesh. Drainage was through an outlet steel pipe (inner diameter 50 mm) placed at the other end 10 cm below the water surface. The hydraulic loading was controlled by a pump (Photo by Lin Shang 2015).

5.4.2 Contribution of water hyacinth to nitrogen removal in phytoremediation

Nitrogen dynamic in eutrophic water is very complex, especially regarding denitrification. However, water hyacinth can promote denitrification that removes large quantities of nitrogen from eutrophic water via enhancing microorganism activities (Yi et al. 2014, Zheng et al. 2015).

The nitrogen removal by water hyacinth varies with different nitrogen concentration in water. Generally, water hyacinth may achieve lower removal percentage at high nitrogen concentration (>15 mg L^{-1}), and higher at low nitrogen concentration (<6.0 mg L^{-1}). For example, water hyacinth removed 100, 83, 46 and 42% of total nitrogen at initial nitrogen concentrations of 2.1, 6.2, 15.1 and 20.1 mg L^{-1}, respectively (Zhang et al. 2010). Another example was from the experiment using high nitrogen concentration (40–300 mg L^{-1}) in indoor simulation test; assimilation by water hyacinth removed 60–85% of total nitrogen (Fox et al. 2008).

5.5 Nitrogen removal by denitrification

5.5.1 Nitrogen balance in aquatic environments as influenced by water hyacinth

Literature confirmed that water hyacinth can remove large amounts of nitrogen and phosphorus from eutrophic waters in ecological engineering projects or natural habitats (Rodríguez et al. 2012), but denitrification processes also remove large quantities of nitrogen from aquatic systems. Although denitrification in natural aquatic systems is relatively well understood, a contribution of water hyacinth to denitrification is less clear. Most studies have focused on absorption of nutrients by water hyacinth, but nitrogen balance showed that the total removed nitrogen is often greater than the amount assimilated by water hyacinth (Moorhead et al. 1988a). In the large-scale phytoremediation using water hyacinth nitrogen removed through harvest of water hyacinth biomass only contributed 64% of the total nitrogen removed from the eutrophic water with initial nitrogen concentration 5.5–14.5 mg L^{-1}. The report showed that during a 7-month experiment period, harvested fresh water hyacinth biomass amounted to 211,000 tonnes and contained 486 tonnes of nitrogen, while nitrogen balance showed that the removed nitrogen was 761 tonnes, suggesting that about 275 tonnes of nitrogen might have been removed by denitrification (Wang et al. 2013).

Most studies in nitrogen dynamic have assumed the missing part of nitrogen (NO_3^-, NH_4^+ and NO_2^-) would be due to denitrification and other biological transformation processes so that the nitrogen is finally released to atmosphere in the form of N_2O and N_2. However, in management decision making regarding eutrophic waters and weed control, assumptions are not enough. More data need to be collected in phytoremediation processes on how much and by what mechanism denitrification may be influenced by macrophytes, especially by water hyacinth. Denitrification depends on a number of complex processes in aquatic environments. The processes involve microorganisms and macrophytes, and are influenced by physical and chemical factors in environment such as dissolved oxygen, pH, temperature, and available organic matter and nutrients. To understand the complex processes, the relationship between macrophytes and microorganisms together with the impacts of physical and chemical environmental factors should be investigated in detail. It should also be borne in mind that N_2O from the process of denitrification is a greenhouse gas that affects global warming (Seitzinger et al. 2006).

5.5.2 Fate of nitrogen traced by ^{15}N

Nitrogen stable isotope tracer technology using ^{15}N is still the most reliable way to accurately track the fate of nitrogen in systems. Moorhead et al. (1988b) used stable isotope ^{15}N-labeled fertilizer ($^{15}NH_4^+$ and $^{15}NO_3^-$) and water hyacinth in purification of sewage effluent with 50 mg L^{-1} BOD, 6 mg NH_4^+-N L^{-1}, total-Kjeldahl N 9 mg L^{-1}, 2 mg TP L^{-1} and they topped N up to 20 mg N L^{-1} with $^{15}NH_4^+$ (10.05% atomic percentage excess) or $^{15}NO_3^-$ (10.0% atomic percentage excess) in the experiment. After adding $^{15}NO_3^-$ to water, water hyacinth took up 57–72% of ^{15}N; in the $^{15}NH_4^+$ treatment, water hyacinth absorbed 70–89% of ^{15}N. In the treatment without water hyacinth, denitrification removed 13–89% of $^{15}NH_4^+$ and 48–96% of $^{15}NO_3^-$. However, the study did not explain whether there was any nitrogen removed by algae because algae play an important role in denitrification in the eutrophic waters. Gao et al. (2012) used ^{15}N trace technology combined with mass balance method in a laboratory simulation test to accurately track the fate of nitrogen in the water purification system by water hyacinth. The simulated eutrophic water contained 5.4 mg L^{-1} of $^{15}NO_3^-$ (9.98% atomic percentage excess) (7.6 mg L^{-1} of TN) or 5.6 mg L^{-1} of $^{15}NH_4^+$ (10.08% atomic percentage excess) (9.1 mg L^{-1} of TN) at the start of the experiment. After excluding the residual in water and algal assimilation of $^{15}NO_3^-$ and $^{15}NH_4^+$, the result revealed that denitrification removed 26.1% of $^{15}NO_3^-$ or 17.7% of $^{15}NH_4^+$ in the treatment without water hyacinth, but 34.4% of ^{15}N-NO_3^- or 20.8% of $^{15}NH_4^+$ in the treatment with water hyacinth. The conclusion was clear that water hyacinth promoted the denitrification processes. Ma et al. (2013) and Gao et al. (2013 and 2014) collected nitrogen (N_2) and nitrous oxide (N_2O) gases *in situ* to directly track denitrification from eutrophic waters as impacted by water hyacinth. The effects of water hyacinth on promoting denitrification were confirmed (1.1–2.7 times that without water hyacinth).

Water hyacinth has a well-developed fibrous root system with an average surface area of 30–60 m^2 per individual macrophyte (Zhou et al. 2012) and an average root diameter of 1 mm (Hadad et al. 2009). The flexible morphology enables the macrophyte to extend its root as long as 2 meters in water (Rodríguez et al. 2012). These biological characteristics make water hyacinth easily intercept detritus to provide favorable micro-environment for microbes in water (Kim and Kim 2000, Yi et al. 2009). Water hyacinth roots can also transfer oxygen to the rhizosphere to provide the required oxygen for microbial nitrification on the root. Ma et al. (2014) investigated the function of water hyacinth roots in releasing oxygen to an ambient environment using titanium (3^+) citrate colorimetric method and found that the single young root of whole water hyacinth plant can release 56 μmol O_2 h^{-1}; at late growth stage, the root of a whole water hyacinth plant released 93 to 106 μmol O_2 h^{-1}. Young water hyacinth roots (< 0.1 g dry weight) released 3.7 g O_2 h^{-1} kg^{-1} (dry weight root) as measured by an oxygen electrode method; when the water hyacinth root dry matter increased to >1 g, the oxygen releasing capacity dropped to 0.11 g O_2 h^{-1} kg^{-1} (dry weight root) (Moorhead and Reddy 1988). The same authors also estimated diffusional capacity of water hyacinth roots to transport O_2 at 0.12–1.3 mg O_2 g^{-1} (dry weight root) h^{-1}. On the other hand, a water hyacinth mat on water surface hampers oxygen exchange between water and atmosphere and, to some extent, decreases dissolved oxygen in water, which may be favorable for anaerobic denitrification processes. Furthermore, water hyacinth roots can also secrete large amounts of organic carbon to provide carbon source for microorganisms in the rhizosphere. The complex interactions among macrophyte, microbes and the environment may influence nitrification and denitrification processes (Ma et al. 2014).

5.5.3 Impact of water hyacinth on denitrifier community

Denitrification is an important pathway in the nitrogen biogeochemical cycle and also in phytoremediation of eutrophic waters or wastewater treatment, and is responsible for about 30–50% of oceanic N_2 production (Bulow et al. 2010). In natural aquatic ecosystems, loss of nitrogen is dominated by heterotrophic denitrification, although other nitrogen losses also occur (Ward et al. 2009). Heterotrophic denitrification is an anaerobic respiratory process in which nitrate is reduced to nitrogen gas (N_2) through the intermediates nitrite, nitric oxide and nitrous oxide. Given that N_2O and N_2 from denitrification escape from water, the impact of denitrification on removal of excess nitrogen from sewage or eutrophic waters becomes important in their treatment and management. Denitrification process consists of four steps: (1) reduction of nitrate (NO_3) to nitrite (NO_2) by membrane-bound respiratory nitrate reductase *narX*[2], periplasmic nitrate reductase *napY*[3] or assimilatory nitrate reductase *nasZ*[4]; (2) reduction of nitrite (NO_2) to Nitric Oxide (NO) by nitrite reductase Cu type, nitrite reductase *nirK* or cytochrome cd_1 type *nirS*; (3) reduction of

[2] X refers to *A, B, C, D, E, G, H, I, J, L.*
[3] Y refers to *A, B, C, D, E.*
[4] Z refers to *A, B, C, D, E, F.*

nitric oxide (NO) to nitrous oxide (N_2O) by nitric acid reductase *norN*[5]; and (4) reduction of nitrous oxide (N_2O) to molecular nitrogen by nitrous oxide reductase *nosM*[6] (Zumft 1997).

A wide variety of bacterial and archaeal genera from different phyla are involved in denitrification. During the wastewater treatment or eutrophic water purification processes dominated by macrophytes, the underlying sediment and other environmental factors impact the abundance and species community structure of denitrifying microbes. Understanding these impacts can provide valuable information to assess the state and strength of denitrification and also a role of management strategies in governing microorganisms to enhance biological denitrification. The abundance, diversity and community structure of denitrifier genes such as *amoA, narG, napA, nirK, nirS* and *nosZ* can be easily detected by Polymerase Chain Reaction (PCR) technology, fluorescence real-time quantitative PCR (Real-time PCR), Denaturing Gradient Gel Electrophoresis (DGGE), Fluorescence *In Situ* Hybridization (FISH), microarray analysis and high-throughput sequencing techniques.

The PCR technique is based on synthesis of new DNA or RNA based on two sets of primers with desired sequences serving as a template for DNA or RNA sample amplification. PCR allows selective amplification of for example *nirS* or *nirK*, can produce high amounts of pure DNA, and enable analysis of DNA samples even from very small amounts of environmental samples. PCR technology is often indispensable in denitrification research.

Real-time PCR is an established technology for DNA quantification that measures the accumulation of DNA product after each round of PCR amplification. It uses fluorescence to monitor the amplification of targeted DNA sequence, i.e., analyze the number (copies) of new DNA produced. Real-time PCR methods allow the estimation of the amount of a given sequence present in a sample and are used to quantitatively determine levels of gene expression in denitrification research.

To get quantitative information on the diversity and abundance of denitrifiers, denaturing gradient gel electrophoresis (DGGE) is often employed. The principle of the DGGC technology is based on different DNA fragments (created by denaturing agents) traveling different distances on a gradient polyacrylamide gel to form bands. Each band theoretically represents a type of DNA, and the intensity of different bands represents the relative quantity of a DNA fragment.

In wastewater treatment or eutrophic water studies, the DNA in environmental samples usually exists in very small quantities that are often not sufficient for identifying specific targets such as *nirS* or *nirK* that are the markers for denitrifiers. These two markers are functionally equivalent (yet structurally divergent) nitrite reductases catalyzing conversion of nitrite to nitric oxide. *nirS* is iron-based cytochrome cd_1 reductase encoded by *nirS*

[5] *N* refers to *B, C, D, E, Q, Z*.

[6] *M* refers to *A, D, F, L, X, Y, Z*.

gene. *nirK* is a copper-based nitrite reductase encoded by *nirK* gene. Both *nirS* and *nirK* have been used as marker genes for denitrifying bacteria (Braker et al. 1998, Hallin and Lingren 1999). In practice, targeted DNA samples from sediments, roots, water and detritus are often amplified using PCR technology (Liang and Zuo 2008, Wang and Hu 2010) to several orders of magnitude, generating thousands to millions of copies of a particular *nirS or nirK* sequence.

The DGGE technique was used in studying microbial community structure and demonstrated the unique advantages in revealing the genetic diversity and abundance of 16S rRNA (Muyzer et al. 1993). A microbial DNA fragment electrophoresis pattern (different bands) reflects the diversity and complexity of gene sequences. so that the DGGE technique can be applied to study microbial community in nitrogen removal systems (Liu et al. 2011). Although water hyacinth roots were hypothesized to promote the nitrification and denitrification processes by providing good adhesion and propagation medium for bacteria (Sooknah 2000), there was a lack of evidence either from microbiology or from nitrification/denitrification products directly collected from water hyacinth roots. Yi et al. (2014) comparatively studied the abundance and diversity of denitrifying bacteria on water hyacinth roots by adding different forms of nitrogen (NO_3^- and NH_4^+) in artificial sewage wastewater and then used DGGE assay to elucidate denitrifying bacteria diversity on the water hyacinth roots and in the ambient water. They found that the richness and Shannon diversity indices of *nirK* and *nirS* were significantly lower in the ambient water than on the water hyacinth roots. The Correspondence Analysis (CA) from the DGGE electrophoresis data revealed that the denitrifying population structures of *nirK* and *nirS* have similar characteristics both in water and on the water hyacinth roots.

Real-time PCR technology was used to investigate the effects of water hyacinth roots on the abundance and diversity of attached denitrifying bacteria in artificial sewage water supplemented with different forms of nitrogen (NO_3^- and NH_4^+). The *nirS* and *nirK* copy numbers were significantly higher in the treatment with water hyacinth than without. The result also showed a positive correlation ($p < 0.01$) between the total number of gene copies (*nirS*, *nirK*, *nosZ*) and gaseous nitrogen loss percentage (Yi et al. 2014). Hence, these results clearly confirmed that water hyacinth roots can increase the diversity and abundance of denitrifying bacteria and improve the nitrogen removal process though gaseous nitrogen losses in sewage treatment or eutrophic water management. Furthermore, *nirS*, *nirK* and *nosZ* have different responses to the presence of water hyacinth under different environmental conditions. For example, among the three genes, *nosZ* had the highest diversity and abundance in the treatment with NO_3^- as nitrogen source (Yie et al. 2014).

Similar results were reported from large scale phytoremediation of eutrophic Lake Dianchi in Kunming, Yunnan Province, China. There are six rivers transporting effluents from different sources, including wastewaters from municipal treatment plants, from industries and from agricultural sources. DGGE profiles and real-time PCR quantitative analysis of these different pollutant sources revealed that the abundance and diversity of *nirS*, *nirK* and *nosZ* varied widely (Yi et al. 2015a). The wastewater from municipal treatment plants and from agricultural sources increased, and wastewater from industrial sources decreased, the abundance and diversity of denitrifiers carrying *nirS*, *nirK* and *nosZ* genes. This finding was in accordance with other reports on industry wastewater decreasing the abundance and diversity of those genes due to increased heavy metal concentration (Mahmoud et al. 2005, Xiong et al. 2012).

By growing water hyacinth in eutrophic or polluted waters, microbial growth on the water hyacinth roots could be promoted, and the abundance and diversity of denitrifying bacteria could be increased significantly, improving the ecological function. Furthermore, redundancy analysis (RDA) between denitrifying bacterial community structure and environmental factors found that *nirS*, *nirK* and *nosZ* genetic diversity and denitrifying bacteria abundance in water as well as on the water hyacinth roots were modulated by both plant and environmental factors such as temperature, dissolved oxygen, pH, and concentrations of nitrate, ammonium and total nitrogen (Yi et al. 2015b). For this reason, during the implementation of phytoremediation in the ecological engineering projects using water hyacinth, the biology and environmental factors need to be taken into consideration in order to promote nutrient assimilation by plants as well as nitrogen removal through denitrification by bacteria.

References cited

Alves, E., L. R. Cardoso, J. Savroni, L. C. Ferreira, C. S. F. Boaro and A. C. Cataneo et al. 2003. Physiological and biochemical evaluations of water hyacinth (*Eichhornia crassipes*), cultivated with excessive nutrient levels. *Planta Daninha* 21(spe): 27–35.

Bernhardt, E. S. 2013. Cleaner lakes are dirtier lakes. *Science* 342(6155): 205–206.

Braker, G., A. Fesefeldt and K. P. Witzel. 1998. Development of PCR primer systems for amplification of nitrite reductase genes (*nirK* and *nirS*) to detect denitrifying bacteria in environmental samples. *Applied and Environmental Microbiology* 64(10): 3769–3775.

Bulow, S. E., J. J. Rich, H. S. Naik, A. K. Pratihary and B. B. Ward. 2010. Denitrification exceeds anammox as a nitrogen loss pathway in the Arabian Sea oxygen minimum zone. *Deep Sea Research Part I: Oceanographic Research Papers* 57(3): 384–393.

Center, T. D. and N. R. Spencer. 1981. The phenology and growth of water hyacinth (*Eichhornia crassipes* (Mart.) Solms) in a eutrophic north-central Florida lake. *Aquatic Botany* 10: 1–32.

Chen, H. G., F. Peng, Z. Y. Zhang, G. F. Liu, W. Da Xue, S. H. Yan et al. 2012. Effects of engineered use of water hyacinths (*Eichhornia crassipes*) on the zooplankton community in Lake Taihu, China. *Ecological Engineering* 38(1): 125–129.

Conley, D. J., C. Humborg, L. Rahm, O. P. Savchuk and F. Wulff. 2002. Hypoxia in the baltic sea and basin-scale changes in phosphorus biogeochemistry. *Environmental Science and Technology* 36(24): 5315–5320.

Cornwell, D. A., J. Zoltek, C. D. Patrinely, T. Furman and J. I. Kim. 1977. Nutrient removal by water hyacinths. *Water Pollution Control Federation* 49(1): 57–65.

DeBusk, T. A. and F. E. Dierberg. 1984. Effect of nitrogen and fiber content on the decomposition of the water hyacinth (*Eichhornia crassipes* [Mart.] Solms). *Hydrobiologia* 118: 199–204.

Fang, Y. 2006. Efficiency and mechanism of uptaking and removing nitrogen from eutrophicated water using aquatic macrophytes. Ph.D. Thesis, College of Environment and Resources. Zhejiang University.

Finlay, J. C., G. E. Small and R. W. Sterner. 2013. Human influences on nitrogen removal in lakes. *Science* 342(6155): 247–50.

Fitzsimons, R. E. and R. H. Vallejos. 1986. Growth of water hyacinth (*Eichhornia crassipes* (Mart) Solms) in the middle Parana River (Argentina). *Hydrobiologia* 131(3): 257–260.

Fox, L. J., P. C. Struik, B. L. Appleton and J. H. Rule. 2008. Nitrogen phytoremediation by water hyacinth (*Eichhornia crassipes* (Mart.) Solms). *Water, Air, and Soil Pollution* 194: 199–207.

Gao, Y., X. Liu, N. Yi, Y. Wang, J. Guo, Z. Zhang et al. 2013. Estimation of N_2 and N_2O ebullition from eutrophic water using an improved bubble trap device. *Ecological Engineering* 57(0): 403–412.

Gao, Y., N. Yi, Y. Wang, T. Ma, Q. Zhou, Z. Zhang et al. 2014. Effect of *Eichhornia crassipes* on production of N_2 by denitrification in eutrophic water. *Ecological Engineering* 68: 14–24.

Gao, Y., N. Yi, Z. Zhang, H. Liu and S. Yan. 2012. Fate of $^{15}NO_3^-$ and $^{15}NH_4^+$ in the treatment of eutrophic water using the floating macrophyte, *Eichhornia crassipes*. *Journal of Environmental Quality* 41(5): 1653–60.

Gopal, B. 1987. *Water hyacinth (Aquatic Plant Studies)*. Amsterdam, Netherlands: Elsevier Science Ltd.

Hadad, H. R., M. A. Maine, M. Pinciroli and M. M. Mufarrege. 2009. Nickel and phosphorous sorption efficiencies, tissue accumulation kinetics and morphological effects on *Eichhornia crassipes*. *Ecotoxicology (London, England)* 18(5): 504–13.

Hallin, S. and P.-E. Lingren. 1999. PCR detection of genes encoding nitrite reductase in denitrifying bacteria. *Applied and Environmental Microbiology* 65(4): 1652–1657.

Kim, Y. and W. Kim. 2000. Roles of water hyacinths and their roots for reducing algal concentration in the effluent from waste stabilization ponds. *Water Research* 34(13): 3285–3294.

Levine, S. N. and D. W. Schindler. 1999. Influence of nitrogen to phosphorus supply ratios and physicochemical conditions on cyanobacteria and phytoplankton species composition in the Experimental Lakes Area, Canada. *Canadian Journal of Fisheries and Aquatic Sciences* 56(3): 451–466.

Li, C. 2012. A feasibility study on blue algae pollution control by water hyacinth in Lake Dianchi. *Environmental Science Survey* 31(3): 64–68 (In Chinese with English Abstract).

Li, J., Y. Zhang, Y. Wang, M. Liu, D. Pan, J. Li et al. 2003. A preliminary study on decontaminating waste water of slaughter industry by planting hyacinth in northern China. *Journal of Shengyang Agricultural University* 34(2): 103–105 (In Chinese with English Abstract).

Liang, L. and J. Zuo. 2008. Review of modern culture independent methods used to study community structure and function of denitrifying microorganisms. *Acta Scientiae Circumstantiae* 28(4): 599–605 (In Chinese with English Abstract).

Liu, D., H. Wang, H. Yang, Y. Zhou and J. Ge. 2011. Progress of the DGGC application in the wastewater biological nitrogen removal system. *Journal of Anhui Agricultural Science* 39(19): 11695–11697 (In Chinese with English Abstract).

López, E. G., R. H. Delgadillo and M. M. Jiménez. 1996. Water hyacinth problems in Mexico and practised methods for control. In: Strategies for Water hyacinth Control - Report of a Panel of Experts Meeting, ed. R. Charudattan, R. Labrada, T. D. Center, and C. Kelly-Begazo, 125–135 pp. Rome, Italy: Food and Agricultural Organization of the United Nations.

Ma, T., N. Yi, Z. Zhang, Y. Wang, Y. Gao and S. Yan. 2014. Oxygen and organic carbon releases from roots of *Eichhornia crassipes* and their influence on transformation of nitrogen in water. *Journal of Agro-Environment Science* 33(10): 2003–2013 (In Chinese with English Abstract).

Ma, T., Z. Zhang, N. Yi, X. Liu, Y. Wang, S. Yan et al. 2013. Nitrogen removal via denitrification from eutrophic water as influenced by *Eichhornia crassipes* and sediment. *Journal of Agro-Environment Science* 32(12): 2451–2459 (In Chinese with English Abstract).

Madsen, J. D. 1993. Growth and biomass allocation patterns during water-hyacinth mat development. *Journal of Aquatic Plant Management*.

Mahmoud, A., M. A. Huda, G. Raymond and G. R. Carvalhi. 2005. The response of epilithic bacteria to different metals regime in two upland streams: assessed by conventional microbiological methods and PCR-DGGE. *Archiv für Hydrobiologie - Hauptbände* 163(3): 405–427.

MEP-PRC. 2002. Environmental quality standards for surface water (GB3838-2002). Baijing, China: Ministry of Environmental Protection of The Peoples's Republic of China.

Mishra, V. K. and B. D. Tripathi. 2009. Accumulation of chromium and zinc from aqueous solutions using water hyacinth (*Eichhornia crassipes*). *Journal of Hazardous Materials* 164(2-3): 1059–1063.

Moorhead, K. K. and K. R. Reddy. 1988. Oxygen transport through selected aquatic macrophytes. *Journal of Environment Quality* 17(1): 138–142.
Moorhead, K. K., K. R. Reddy and D. A. Graetz. 1988a. Water hyacinth productivity and detritus accumulation. *Hydrobiologia* 157(2): 179–185.
Moorhead, K. K. 1988b. Nitrogen transformations in a waterhyacinth-based water treatment system. *Journal of Environment Quality* 17(1): 71–76.
Muyzer, G., E. C. De Waal and A. G. Uitterlinden. 1993. Profiling of complex microbial populations by denaturing gradient gel electrophoresis analysis of polymerase chain reaction-amplified genes coding for 16S rRNA. *Applied and Environmental Microbiology* 59(3): 695–700.
Niu, J. 2012. Effects on the growth and tillering of water hyacinth at different nitrogen and phosphorus levels and the basis for scientific salvage. MSc. Thesis, Department of Botany, Soochow University, Suzhou, China.
Petrucio, M. M. and F. A. Esteves. 2000. Uptake rates of nitrogen and phosphorus in the water by *Eichhornia crassipes* and *Salvinia auriculata. Revista brasileira de biologia* 60(2): 229–236.
Polomski, R. F., M. D. Taylor, D. G. Bielenberg, W. C. Bridges, S. J. Klaine and T. Whitwell. 2009. Nitrogen and phosphorus remediation by three floating aquatic macrophytes in greenhouse-based laboratory-scale subsurface constructed wetlands. *Water, Air, and Soil Pollution* 197(1-4): 223–232.
Qi, W., G. Chen, Z. Sun, G. Wang, J. Xi and Q. Ma. 2005. Questions of TN and TP monitoring. *Environmental Monitoring in China* 21(2): 31–35 (In Chinese with English Abstract).
Qin, B. and C. Fan. 2002. Exploration of conceptual model of nutrient release from inner source in large shallow lake. *China Environmental Science* 22(2): 150–153 (In Chinese with English Abstract).
Qin, H., Z.Y. Zhang, Z.H. Zhang, X. Wen, H. Liu and S. Yan. 2015. Analysis of the death causes of water hyacinth planted in large-scale enclosures in Dianchi Lake. *Resources and Environment in the Yangtze Basin* 24(4): 594–602 (In Chinese with English Abstract).
Reddy, K. R., M. Agami and J. C. Tucker. 1989. Influence of nitrogen supply rates on growth and nutrient storage by water hyacinth (*Eichhornia crassipes*) plants. *Aquatic Botany* 36(1): 33–43.
Reddy, K. R. and J. C. Tucker. 1983. Productivity and nutrient uptake of water hyacinth, *Eichhornia crassipes* I. effect of nitrogen source. *Economic Botany* 37(2): 237–247.
Ritchie, R. J. 2006. Estimation of cytoplasmic nitrate and its electrochemical potential in barley roots using $^{13}NO_3^-$ and compartmental analysis. *The New Phytologist* 171(3): 643–55.
Rodríguez, M., J. Brisson, G. Rueda and M. S. Rodríguez. 2012. Water quality improvement of a reservoir invaded by an exotic macrophyte. *Invasive Plant Science and Management* 5(2): 290–299.
Rommens, W., J. Maes, N. Dekeza, P. Inghelbrecht, T. Nhiwatiwa, E. Holsters et al. 2003. The impact of water hyacinth (*Eichhornia crassipes*) in a eutrophic subtropical impoundment (Lake Chivero, Zimbabwe). I. water quality. *Archiv für Hydrobiologie* 158(3): 373–388.
Seitzinger, S., J. A. Harrison, J. K. Böhlke, A. F. Bouwman, R. Lowrance, B. Peterson et al. 2006. Denitrification across landscapes and waterscapes: a synthesis. *Ecological Applications* 16(6): 2064–2090.
Shiralipour, A., L. A. Garrard and W. T. Haller. 1981. Nitrogen source, biomass production, and phosphorus uptake in waterhyacinth. *Journal of Aquatic Plant Management* 19: 40–43.
Snow, A. M. and A. E. Ghaly. 2008. A comparative study of the purification of aquaculture wastewater using water hyacinth, water lettuce and parrot's feather. *American Journal of Applied Sciences*.
Sooknah, R. 2000. A review of the mechanism of pollutant removal in water hyacinth systems. *Science and Technology* 6: 49–57.
Temponeras, M. and J. Kristiansen. 2000. Seasonal variation in phytoplankton composition and physical-chemical features of the shallow Lake Doïrani, Macedonia, Greece. *Hyd* 424: 109–122.
Trimmer, M., J. Grey, C. M. Heppell, A. G. Hildrew, K. Lansdown, H. Stahl et al. 2012. River bed carbon and nitrogen cycling: state of play and some new directions. *The Science of the Total Environment* 434: 143–58.
Tucker, C. S. 1981. The effect of ionic form and level of nitrogen on the growth and composition of *Eichhornia crassipes* (Mart.) Solms. *Hydrobiologia* 83: 517–522.

Wang, C., X. Yan, P. Wang and C. Chen. 2008. Interactive Influence of N and P on their uptake by four different hydrophytes. *African Journal of Biotechnology* 7(19): 3480–3486.
Wang, Y. and C. -S. Hu. 2010. Research advances on community structure and function of denitrifiers. *Chinese Journal of Eco-Agriculture* 18(6): 1378–1384.
Wang, Z., Z. Zhang, Y. Han, Y. Zhang, Y. Wang and S. Yan. 2012. Effects of large-area planting water hyacinth (*Eichhornia crassipes*) on water quality in the bay of Lake Dianchi. *Chinese Journal of Environmental Engineering* 6(11): 3827–3832 (In Chinese with English Abstract).
Wang, Z., Z. Zhang, Y. Zhang, J. Zhang, S. Yan and J. Guo. 2013. Nitrogen removal from Lake Caohai, a typical ultra-eutrophic lake in China with large scale confined growth of *Eichhornia crassipes*. *Chemosphere* 92(2): 177–183.
Ward, B. B., A. H. Devol, J. J. Rich, B. X. Chang, S. E. Bulow, H. Naik et al. 2009. Denitrification as the dominant nitrogen loss process in the Arabian Sea. *Nature* 461(7260): 78–81.
Wolverton, B. C. and R. McDonald. 1979. Water hyacinth (*Eichhornia crassipes*) productivity and harvesting studies. *Economic Botany* 33(1): 1–10.
Wolverton, B. C. and R. C. McDonald. 1978. Nutritional composition of water hyacinths grown on domestic sewage. *Economic Botany* 32(4): 363–370.
Wooten, J. W. and J. D. Dodd. 1976. Growth of water hyacinths in treated sewage effluent. *Economic Botany* 30(1): 29–37.
Wu, Y., H. Xu, G. Yang, G. Zhu and Boqiang Qin. 2014. Progress in nitrogen pollution research in Lake Taihu. *Journal of Lake Sciences (China)* 26(1): 19–28 (In Chinese with English Abstract).
Xiong, J., Z. He, J. D. Van Nostrand, G. Luo, S. Tu, J. Zhou et al. 2012. Assessing the microbial community and functional genes in a vertical soil profile with long-term arsenic contamination. *PloS One* 7(11): e50507.
Xu, H., H. W. Paerl, B. Qin, G. Zhu and G. Gao. 2010. Nitrogen and phosphorus inputs control phytoplankton growth in eutrophic Lake Taihu, China. *Limonology and Oceanography* 55(1): 420–432.
Yi, N., Y. Gao, X. Long, Z. Zhang, J. Guo, H. Shao et al. 2014. *Eichhornia crassipes* cleans wetlands by enhancing the nitrogen removal and modulating denitrifying bacteria community. *CLEAN–Soil, Air, Water* 42(5): 664–673.
Yi, N., Y. Gao, Z. Zhang and S. Yan. 2015a. Response of spatial patterns of denitrifying bacteria communities to water properties in the stream inlets at Dianchi Lake, China. *International Journal of Genomics* 2015: 11 pp. (online publication).
Yi, N. 2015b. Water properties influencing the abundance and diversity of denitrifiers on *Eichhornia crassipes* roots: a comparative study from different effluents around Dianchi Lake, China. *International Journal of Genomics* 2015: 12 pp. (online publication).
Yi, Q., Y. Kim and M. Tateda. 2009. Evaluation of nitrogen reduction in water hyacinth ponds integrated with waste stabilization ponds. *Desalination* 249: 528–534.
Yuan, C., Q. Zhao and Z. Wu. 1983. Potential of nutrient recycle in agro-ecosystem using three macrophyes. *Jiangsu Agricultural Sciences* 9: 27–29, 23 (In Chinese).
Zhang, Y., Z. Zhang, Y. Wang, H. Liu, Z. Wang, S. Yan et al. 2011. Research on the growth characteristics and accumulation ability to N and P of *Eichhornia crassipes* in different water areas of Dianchi Lake. *Journal of Ecology and Rural Environment* 27(6): 73–77 (In Chinese with English Abstract).
Zhang, Z., Y. Gao, J. Guo and S. Yan. 2014. Practice and reflections of remediation of eutrophicated waters: a case study of haptophyte remediation of the ecology of Dianchi. *Journal of Ecology and Rural Environment* 30(1): 15–21 (In Chinese with English Abstract).
Zhang, Z. Y., J. C. Zheng, H. Q. Liu, Z. Z. Chang, L. G. Chen and S. H. Yan. 2010. Role of *Eichhornia crassipes* uptake in the removal of nitrogen and phosphorus from eutrophic waters. *Chinese Journal of Eco-Agriculture* 18(1): 152–157 (In Chinese with English Abstract).
Zhang, Z., J. Zhang, H. Liu, L. Chen and S. Yan. 2011. Apparent removal contributions of *Eichhornia crassipes* to nitrogen and phosphorous from eutrophic water under different hydraulic loadings. *Jiangsu Journal of Agricultual Sciences* 27(2): 288–294 (In Chinese with English Abstract).
Zheng, J., Z. Chan, L. Chen, P. Zhu and J. Shen. 2008. Feasibility studies on N and P removal using water hyacinth in Taihu Lake region. *Jiangsu Agricultural Science* 3: 247–250 (In Chinese).

Zheng, Y. K., K. Liu, Z. J. Xiong, C. P. Miao, Y. W. Chen, L. H. Xu et al. 2015. Effect of large-scale planting water hyacinth on cultivable bacterial community structure in the eutrophic lake. *Microbiology China* 42(1): 42–53 (In Chinese with English Abstract).

Zhou, Q., S. Q. Han, S. H. Yan, W. Song and J. P. Huang. 2012. The mutual effect between phytoplankton and water hyacinth planted on a large scale in the eutrophic lake. *Acta Hydrobiologica Sinica* 36(4): 873–791 (In Chinese with English Abstract).

Zhou, Q., S. Han, S. Yan, W. Song and G. Liu. 2014. Impacts of *Eichhornia crassipes* (Mart.) Solms stress on the growth characteristics, microcystins and nutrients release of *Microcystis aeruginosa*. *Environmental Science* 35(2): 597–604 (In Chinese with English Abstract).

Zhu, J. and X. Zhu. 1998. Treatment and utilization of wastewater in the Beijing Zoo by an aquatic macrophyte system. *Ecological Engineering* 11(1–4): 101–110.

Zimmels, Y., F. Kirzhner and A. Malkovskaja. 2007. Advanced extraction and lower bounds for removal of pollutants from wastewater by water plants. *Water environment research: a research publication of the Water Environment Federation* 79(3): 287–96.

Zumft, W. G. 1997. Cell biology and molecular basis of denitrification. *Microbiology and molecular biology reviews: MMBR* 61(4): 533–616.

CHAPTER 6

Mechanism and Implications of Phosphorus Removal

Y. Y. Zhang and *S. H. Yan**

6.1 Phosphorus in aquatic ecosystem

6.1.1 Phosphorus as an important element in eutrophication

Eutrophication is a natural process but has been greatly accelerated with rapid population growth, economic development and industrialization at the global scale (Dokulil and Teubner 2011). The acceleration of eutrophication has reduced drinking water resources and degraded water quality in many parts of the world (Meybeck 2003). Eutrophication is mainly assessed or classified by primary production in aquatic environment and described by eutrophic status (Carlson and Simpson 1996). Although different classification methods have varied definitions, the difference is in concentration of various nutrients in aquatic ecosystem. For example, North America and Europe usually summarize eutrophic status by an index in a multi-parameter classification system, representing a combination of nutrient concentrations, phytoplankton biomass (chlorophyll-*a*), transparency, etc. (Table 6.1.1-1). China has a similar classification system, but added total nitrogen and chemical oxygen demand (COM_{Mn}) to the calculation parameter list, and has more levels on the eutrophic status index above 50 (China Environmental Monitoring Station 2001).

Phosphorus is a critical nutrient (Lapointe et al. 1994) for growth of algae and other aquatic organisms, accumulation of organic matter and the functionality of nutrient cycling and energy flow in an aquatic ecosystem. The literature suggested that many lakes, streams and estuaries

50 Zhong Ling Street, Nanjing, China.
Email: fly8006@163.com
* Corresponding author: shyan@jaas.ac.cn

Table 6.1.1-1. Eutrophic status index represented by chlorophyll-*a*, total phosphorus, transparency and classification (Carlson and Simpson 1996, China Environmental Monitoring Station 2001).

Eutrophic status index	Chlorophyll-*a* (μg L^{-1})	Total Phosphorus (μg L^{-1})	Transparency (m)	Carlson and Simpson classification	China classification
<30–40	0–2.6	0–12	>8–4	Oligotrophic	Oligotrophic
40–50	2.6–20	12–24	4–2	Mesotrophic	Mesotrophic
51–60	20–56	24–96	2–0.5	Eutrophic	Light Eutrophic
61–70					Moderate Eutrophic
70–100	56–155 +	96–384 +	0.5<0.25	Hypereutrophic	Hypereutrophic

showed excessive nutrient concentrations that have been attributed to human-induced accumulation, especially of phosphorus (Lee 1973). Phosphorus does not function alone but must be in balance with nitrogen (Bernhardt 2013). In a phosphorus-overloaded ecosystem, the sources of phosphorus may be classified as endogenous and exogenous. To control the overload of phosphorus, the two sources must be controlled at the same time; otherwise, the management strategies may not reach the target, especially in shallow lakes (Wang et al. 2008, Xie et al. 2003, Coveney et al. 2005, Mehner et al. 2008). The chemical and biological characteristics of phosphorus determine it may not be as active as nitrogen that involves air-water exchanges in gaseous form via denitrification and nitrogen fixation. Removal of phosphorus from eutrophic waters and maintaining the nutrient balance can help mitigate eutrophication (Edmondson 1994). For example, research in ~40 Yangtze lakes revealed that phosphorus may be the main element to be controlled due to the phytoplankton always being linked to the concentration of phosphorus (Wang and Wang 2009). Another example was the 37-year monitoring results in #227 lake of Lake Ontario, Canada. Nitrogen fixation allowed phytoplankton to grow proportionally to the phosphorus concentration during 37 years to keep the lake in hypereutrophication state (Schindler et al. 2008). Lake restoration practices in Europe and the United States also confirmed that decreasing phosphorus concentration can effectively alleviate eutrophication (Edmondson and Lehman 1981). Lake Washington in the United States had an overload of exogenous phosphorus causing blooms of cyanobacteria. Phosphorus removal started in 1936, and by 1960–1970 there were decreases of about 50% in total nitrogen and about 80% in total phosphorus. The chlorophyll-*a* was always proportional to the concentration of total phosphorus, and dominant phytoplankton (cyanobacteria) was reduced from 90% in 1960 to 20% in 1970 (Edmondson and Lehman 1981, Edmondson 1994).

6.1.2 Phosphorus forms in aquatic ecosystems

The forms of phosphorus in waters can be categorized into three groups: orthophosphate, polyphosphates and organic phosphorus according to their chemical properties, or into two groups: particulate phosphorus and dissolved phosphorus according to their physical characteristics. In general, dissolved orthophosphate is easily assimilated and utilized by phytoplankton and macrophytes. The US Environmental Protection Agency noted that amount of wastewater discharged into lakes and reservoirs needs to be linked with the phosphorus concentration of the location based on the trend of phosphorus concentration changes in a particular water body. In 2002, the general guidelines on the total phosphorus in standard discharge wastewater were 0.008–0.0375 mg L^{-1} (EPA 2007). For example, in Wisconsin, the municipal sewage treatment plant can only discharge effluent with total phosphorus concentration less than 1 mg L^{-1} by controlling the amount discharged to ensure sufficient dilution to keep the final concentration of phosphorus lower than 0.0375 mg L^{-1} in the receiving lakes and reservoirs (Department of Natural Resources 2013). China has different guidelines for total phosphorus concentration in effluent from municipal sewage treatment plants (less than 0.5–1.0 mg L^{-1} (Ministry of Environmental Protection of the People's Republic of China 2003) (Table 6.1.2-1).

In aquatic ecosystems, endogenous and exogenous phosphorus are in dynamic equilibrium. Exogenous phosphorus from effluent, nonpoint pollution, leaching from farmland and atmospherically deposition can be accumulated in sediment via biological or physical processes. The phosphorus in vegetation biomass can be released via mineralization and P from sediment by a chemical process to become soluble in water, and thus take part in cyanobacteria and algal blooms causing water quality degradation (Cui 2013). When the exogenous source of phosphorus is limited, the release of endogenous phosphorus can maintain the concentration of soluble phosphorus to keep the eutrophic state (Berelson et al. 1998). For example, Pitkänen et al. (2001) reported the nutrient dynamics in eastern Gulf of Finland when the external input of phosphorus was reduced by 30%, the phosphate concentration of the water was still rising due to the release of endogenous phosphorus.

The accumulated endogenous phosphorus in sediments originated mainly from organic matter deposition being converted to inorganic phosphorus via mineralization. The inorganic phosphorus is easily released from sediment to

Table 6.1.2-1. Maximum total phosphorus concentration in effluent discharged from water sewage treatment plants (mg L^{-1}).

Basic Control Project		Standard		Secondary[3]	Tertiary[4]
		Standard A[1]	Standard B[2]		
Total phosphorus (as mg P L^{-1})	Before 31 December 2005	1	1.5	3	5
	After 1 January 2006	0.5	1	3	5

Note: [1] Effluent suitable for reuse after dilution; [2] Effluent for direct discharge to non-drinking water resources and non-tourist water bodies; [3] Effluent only for industry and non-tourist utilization without direct body contact; [4] Irrigation and landscape utilization.

overlying water. This process is greatly impacted by aquatic environmental factors such as pH, oxygen concentration, temperature and other biological, chemical and physical properties; it is particularly prominent in shallow waters and characterized by seasonal pattern (Ren et al. 2010).

Phosphorus in natural aquatic ecosystems is often described as Total Phosphorus (TP), Total Soluble Phosphorus (TSP), Particulate Phosphorus (PP), Dissolvable Organic Phosphorus (DOP) and Soluble Reactive Phosphorus (SRP) or biologically available orthophosphate. Total phosphorus represents all phosphorus in water, whereas total soluble phosphorus represents dissolvable organic phosphorus and soluble reactive phosphorus. The concepts of soluble reactive phosphorus and biologically available orthophosphate are overlapping, but not all orthophosphate may be biologically available. Dissolved organic phosphorus and insoluble forms of organic and inorganic phosphorus are generally not biologically available for macrophytes if not transformed into soluble reactive phosphorus. Although the terminology is complicated to describe the relationship among these forms of phosphorus in aquatic ecosystem, in phytoremediation, only biologically available and transformation of biologically non-available phosphorus need to be investigated.

6.1.3 Phosphorus transformation and migration in aquatic ecosystems

The important processes in aquatic ecosystems are the interchange of inorganic and organic phosphorus and particulate phosphorus adsorption and deposition. Conversion of inorganic to organic phosphorus is referred to organic phosphorus fixation, whereas conversion of organic to inorganic phosphorus is mineralization. During organic phosphorus fixation (e.g., plant assimilation), the inorganic phosphorus is incorporated into biological components and stored in tissue. During organic biomass decomposition phosphorus is released. Transformation of organic to inorganic phosphorus depends mainly on the source of organic matter and environmental factors such as pH, temperature, oxygen concentration and microorganism activities (Ruttenberg 2003).

Sediment-water interface is a very active layer of lake biogeochemistry and important physical and chemical interface in aquatic ecosystems for delivery and exchange of phosphorus and biogeochemical cycling of nutrients (Song 2009). A dynamic balance of adsorption-desorption of phosphorus on the sediment-water interface depends on the relative phosphorus concentrations between interstitial (pore) water and overlying water. The concentration of phosphorus in overlying water is referred to as an equivalent when the balance is established. When the concentration of phosphorus in overlying water is higher than the equivalent, sediment adsorption occurs; in contrast, endogenous phosphorus in sediment is released to overlying water through a series of physical, chemical and biological processes when the concentration of the phosphorus is reduced below the equivalent (Jin et al. 2008). The processes of phosphorus interchange at the sediment-water interface involve particle re-suspension, the phosphorus concentration gradient, dissolution and desorption of iron oxide, degradation and mineralization of organic matter or organic phosphorus release between interstitial water and overlying water (Guo 2007). The most important pathway is the exchange of phosphorus between interstitial water and overlying water, with the concentration gradient and molecular diffusion being the main mechanisms (Zhang 2004). The above discussion implied that the management target for remediation of eutrophic waters must consider both the short-term water quality improvement and long-term endogenous pollutant removal. Given that many factors control the endogenous phosphorus at the sediment-water interface, such as hydraulic characteristics, physical and chemical as well as biological factors, it is difficult to predict the amount of phosphorus releasing from endogenous pool. Literature suggested that a multi-target management strategy to understand and control loading of phosphorus (both exogenous and endogenous) as well as removal is needed to mitigate eutrophication, improve water quality and maintain balanced functioning of an aquatic ecosystem. To achieve this target, the first priority is to employ the appropriate methods to remove total phosphorus from water at a rate that exceeds the rate of endogenous phosphorus release and exogenous phosphorus loading (Equation 6.1.3-1).

$$P_r - P_e - P_x > 0 \qquad [6.1.3\text{-}1]$$

Where P_r is total phosphorus removal rate (mg L^{-1} m^{-2} d^{-1}); P_e is endogenous phosphorus release rate (mg L^{-1} m^{-2} d^{-1}); P_x is exogenous phosphorus loading rate (mg L^{-1} m^{-2} d^{-1}).

During the planning stage of a phytoremediation program for phosphorus control, the above equation must be used to estimate the final output of the program, and then detailed operation procedures need to be determined to achieve the target (Rodríguez et al. 2012, Wang et al. 2012).

6.2 Impact of phosphorus on water hyacinth growth

6.2.1 Phosphorus content in water hyacinth organs

The phosphorus content varies in different parts of the macrophyte depending on the growth and development stages and phosphorus concentration in the habitat. Literature reported that the phosphorus content was higher in blades than stem and roots (Haller and Sutton 1973, Polomski et al. 2009). However, with continuous nutrient renewal to mimic natural habitats and harvest to keep macrophyte in the early growth stage, the content of phosphorus may be lower in blades than roots (Table 6.2.1-1).

The variable distribution of phosphorus together with other plant nutrients among water hyacinth organs demonstrated strong biological flexibility of

Table 6.2.1-1. Phosphorus content of blades, petioles, roots and the whole plant of water hyacinth grown in nutrient solutions with different phosphorus (P) concentrations.

Treatment concn ($mg\ P\ L^{-1}$)	Phosphorus content ($g\ kg^{-1}$ Dry Weight) in water hyacinth				Ref.
	Blades	Petioles	Roots	Whole plants	
0	1.17^{a}	0.71^{a}	0.96^{a}	0.98^{a}	(Haller and Sutton 1973)
5	4.96^{b}	3.00^{b}	1.97^{b}	3.77^{b}	
10	6.77^{c}	4.80^{c}	3.12^{c}	5.52^{c}	
20	8.16^{d}	6.73^{d}	6.05^{d}	7.22^{d}	
40	8.80^{d}	9.30^{c}	9.26^{c}	9.07^{c}	
0.07	1.54^{a}	—	1.29^{b}	—	(Polomski et al. 2009)
0.18	1.30^{a}	—	1.18^{a}	—	
1.86	1.45^{a}	—	1.26^{b}	—	
3.63	1.78^{a}	—	1.67^{a}	—	
6.77	2.53^{a}	—	1.71^{b}	—	
0.14	3.46		6.02	4.17	(Zhang et al. 2010)[1]
0.34	5.45		7.43	5.96	
1.07	6.43		7.70	6.77	
1.43	6.90		8.50	7.30	
8.50	3.00^{c}		6.27^{a}	4.20^{b}	(Zhang et al. 2011)[1]
11.7	3.70^{bc}		6.09^{a}	4.50^{b}	
19.2	4.64^{b}		6.37^{a}	5.50^{b}	
56.5	6.41^{a}		7.82^{a}	6.69^{ab}	

Note: conc. = concentration; Different superscript letters represent the significant differences among the treatments in a given study; [1] Experiment with nutrient renewal and macrophyte harvest at 21-day interval; the data were averaged over 15 intervals.

water hyacinth to adapt to different habitats for survival and growth. Below 31 mg P L^{-1} of culture solution, the phosphorus content in the macrophyte showed a positive relationship with phosphorus concentration in the media (Gossett and Norris Jr 1971); at 40 mg P L^{-1} of culture solution, the phosphorus content in water hyacinth reached the maximum (Haller and Sutton 1973).

Under natural conditions, the phosphorus content in water hyacinth varied widely (1.4–8.0 g kg^{-1} dry matter) with an average of 5.4 g kg^{-1} dry matter (Boyd and Vickers 1971). Water hyacinth can yield 65.2 tons (dry matter) ha^{-1} yr^{-1} with 15.4–20.5 g N kg^{-1} dry matter and 1.6–2.9 g P kg^{-1} dry matter (Table 6.2.1-2).

The wide range of phosphorus content in water hyacinth growing in different habitats reflects variable nutrient conditions such as concentration and balance of nutrients as well as other environmental factors such as pH, incident light and temperature. For example, phosphorus content was greater in water hyacinth float and blades from the mat center than the mat edge (Table 6.2.1-3).

This result was consistent with other investigations (Pinto-coelho et al. 1999) and indicated that phosphorus was responsible for resource allocation among plant parts for their growth and development (Xie et al. 2004).

The phosphorus content in different water hyacinth parts may represent biological adaptation to specific habitats. Regarding phytoremediation management, it is more important to examine the total amount of phosphorus accumulated aerial and under-water parts. The general principle is that water hyacinth accumulates more phosphorus in aerial parts with an increase in concentration of available phosphorus in water (Table 6.2.1-4).

The data shown in Table 6.2.1-4 are expected for a noxious invasive species because after concentration of available phosphorus in water reaches sufficient levels to support basic functions of water hyacinth, the assimilated phosphorus is allocated predominantly to aerial parts to allow a plant to compete with others for survival and expansion (production). The biological characteristics of phosphorus allocation in water hyacinth plant biomass may imply that in

Table 6.2.1-2. Phosphorus content (g kg^{-1} dry weight) in water hyacinths cultivated in nutrient-enriched ponds (Boyd 1976).

Ponds	Sampling date			
	20 June	**6 August**	**5 September**	**28 September**
1	2.4	2.0	1.6	2.4
2	2.6	2.5	2.4	2.9
3	2.8	1.6	1.8	1.9
Average	2.6	2.0	1.9	2.4

Table 6.2.1-3. Phosphorus content (g kg^{-1} dry weight) of water hyacinth parts at a different location of a dense mat.

Ref.	Sample location	Plant parts		
		Root	**Petiole**	**Blade**
(Musil and Breen 1977)	mat edge	0.311	0.321	0.547
	mat center	0.308	0.525	0.588

Table 6.2.1-4. Phosphorus allocation in different water hyacinth parts influenced by different nutrient concentrations in the growing habitat (after Polomski et al. 2009).

Phosphorus concn in water (mg P L^{-1})	Ratio (aerial part/ under-water part)	Phosphorus content (mg P/whole plant)	
		Aerial part	**Under-water part**
0.07	7.9	17.954 **	2.287
0.18	6.9	18.804 **	2.708
1.86	7.5	20.647 **	2.763
3.63	11.3	38.617 **	3.429
6.77	22.9	80.134 **	3.502

** represent significant differences; X refers to phosphorus concentration in culture solution (mg P L^{-1}); Ratio = $7.397–0.545X + 0.420X^2$, $R^2 = 0.997$, $0.07 \leq X \leq 6.77$.

case of phytoremediation, different management and biomass end-use may be appropriate in different eutrophic situations. At high phosphorus concentration (>1 mg L^{-1}) in eutrophic water and early growth stage, harvested water hyacinth blades may be more suitable for animal feed if other pollutants are below the limits, whereas at low phosphorus concentration in eutrophic water and mature growth stage, water hyacinth may be more suitable for methane and fertilizer fermentation due to low nutrient content in the plant. At different eutrophic zones of a eutrophic water body, water hyacinth population needs to be managed for the specific water quality control targets; otherwise, the efficiency of phosphorus removal may be decreased.

6.2.2 *Impact of available phosphorus in water on the growth of water hyacinth*

The important biological characteristics of water hyacinth with implications for phytoremediation are flexible morphology and hyperaccumulation of nutrients. Flexible morphology refers to water hyacinth having the capability to easily adapt to various habitats, with changes in root length, petiole length and shape and shoot-root ratio, and especially the properties of roots. In very low nutrient concentration, water hyacinth can increase root length to 2 meters to enhance nutrient acquisition (Rodríguez et al. 2012) or strike root into sediment and survive at least long enough to flower and produce seeds. Hyperaccumulation refers to water hyacinth having the capacity to absorb phosphorus and other nutrients in excess of its physiological needs and store phosphorus in tissues (Reddy et al. 1989).

The highest concentration of phosphorus together with the balanced concentration of other nutrients in eutrophic waters are critical basic information that is required to estimate water hyacinth population growth, harvesting schedule and water hyacinth biomass quantity and quality for post-harvest processing and utilization. The reported highest phosphorus concentration to support maximum growth of water hyacinth in eutrophic water was 20 mg L^{-1} (Haller and Sutton 1973), although concentrations higher than 1.06 mg L^{-1} may not further increase the production of water hyacinth biomass (Reddy et al. 1990). The difference in the two separate experiments may be due to different methods used in the experiment. Haller and Sutton's experiment used a 11-liter water tank for three plants growing for 4 weeks, whereas Reddy et al. used a 1000-liter tank with the temperature (5–33°C) and irradiance (5–20 MJ m^{-2} d^{-1}) during the experiment; static nutrient solution was used in both studies. Although Haller and Sutton did not mention the surface area of the tank, presumably three plants might have increased several folds during the experiment so that the density to surface area and phosphorus supply below 20 mg L^{-1} treatments were unclear; on the other side, the temperature and irradiance during the Reddy et al. experiment were not at the optimum level plus the nitrogen/phosphorus ratio was unbalanced at the end of the experiment in treatments with phosphorus concentration at high range so that water hyacinth growth might have been effected. Although both experiments seemed less than optimal, it can be said with confidence that phosphorus concentrations from 1.06 mg L^{-1} up to 20 mg L^{-1} may be regarded as the maximum range of P concentration to support water hyacinth growth so that the expected production of biomass may be planned for scheduling harvests and post-harvest processing. The second conclusion from both experiments was hyperaccumulation of phosphorus by water hyacinth growing in eutrophic waters at high phosphorus concentration. Experiments suggested that when the plant tissue P reached a steady-state of around 1.5 g kg^{-1} dry matter, the internal P cycling within water hyacinth mats is adequate to maintain this tissue P level. At tissue phosphorus content of 0.5 g kg^{-1}, the net productivity approached zero (Reddy et al. 1990). The highest whole plant tissue P content was 13.5 g kg^{-1} at phosphorus supplied at 761 mg m^{-2} d^{-1}, and other literature suggested tissue P content 5.5–9.1 g kg^{-1} at phosphorus supply from 1–40 mg L^{-1} in culture solution (Ornes and Sutton 1975, Gossett and Norris Jr. 1971, Haller and Sutton 1973). The wide variations indicated the measurements obtained at different nitrogen concentrations or growing stage or other environmental influence. These results revealed the fact that water hyacinth can luxuriously accumulate phosphorus in plant tissue at concentrations nine-fold higher than its physiological need (Reddy et al. 1990).

Literature suggested that phosphorus concentration in water may promote water hyacinth tillering and lateral growth. The survey of water hyacinth population growth and response to water quality in the Minjiang River in Fujian province, China, showed that, when the total phosphorus concentration in the river was between 0.18–0.25 mg L^{-1}, and phosphorus/nitrogen at between 12–17, the numbers of water hyacinth stolons and petioles were related positively, and the lengths of stolon and leaves were related negatively, with total phosphorus concentration in water, whereas water hyacinth biomass was not related significantly (Zhou 2008). The amount of available phosphorus also influences water hyacinth growth and the morphology of its lateral roots. At 0.6 g P m^{-2} yr^{-1}, water hyacinth lateral root length and density were higher than at 4.8 g P m^{-2} yr^{-1}; but root diameter was lower at high P concentration. Further study revealed that at low concentration the lateral root biomass was 85% of the total root biomass, but represented 99.8% of root surface (Xie 2003). This type of root morphological change assisted water hyacinth in acquiring as much as available phosphorus from the environment as possible to adapt to low phosphorus condition. Indeed, root morphological flexibility is very important for plant nutrient intake.

The flexibility of plant morphology may be influenced by growth hormone contents, although nitrogen also played an important role. Tillering number is controlled by endogenous hormone contents that respond to nitrogen and phosphorus contents in plant tissues. When tillering occurs, indoleacetic acid (IAA) content decreased and Zeatin Riboside and Zeatin (ZR + Z) contents increased; IAA/(ZR + Z) ratio was inversely related to tiller number and negatively correlated with phosphorus concentration in culture solution ($p<0.001$). Analysis of the impacts of both phosphorus and nitrogen on the hormone contents, phosphorus played a more important role than nitrogen (Niu et al. 2012).

6.2.3 Influence of available nitrogen on phosphorus uptake and utilization by water hyacinth

Below the optimum levels of available nitrogen and phosphorus, the growth of water hyacinth and nutrient uptake were positively related to concentrations of available nitrogen and phosphorus in the habitat (Haller et al. 1970, Haller and Sutton 1973, Shiralipour et al. 1981, Reddy and Tucker 1983) and were dependent on the N/P ratio (Reddy et al. 1989). An amount of available nutrient and nutrient concentration are different concepts, especially in moving waters such as rivers, large lakes and reservoirs. In static waters, the differences of the two may depend on the chemical forms because inorganic and soluble forms can be assimilated easily by macrophytes. In moving waters, both nitrogen and phosphorus concentrations can be very low, but water hyacinth may grow well and may have high yield because water passing through the root zone of water hyacinth supplies adequate nutrients to the macrophyte.

Reddy and Tucker (1983) investigated the effects of six nitrogen sources on water hyacinth growing in greenhouse containers with adequate nutrient supply by changing culture solution every week and reported no significant effect on phosphorus content in the biomass, but significant effect on the net plant production. The nitrogen sources tested were potassium nitrate (KNO_3), ammonium chloride (NH_4Cl), ammonium chloride + nitrapyrin[1] ($NH_4Cl + C_6H_3Cl_4N$), ammonium chloride + nitrapyrin + potassium nitrate ($NH_4Cl + C_6H_3Cl_4N + KNO_3$), urea (all treatments with 20 mg N L^{-1}) and methane digester effluent at 6 mg N L^{-1}. The results were expected because the N/P ratio was 2.5–5 (in the first five treatments), which was in the range required for maximum biomass yield (Reddy and Tucker 1983). This ratio can be used in planning phytoremediation for removal of both nitrogen and phosphorus. However, it should be borne in mind that manual adjustment of the N/P ratio in practice is not realistic, except in rare situations.

In practice of phytoremediation aimed at removing phosphorus from eutrophic waters, nutrient loading should be considered in planning and management. Nutrient loading is defined by nutrient concentration in water multiplied by an amount of water passing the cross section of a unit area (expressed as mg m^{-2} d^{-1}). In water hyacinth phytoremediation system, the cross section may refer to the cross section of the root zone (usually 200–500 mm vertically, although 100–2000 mm is also quite common depending

[1] Nitrapyrin [2-Chloro-6-(trichloromethyl) pyridine] mainly functions as nitrification inhibitor.

on the amount of available nutrients in water). For example, Reddy et al. (1989) designed experiment using 900 liters of culture solution in a container of 1.7 m^2 surface area (530 mm in depth) with ammonium nitrate concentrations 0.5, 2.5, 5.5, 10.5, 25.5 and 50.5 mg N L^{-1} and replacing the culture solution every week. Based on the above definition, the corresponding nitrogen loadings were 38, 189, 416, 794, 1930 and 3820 mg N m^{-2} d^{-1} (concentration in mg N L^{-1} multiplied by 900 liters divided by 1.7 m^2 divided by 7 days). In the experiment, the phosphorus concentration was 3 mg P L^{-1} (loading 227 mg P m^{-2} d^{-1}). The results showed that the net production of water hyacinth increased as nitrogen loading increased to 416 mg N m^{-2} d^{-1} at N/P ratio 1.83; higher nitrogen loadings did not significantly increase the net production, but increased nitrogen concentration in water hyacinth biomass up to the highest loading at 3820 mg N m^{-2} d^{-1}. The highest nitrogen concentration in water hyacinth biomass was 26.7 g kg^{-1} dry weight (equivalent to crude protein level of 167 g kg^{-1} dry biomass). The phosphorus content of water hyacinth was influenced by available nitrogen and was as high as 6.7 g P kg^{-1} dry biomass at nitrogen loading of 416 mg N m^{-2} d^{-1}. Further analysis indicated (i) that the standing crop may be above the optimum level (1–2 kg dry weight m^{-2}) for free growth of water hyacinth and (ii) the nitrogen loading below 416 mg N m^{-2} d^{-1} may not be adequate.

In practice of phytoremediation, project design and planning often need to be informed by (in the order of priority): (i) final water quality, (ii) amount of biomass produced, and (iii) the quality of biomass. Final water quality is the main target of phytoremediation that needs to be designed based on the source of the pollutant. Amount of biomass produced is important for planning the harvesting schedule and post-processing. The quality of the biomass is related to the possible ways of biomass utilization, which may involve a financial return. The pollution from non-point sources (e.g., agricultural land) and effluent from municipal waste treatment plants usually have higher nitrogen/phosphorus ratio (7–13) (Wang et al. 2013) than optimum ratio of 4 for water hyacinth growth. Nevertheless, phytoremediation reliant on water hyacinth growth may still achieve good removal of both phosphorus and nitrogen because surplus nitrogen may be dissipated by natural nitrification and denitrification processes.

The above discussion indicated that the final water quality and nutrient removal are closely linked with nutrient loading level. In large lakes and reservoirs, the practical application was to control exogenous phosphorus loading and to gradually reduce endogenous phosphorus source. Removal of exogenous phosphorus usually can be achieved by growing water hyacinth at or near river mouths (high loading sites). Removal of endogenous phosphorus can be achieved by growing water hyacinth at lee site where algae would accumulate and die, with released nutrients absorbed by water hyacinth.

6.3 Phosphorus removal by water hyacinth

6.3.1 Phosphorus removal by absorption

Literature reported that in both static simulation laboratory tests and practical wastewater treatment systems, water hyacinth exhibited excellent capacity to remove soluble reactive phosphorus (PO_4^{3-}). Under natural sunlight and air temperature between 27 and 31°C water hyacinth was cultivated in 700 mL of nutrient solution for 24 hours, removing 19 to 97% of total phosphorus and 26 to 99% of soluble reactive phosphorus (Petrucio and Esteves 2000). It was understandable that the soluble reactive phosphorus was removed more effectively by water hyacinth because it is immediately available, whereas some other phosphorus forms that contribute to total P need to go through processes before becoming available to plants. The wide range of removal percentages depended on the nitrogen balance and other nutrient concentrations as influenced by hydraulic retention time. For example, the treatment with high nutrient concentration (8.0 mg NO_3^- L^{-1}, 10.0 mg NH_4^+ L^{-1}, 6.0 mg PO_4^{3-} L^{-1}) had the lowest removal of total phosphorus (19%) and soluble reactive phosphorus (26%) due to short hydraulic retention time of only 24 hours. At low initial nutrient concentration (SRP 0.6–3.0 mg PO_4^{3-} L^{-1}) the removal was from 87–99%. Due to different initial concentration, the low removal of 26% equaled the actual removed amount of 1.62 mg L^{-1}, and the high removal of 99% represented the actual removed amount of 0.63 mg L^{-1}. The amount of phosphorus removal was highest in the treatment with 3.0 mg PO_4^{3-} L^{-1} and lowest in the treatment with 0.6 mg PO_4^{3-} L^{-1} (Petrucio and Esteves 2000). This example demonstrated that the effectiveness and efficiency of phosphorus removal are different concepts. The effectiveness refers to the percentage of phosphorus removed or the final water quality the phytoremediation project can achieve. The efficiency of phosphorus removal refers to the amount of phosphorus being removed per units of area, time and fresh weight of water hyacinth. These two concepts can be used in different phytoremediation stages or to target management. In the situation of high pollution loads, the criterion of efficiency may be the priority. However, when water reclamation is required for drinking water resources, the criterion of effectiveness may be the priority. Usually, effectiveness is more difficult to achieve and needs longer hydraulic retention time and careful management of nutrient balance during the phytoremediation.

Apart from the concepts of effectiveness and efficiency of phosphorus removal in phytoremediation, the interaction between algae and macrophyte may also greatly influence the process and the final target of a phytoremediation project. In natural waters, regardless of whether macrophyte exists or not, algae are always important primary production contributors. During phytoremediation, algal community may significantly change the amount and availability of phosphorus by transforming soluble reactive phosphorus into organic phosphorus. Because algae have a relatively short life cycle of a few days to weeks and can be moved by hydraulic and wind forces, the organic phosphorus in tissues of algae can be transported to and released in different locations, influencing the water quality inside and outside of water hyacinth mat. Literature reported that the algal abundance was 1.7–11.2 times higher inside the water hyacinth mat than outside (Zhou et al. 2012). Deposition of large algal populations inside the water hyacinth mat may significantly increase phosphorus concentration (by up to 350 times) compared with the area outside of water hyacinth mat (Qin 2015).

6.3.2 Phosphorus removal by adsorption

Particulate phosphorus includes insoluble inorganic phosphorus particles such as ferric, magnesium, calcium and aluminum phosphate in detritus and organic phosphorus in planktonic alga in water and sediment. The most phosphorus in natural waters may be particulate phosphorus. For example, particulate phosphorus as the fraction of total phosphorus was as high as 66% in Lake Taihu in 2013 (Jin et al. 2015), 67% in Conestogo River and 77% in Grand River, Ontario, Canada in 2012 (Hu 2013). Although particulate phosphorus is not biologically active, it is the most active form of phosphorus in biogeochemical cycling in aquatic ecosystems because it is influenced strongly by environmental and biological factors in space and time. The specific forms of inorganic particulate phosphorus are mainly determined by the parent material in the sediment and water chemistry.

Spatial and seasonal distributions of particulate phosphorus generally follow the changes in the hydraulic, temperature and water chemistry patterns as well as the influence of exogenous non-point and point source loading. Temperature (seasonal variation), pH and biological activity in aquatic ecosystems may determine the amount of particulate P being dissolved. The spatial distribution pattern may follow the hydraulic and wind forces. In the eutrophic and shallow Lake Taihu, Jiangsu Province, China, the concentration of particulate phosphorus showed a clear spatial pattern in four sections of the lake, being highest in the north-west section and lowest in the south-east section due to monsoon phenomena and cyanobacteria migration and accumulation patterns (Jin et al. 2015). Seasonal distributions of particulate phosphorus may be impacted by the variations in temperature, pH and algal activities. For example, in Lake Taihu region, summer and autumn were the seasons when phosphorus desorption prevails, and spring and winter were the seasons when adsorption predominated (Jin et al. 2008).

The source of particulate phosphorus in sediment was mainly from phosphorus in the overlying water. The concentration of particulate phosphorus was negatively correlated with the distance to the sediment surface (Hong et al. 1989, Yuan et al. 2009). Acidic pH can provide conditions for solubilization of calcium phosphate and alkaline pH for ferromanganese and ferric-aluminum phosphate (Cui 2013). Particulate phosphorus can be a potential source of soluble reactive phosphorus influenced by a concentration gradient, pH and temperature variation and physical disturbance by hydraulic and wind forces as well as biological activities including macrophytes, algae and microorganisms. The potential bioavailable phosphorus may be as high as 358–448 mg kg^{-1} sediment solid, or 87% of total phosphorus (Jiang et al. 2007). Although the mechanisms of dynamic changes in particulate phosphorus are not fully understood, the transformations between soluble reactive phosphorus and particulate phosphorus are mainly dominated by the SRP concentration gradients that are controlled by the interactions among macrophytes, algae and microorganisms.

Macrophytes can decrease SRP by assimilation and adsorption of particulate phosphorus on the surface of leaves in case of submerged aquatic plants or on the surface of roots in case of floating aquatic plants such as water hyacinth. The SRP assimilation and adsorption create a concentration gradient in overlying water. Adsorption is only a physical process, when both inorganic and organic phosphorus (including phosphorus assimilated by planktonic algae) can be transformed to SRP by extracellular enzymes or microbes, although the pH and temperature changes on a micro-scale may also help solubilization of phosphorus from particulate phosphorus (Pan et al. 2004). The algae assimilated phosphorus can only be released through algal decomposition that is more complex than a physical desorption process.

Chen et al. (2015) investigated the interaction between water hyacinth and algae, including phosphorus released from algae in a simulation test. The experiment was designed using a 215-L container with surface area of 0.44 m^2 and 100 mm thick sediment. The concentration of phosphorus in test water was total dissolved phosphorus 0.173 mg L^{-1} and 0.760 mg L^{-1} P. The water hyacinth, at an average starting density of 2.25 kg m^{-2}, was cultivated under an outdoor canopy made of transparent plastic sheet (temperature 17–24°C) for 80 days with an average final density of 14.91 kg m^{-2}. Water hyacinth removed 95% of total phosphorus in test water, including Total Dissolved Phosphorus (TDP) 77 mg m^{-2} and organic phosphorus 356 mg m^{-2} (Table 6.3.2-1).

Although the above experiment was not perfect due to the analytical method used for the assessment of organic phosphorus in alga, the results still can reflect the effects of water hyacinth on the removal of total phosphorus including organic phosphorus. Two most interesting results were: (1) the TDP was increased to 0.369 mg L^{-1} on day 14 and reduced to 0.016 mg L^{-1} at the end of the experiment, which may indicate the effects on the dissolve of particulate phosphorus in the system; unfortunately, the experiment did not give any further analysis and data; and (2) the total phosphorus reduced to 0.047 mg L^{-1} on day 21. The results may imply that the treatment of eutrophic waters with high particulate phosphorus concentration needs to be prolonged for good results. The low total phosphorus result implies that the eutrophic water with high concentration of particulate phosphorus can be treated to the drinking water quality standard.

Table 6.3.2-1. Fate of phosphorus in water hyacinth purification after 80 days (mg P m^{-2}).

Treatment	Phosphorus from sediment	TDP removed	Phosphate from algae	Phosphorus from water[a]	Phosphorus in plant	Phosphorus lost[b]
Water hyacinth	955	77	356	433	1355	33
Control	28	38.	298	336	-	363

[a] Phosphorus from water represents the removal of phosphorus from water in forms of TDP plus phosphate from algae. [b] Phosphorus lost means phosphorus on the wall of container.

The organic particulate phosphorus can mainly be adsorbed onto the root surface of water hyacinth and then decomposed to release soluble phosphorus that can be absorbed by algae and water hyacinth. The process can greatly affect not only the particulate phosphorus removal from eutrophic waters but also the removal of endogenous phosphorus that is the principal purpose of the management of aquatic ecosystems. Cyanobacteria can move on water surface at speed of 0.04–1.2 km h^{-1} (Deng et al. 2014) when wind speed is less than 11–14 km h^{-1} (Zhu and Cai 1997). However, when wind speed is higher, cyanobacteria movement is more vertical than horizontal (Chen et al. 2012). Algae can accumulate on leeward site or following the water current because light breeze and the buoyancy of 0.7 to 360 μm s^{-1} enable cyanobacteria to float on water surface and move during clear and sunny days (Reynolds et al. 1987). Literature reported observations of cyanobacteria cell density at the north shoreline of Lake Caohu, Anhui Province, China, being 10 to 20 times greater than that at 30 km south of the shoreline on strong south wind and clear sunny weather, but being evenly distributed on north wind (Li et al. 2012). Also, the alga population was 1.7–30 times higher in the root zones of water hyacinth than the open area, nutrient release from algae was also evident (Zhou et al. 2012). This implies that the algae carrying assimilated nutrients at other locations in a lake or reservoir could be naturally transported to the lee site by the wind or water currents, as floating macrophytes would also be naturally located there by the same natural forces. When algae were trapped on the roots of water hyacinth, cell death of cyanobacteria was promoted by the physiological changes in the phycocyanin and phycocyanin/allophycocyanin ratio in the cells within 8 days (Zhou et al. 2014a). Following the decay and lysis of the algal cells, the total dissolved phosphorus concentration was tripled in 12 days (Fig. 6.3.2-1, from Zhou et al. 2014a).

Cell death of cyanobacteria occurred very quickly under the water hyacinth mat (Zhou et al. 2014b); toxin production and release from cyanobacteria was suppressed because the caspase-3 activity in algae, as the proxy for programmed cell death, was reduced significantly by water hyacinth. Although photosystem (PS) II-Hill reaction in the alga was not significantly interrupted, the energy harvest and electron transfer processes in the photosystem of alga might be disturbed due to the damage of phycocyanin

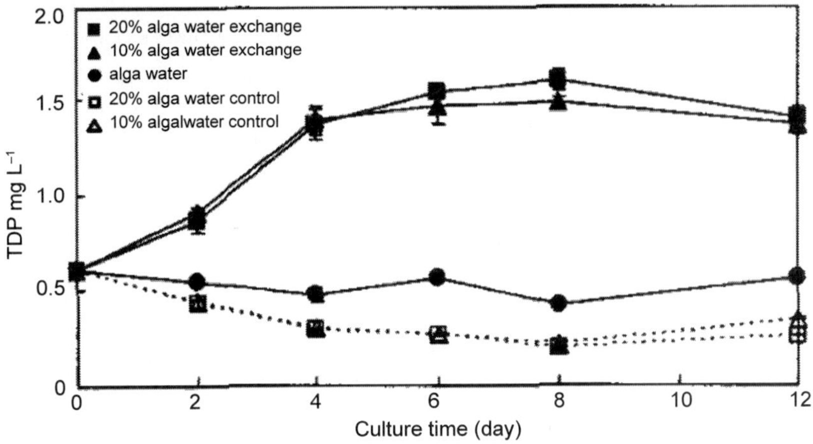

Fig. 6.3.2-1. Changes of the total dissolved phosphorus concentration in *Microcystis aeruginosa* cultured under the impact of water hyacinth. Twenty percent alga water exchange refers to 20% of water hyacinth culture media being replaced by alga culture media daily. The same refers as to 10% alga water exchange only 10% replaced daily.

and a change in the phycocyanin/allophycocyanin ratio. More importantly, the level of extracellular microcystin-LR was significantly reduced from 213 to 19 μg L^{-1} in 6 days and then remained stable in the presence of water hyacinth, but increased from 213 to 1174 μg L^{-1} in control in 8 days. The microcystin-LR level in the whole plant of water hyacinth was 3.9 ng g^{-1} fresh biomass (Zhou et al. 2014b). This result with the presence of water hyacinth was in contrast with the report that microcystin can persist for as long as 9 days at high concentration of 1300–1800 μg L^{-1} following the cyanobacteria decay and lysis in natural water without the presence of water hyacinth (Jones and Orr 1994). These data suggested that the interaction between cyanobacteria and water hyacinth can reduce microcystin concentration in aquatic environment.

The decay and lysis of algal cells are greatly enhanced by the microbes associated with the root surface of water hyacinth; soluble reactive phosphorus can be released from dissolved organic phosphorus by extracellular alkaline phosphatase naturally existing in water. Soluble reactive phosphorus of 1.23–2.42 μg L^{-1} h^{-1} can be released from dissolved organic phosphorus via both extracellular alkaline phosphatase naturally present in water and intracellular alkaline phosphatase during winter with water temperature 4–5°C (Hu 2013).

Migration of cyanobacteria can lead to an increase in concentrations of dissolved nitrogen and phosphorus below and within the root zone of the water hyacinth mat. Estimated from the yields of 15–25 tonnes ha^{-1} yr^{-1} of dry microalgae (Lam and Lee 2012) with an average phosphorus content of 8.06 g kg^{-1} dry biomass (Han et al. 2009), the amount of phosphorus being transported from other locations to leeward site may be huge, although there was no method to accurately quantify the amount. However, this is primarily a way to decrease endogenous phosphorus in large lakes and reservoirs because the areas of water hyacinth population may be negligible compared to the huge area of surface water. Of course, for effective management strategy and target, the amount of algae transportation and the areas of water hyacinth population need to be balanced so that the released nutrients can be totally absorbed.

6.3.3 Effect of water hyacinth on phosphorus removal during effluent purification

Effluent from municipal waste treatment plants and animal production systems represent the major phosphorus load to natural waters and contained mainly the particulate phosphorus, usually have high concentrations of both nitrogen and phosphorus in the balanced ratio to favor phosphorus removal. DeBusk et al. (1995) investigated phosphorus removal from dairy farm wastewater using water hyacinth under optimum temperature (21–35°C) and reported total phosphorus concentration being reduced to ~0.2 mg P L^{-1} by the end of the first week, to ~0.1 mg P L^{-1} by the end of the second week and to undetectable level by the end of the third week from the wastewater that initially contained inorganic nitrogen 21.8 mg N L^{-1} and soluble reactive phosphorus 1.2 mg P L^{-1} (N/P ration 18.2), total nitrogen 28.2 mg N L^{-1} and total phosphorus 7.4 mg P L^{-1} (N/P ration 3.8). During the experiment, the inorganic nitrogen was almost all removed in the first 3 days, soluble reactive phosphorus concentration was unchanged, but total phosphorus was reduced to 2 mg P L^{-1}. This phenomenon can only be interpreted by rapid mineralization of organic phosphorus in the first 3 days. In the later days of the experiment, the inorganic nitrogen was estimated from mineralization of organic nitrogen and was adequate due to high temperature. The interesting phenomenon was an increase in organic nitrogen by the end of the first and the second week when the phosphorus concentration in culture solution was below 0.2 mg P L^{-1}, but inorganic nitrogen was almost zero. However, this phenomenon was unexplainable since the experiment did not mention the influence of algae and possible nutrient deposition in the container. The final phosphorus control effect was excellent.

Literature suggested that a variety of macrophytes have the capacity to absorb excessive phosphorus; and water hyacinth exhibited higher capacity than other plants (Haller and Sutton 1973). High phosphorus absorption capacity and proliferation of water hyacinth are two preferred biological behaviors for phytoremediation in eutrophic waters, but not perfect to use in wastewater treatment system.

For removal of phosphorus from other wastewater treatment systems, depending on the waste treatment technology, some facilities employed deep oxidation of organic matter and anaerobic treatment in nitrogen removal so that the ratio of available nitrogen and phosphorus in effluent from those facilities may be as low as 1–2. The purification of effluent with high phosphorus loading but low available nitrogen may necessitate designing proper procedures of hydraulic retention time to maximize phosphorus accumulation in water hyacinth biomass; otherwise, the system may not work effectively. Even for effluent with balanced nitrogen/phosphorus ratio, cold weather may prevent good final quality of water leaving the system; a removal rate reduction of 30–40% compared to hot weather was quite common (Chen et al. 2010). The seasonal variation in using water hyacinth for waste treatment plants or animal production systems required integration of other technologies in the treatment system to have a full season and reliable phosphorus removal.

6.4 Strategies of phosphorus removal by water hyacinth

Phosphorus removal from eutrophic waters is critical for aquatic ecosystem management and environmental and water resources maintenance. From the previous discussion, removal effectiveness and efficiency are two different concepts and can impact greatly on management strategies and final targets. Effectiveness is defined by the final water quality a phytoremediation project can reach, whereas efficiency is defined by the amount of pollutants that can be removed per unit area and unit time or per unit fresh biomass and unit time. When using water hyacinth in phytoremediation, higher efficiency means lower effectiveness and vice versa. There is no general strategy to apply in a specific phytoremediation project. The final targets are the relevant water types (e.g., static or flowing water bodies), management purposes (e.g., water resources or recreational resources or general environmental protection), nutrient loading (spatial and temporal characteristics) and pollution levels.

6.4.1 Phosphorus removal from static water bodies

Static water bodies may be recognized as ponds without continuous inflow and outflow. These water bodies are seldom drinking water resources, except in large static reservoirs specially created for storing drinking water. This type of water body may have management targets for maintaining moderate primary production and removal of endogenous phosphorus to prevent algal bloom and establish diversified aquatic vegetation. Zhang et al. (2009, 2010) investigated the effectiveness and efficiency of phosphorus removal from static water, concluding that maintenance of good quality water in static water body was a relatively easy task with a need to estimate the endogenous phosphorus loading and the population of water hyacinth (Table 6.4.1-1).

Another report confirmed that water hyacinth was a good candidate for removal of endogenous phosphorus, effectively removing phosphorus released from sediment (Table 6.4.1-2) in shallow water (Zhang et al. 2016).

The result showed that 80–86% of total phosphorus removed by water hyacinth originated mainly from desorption from sediment through plant uptake. Root adsorption contributed about 6.2–7.8% while 1.4–1.9% may be

Table 6.4.1-1. Total phosphorus removal from static water by water hyacinth at temperatures ranging from 5 to 37°C.

Starting total P concn (mg L^{-1})	Total P concn after 21 days (mg L^{-1})	Removal rate (%)	Efficiency (mg kg^{-1} d^{-1})[a]	Total removal in 21 days (g m^{-2})	Total absorption in 21 days (g m^{-2})
0.14	0.03	76	1.2	1.6	8.7
0.34	0.04	87	3.2	4.5	12.3
1.07	0.08	92	10.3	14.8	14.2
1.43	0.11	93	13.8	19.9	16.6
Ref.	1	1	1	2	2

[a] Efficiency is defined as mg phosphorus removed by per kg fresh biomass of water hyacinth per day; Ref. 1: (Zhang et al. 2009): Ref. 2: (Zhang et al. 2010).

Table 6.4.1-2. Pathways of phosphorus removal from static water by water hyacinth at temperature 17.7–25.2°C within 30 days (Zhang et al. 2016).

Coverage %	Total phosphorus removed from			Different pathways of phosphorus removal					
	Water (mg P)	Sediment (mg P)	Total (mg P)	Plants absorption		Root adsorption		Root shedding	
				(mg P)	%	(mg P)	%	(mg P)	%
60	29.8	222	252	201	80	15.6	6.2	3.6	1.4
100	29.2	236	265	229	86	20.7	7.8	4.9	1.9

returned to the sediment via root shedding. Total phosphorus concentration in water decreased to 0.06 mg P L^{-1} from initial concentration of 0.25 mg P L^{-1} within 5 days (Zhang et al. 2016).

The above discussion showed that good water quality could be achieved within 5–21 days in static water body with relatively low concentration of phosphorus in eutrophic water, whereas a longer time may be required if the concentration was above 1.4 mg L^{-1}. Efficiency of phosphorus removal was positively correlated with the initial phosphorus concentration. However, high efficiency (63.1 mg P m^{-2} d^{-1}) was associated with lower final water quality (0.11 mg P L^{-1}), and low efficiency (5.0 mg P m^{-2} d^{-1}) was associated with higher final water quality (0.03 mg P L^{-1}). The results implied that the phytoremediation targets need to be set based on the multidimensional strategies for final water quality (e.g., schedule of harvest intervals), scale of the removal of endogenous phosphorus (e.g., initial water hyacinth population to maintain acceptable phosphorus concentration for a functional aquatic ecosystem) and phosphorus concentration rebound due to water hyacinth underperforming in cold weather.

6.4.2 Phosphorus removal from flowing water bodies

Flowing water bodies receive inflows for most of the year and have outflows to keep water surface within a certain range. Rivers, lakes and most reservoirs are flowing water bodies and usually have relatively low phosphorus loading as either soluble reactive phosphorus or particulate phosphorus. Phosphorus removal from this type of water body needs to consider the exogenous nutrient loading, endogenous phosphorus release and confined growth of water hyacinth. More important characteristics of this type of phytoremediation are seasonal variation in the pollutants together with all the characteristics of phosphorus loading in static water bodies mentioned above. The quality of water leaving the water hyacinth mat reflects the efficiency of the phytoremediation system. Relatively low phosphorus concentration in the inflow water may not reflect the total production of water hyacinth biomass due to nutrients being supplied continuously but depending on the hydraulic retention time. Zhang et al. (2010, 2011) investigated the effects of water hyacinth in phytoremediation of eutrophic water with simulated water inflows (8-hours inflow and 16-hours break daily). Water hyacinth was initially stocked at 3.0 kg m^{-2}; conditions were: ambient temperature 18.5–35.5°C, total phosphorus concentration in the inflow 0.50 mg P L^{-1}, and hydraulic loading at 0.14, 0.20, 0.33 or 1.00 m^3 m^{-2} d^{-1}. These hydraulic loadings are equivalent to hydraulic retention times of 7, 5, 3 and 1 day, respectively. The results showed that although the phosphorus concentration in inflow water was the same, the phosphorus removal efficiency and final water quality obtained were significantly different (Zhi-yong Zhang et al. 2010). Generally, higher hydraulic loading was associated with higher removal efficiency but lower effectiveness (Table 6.4.2-1).

The experiment concluded that under the temperature and nutrient balance of the eutrophic water, the suitable hydraulic loadings were 0.14–0.33 m^3 m^{-2} d^{-1}. The final total phosphorus concentration after phytoremediation was 0.09–0.19 mg P L^{-1} (quality levels from drinking water to aquatic recreational use) (MEP-PRC 2002).

In 2014 at Jiangsu Academy of Agricultural Sciences, Nanjing, China, a remediation system with 24-hour continuous water flow was constructed using stainless steel containers (length, width and height of 20.0 by 1.0 by 0.5 m) and hydraulic loadings of 0.07, 0.20 and 0.42 m^3 m^{-2} d^{-1} with water hyacinth initially stocked at 20 kg m^{-2} (Fig. 6.4.2-1).

The results (Table 6.4.2-2) were consistent with the report by Zhang et al. (2010, 2011) and indicated the reverse relationship of effectiveness and efficiency.

The results showed that hydraulic retention time of 5 days for initial phosphorus concentration 0.84 mg P L^{-1} was sufficient to achieve water quality for aquatic recreational use.

A similar experiment was set up in 2015 at Jiangsu Academy of Agricultural Sciences in a natural pond with continuous flow of municipal wastewater and conducted for 6 days with stainless steel containers (length 10.0 m, width 1.0 m and depth 0.5 m) with initial stocking of water hyacinth at 20 kg m^{-2} (Fig. 6.4.2-2).

Table 6.4.2-1. Effects of phytoremediation involving water hyacinth on total phosphorus removal at different hydraulic loadings in 112 days with initial average total phosphorus concentration of 0.5 mg P L^{-1}.

Hydraulic loading (m^3 m^{-2} d^{-1})	Final TP concn (mg P L^{-1})	Apparent TP removal (%)	TP loading (g m^{-2} d^{-1})	TP removal (g m^{-2})	Plant absorption of P (g m^{-2})	TP removal efficiency (g m^{-2} d^{-1})
0.14	0.09	81	0.08	6.9	6.5	0.06
0.20	0.12	73	0.10	8.6	7.9	0.07
0.33	0.18	64	0.17	12 .3	8.4	0.11
1.00	0.27	48	0.50	27.0	13.5	0.23

Fig. 6.4.2-1. Remediation experiment set up for 24-hours continuous water flow at hydraulic loadings of 0.07–0.42 m^3 m^{-2} d^{-1} (photo by Lin Shang 2014).

Table 6.4.2-2. Effects of phytoremediation with water hyacinth on total phosphorus removal at different hydraulic loadings, initial average total phosphorus concentration 0.84 mg P L^{-1} and ambient temperature 15–29°C after 14 days (unpubl. data).

Hydraulic load (m^3 m^{-2} d^{-1})	Hydraulic retention time (day)	TP loading (g m^{-2} d^{-1})	Final TP concn (mg P L^{-1})	TP removal rate (%)	TP removal efficiency (g m^{-2} d^{-1})
0.07	14	0.06	0.02	97	0.06
0.20	5	0.17	0.17	79	0.14
0.42	2.4	0.35	0.33	59	0.24

The results showed that efficiency of phosphorus removal ranged from 0.16 to 0.36 g m^{-2} d^{-1} (Table 6.4.2-3). Water quality at the exit was at the level for aquatic recreational use.

The data from flowing water experiments were obtained with fresh biomass of water hyacinth at ~20 kg m^{-2} and under the ambient temperature 15–37°C. The results suggest that the total phosphorus loading of 0.17 g m^{-2} d^{-1} in the phytoremediation system may be treated to the moderate water quality with retention time of 1 day; higher total phosphorus loading may require longer retention times.

Fig. 6.4.2-2. *In situ* experimental layout with stainless steel flow sinks welded into a rectangular frame (10.0 m by 1.0 m by 0.5 m: length, width and depth), with the depth of the flow sinks adjusted by foam floats on both sides of the frame. The flow sinks were fixed to steel posts in the wastewater treatment pond. Wastewater was continuously flowing into each sink through a steel pipe (inner diameter 50 mm, covered by a 1.3-mm nylon mesh to filter out detritus) placed at one end 10 cm below the water surface. Drainage was via an outlet steel pipe (inner diameter 50 mm) placed at the other end 10 cm below water surface. The hydraulic loading was controlled by a precise quantitative pump at 0.5 m^3 m^{-2} d^{-1} (photo by Lin Shang 2015).

Table 6.4.2-3. Effects of phytoremediation with water hyacinth on total phosphorus removal at hydraulic loading of 0.5 m^3 m^{-2} d^{-1} (equivalent to hydraulic retention time of 1 day) at 27–37°C during 6 days (unpubl. data).

Sample date in 2015	Inflow TP concn (mg P L^{-1})	Outflow TP concnn (mg P L^{-1})	TP removal rate (%)	TP removal efficiency (g m^{-2} d^{-1})
1 August	0.70	0.19	73	0.25
2 August	0.95	0.23	76	0.36
3 August	0.57	0.20	64	0.18
4 August	0.58	0.20	66	0.19
5 August	0.50	0.17	65	0.16
6 August	0.48	0.15	69	0.16

6.5 Summary

Water hyacinth can grow in aquatic environments with a wide range of phosphorus concentrations, but with variable effects on growth rate, phosphorus accumulation, phosphorus distribution in plant parts, and efficiency of phosphorus absorption. Water hyacinth can utilize both soluble reactive phosphorus and particulate phosphorus. The amount of phosphorus absorbed and accumulated by water hyacinth was positively related to the initial concentration of the total phosphorus in water or to phosphorus loading in flowing water. When growing in water with low concentration of phosphorus, water hyacinth forms a large root surface for nutrient uptake. The hyperaccumulation properties enable water hyacinth to grow in phosphorus-rich water as well. The phosphorus content in water hyacinth tissue can be 1.5–13.5 g kg^{-1} dry biomass. Phytoremediation with water hyacinth enables water purification to good quality, even producing water with drinking water quality for resource reclamation. In water bodies with phytoremediation-occurring and naturally-occurring water hyacinth, a suitable harvesting schedule may keep both good quality of water and produce biomass for feed, organic fertilizer or bio-energy.

References cited

Berelson, W. M., D. Heggie, A. Longmore, T. Kilgore, G. Nicholson and G. Skyring. 1998. Benthic nutrient recycling in Port Phillip Bay, Australia. *Estuarine, Coastal and Shelf Science* 46(6): 917–934.

Bernhardt, E. S. 2013. Cleaner lakes are dirtier lakes. *Science* 342(6155): 205–206.

Boyd, C. E. 1976. Accumulation of dry matter, nitrogen and phosphorus by cultivated water hyacinths. *Economic Botany* 30(1): 51–56.

Boyd, C. E. and D. H. Vickers. 1971. Variation in the elemental content of *Eichhornia crassipes*. *Hydrobiologia* 38(3-4): 409–414.

Carlson, R. E. and J. Simpson. 1996. A coordinator's guide to volunteer lake monitoring methods. Madison, WI, USA: North American Lake Management Society.

Chen, L., C. Wang and T. Li. 2012. Study on movement of cyanobacteria bloom with special wind-field condition in Taihu Lake. The Administration and Technique of Environmental Monitoring 24(3): 24–34 (In Chinese with English Abstract).

Chen, X., X. Chen, X. Wan, B. Weng and Q. Huang. 2010. Water hyacinth (*Eichhornia crassipes*) waste as an adsorbent for phosphorus removal from swine wastewater. *Bioresource Technology* 101(23): 9025–30.

Chen, Z., Z. Zhang, H. Liu, X. Wen, H. Qin, S. Yan et al. 2015. Research on removal efficiency of phosphorus by four aquatic macrophytes and rule of phosphorus migration in systems. *Journal of Nanjing Agricultural University* 38(1): 107–112 (In Chinese with English Abstract).

China Environmental Monitoring Station. 2001. Guide to lake (reservoir) eutrophication assessment method. Beijing, China: China Environmental Monitoring Station (In Chinese).

Coveney, M. F., E. F. Lowe, L. E. Battoe, E. R. Marzolf and R. Conrow. 2005. Response of a eutrophic, shallow subtropical lake to reduced nutrient loading. *Freshwater Biology* 50(10): 1718–1730.

Cui, F. 2013. Study on phosphorus forms and the releasing characteristics in sediment-water interface of Wuliangsuhai Lake. MSc. Thesis, Department of Hydrology and Water Resources. Inner Mongolia Agricultural University.

DeBusk, T. A., J. E. Peterson and K. Ramesh Reddy. 1995. Use of aquatic and terrestrial plants for removing phosphorus from dairy wastewaters. *Ecological Engineering* 5(2-3): 371–390.

Deng, J., X. Liu, H. Zhang, F. Chen, B. Xu and A. Shen. 2014. Characteristics and impact factors of algal horizontal drifting in Lake Taihu. *Journal of Lake Sciences (China)* 26(3): 358–364 (In Chinese with English Abstract).

Department of Natural Resources. 2013. Chapter NR 217: effluent standards and limitations for phosphorus. *Effluent Standards and Limitations*. State of Wisconsin, US. http://water.epa.gov/scitech/swguidance/standards/wqslibrary/upload/wiwqs_nr217.pdf.

Dokulil, M. T. and K. Teubner. 2011. Eutrophication and climate change: present situation and future scenarios. In: Eutrophication: Causes, Consequences and Control, ed. A. A. Ansari, S. S. Gill, G. R. Lanza, and W. Rast, 1–16. Dordrecht, Netherlands: Springer Netherlands.

Edmondson, W. T. 1994. Sixty years of Lake Washington: a curriculum vitae. *Lake and Reservoir Managemant* 10(2): 75–84.

Edmondson, W. T. and J. T. Lehman. 1981. The effect of change in the nutrient income on the condition of Lake Washington. *Limnology and Oceanography* 26(1): 1–29.

EPA. 2007. Summary table for the nutrient criteria documents. *Environmental Protection Agency*. Washington DC, US: Office of Science and Technology.

Gossett, D. R. and W. E. Norris Jr. 1971. Relationship between nutrient availability and content of nitrogen and phosphorus in tissues of the aquatic macrophyte, *Eichornia crassipes* (Mart.) Solms. *Hydrobiologia* 38(1): 15–28.

Guo, Z. 2007. Distribution and transformation of phosphorus fractionations in the sediments of three typical urban shallow lakes - Xuanwu Lake, Damning Lake and Mochou Lake. MSc. Thesis, Department of Environment and Engineering. Hohai University.

Haller, W. T., E. B. Knifling and S. H. West. 1970. Phosphorus absorption by and distribution in water hyacinths. *Proceedings. Soil and Crop Science Society of Florida* 30: 64–68.

Haller, W. T. and D. L. Sutton. 1973. Effect of pH and high phosphorus concentrations on growth of waterhyacinth. *Hyacinth Control Journal* 11: 59–61.

Han, S., S. Yan, Z. Wang, W. Song, H. Liu, J. Zhang et al. 2009. Harmless disposal and resources utilizations of Taihu Lake blue algae. *Journal of Natural Resources* 24(3): 431–438 (In Chinese with English Abstract).

Hong, H., L. Guo, J. Chen and C. Jin. 1989. Characteristics of particulate phosphate at Jiulong River estuary and Xiamen Harbour. *Journal of Xiamen University (Natural Science)* 28(1): 74–78 (In Chinese with English Abstract).

Hu, Z. 2013. Seasonal variation of phosphorus forms and mutual transformation in Grand River watershed, Canada. Ph.D. Thesis, Department of Soil and Environmental Chemistry. Southwest University.

Jiang, Z., J. Fang, J. Zhang, Y. Mao and W. Wang. 2007. Forms and bioavailability of phosphorus in surface sediment from Sungo Bay. *Environmental Science* 28(12): 2783–2788 (In Chinese with English Abstract).

Jin, X., X. Jiang, Q. Wang and D. Liu. 2008. Seasonal changes of P adsorption-desorption characteristics at the water-sediment interface in Meiliang Bay, Taihu Lake, China. *Acta Scientiae Circumstantiae* 28(1): 24–30 (In Chinese with English Abstract).

Jin, Y., G. Zhu, H. Xu and M. Zhu. 2015. Spatial distribution pattern and stock estimation of nutrients during bloom season in Lake Taihu. *Environmental Science* 36(3): 936–945 (In Chinese with English Abstract).

Jones, G. J. and P. T. Orr. 1994. Release and degradation of microcystin following algicide treatment of a *Microcystis aeruginosa* bloom in a recreational lake, as determined by HPLC and protein phosphatase inhibition assay. *Water Research* 28(4): 871–876.

Lam, M. K. and K. T. Lee. 2012. Microalgae biofuels: a critical review of issues, problems and the way forward. *Biotechnology Advances* 30(3): 673–90.

Lapointe, B. E., D. A. Tomasko and W. R. Matzie. 1994. Eutrophication and trophic state classification of seagrass communities in the Florida Keys. *Bulletin of Marine Science* 54(3): 696–717.

Lee, G. F. 1973. Role of phosphorus in eutrophication and diffuse source control. *Water Research* 7(1): 111–128.

Li, Y., B. Rao, Z. Wang, H. Qin, L. Zhang and D. Li. 2012. Spatial-temporal distribution of phytoplankton in bloom-accimulation area in Lake Chaohu. *Resources and Environment in the Yangtze Basin* 21(Z2): 25–31 (In Chinese with English Abstract).

Mehner, T., M. Diekmann, T. Gonsiorczyk, P. Kasprzak, R. Koschel, L. Krienitz et al. 2008. Rapid recovery from eutrophication of a stratified lake by disruption of internal nutrient load. *Ecosystems* 11(7): 1142–1156.

MEP-PRC. 2002. Environmental quality standards for surface water (GB3838-2002). Baijing, China: Ministry of Environmental Protection of The Peoples's Republic of China.

MEP-PRC. 2003. Discharge standard of pollutants for municipal wastewater treatment plant (GB18918-2002). Beijing, China: China Environmental Science Press.

Meybeck, M. 2003. Global analysis of river systems: from Earth system controls to anthropocene syndromes. *Philosophical transactions of the Royal Society of London. Series B, Biological Sciences* 358(1440): 1935–1955.

Musil, C. F. and C. M. Breen. 1977. The influence of site and position in the plant community on the nutrient distribution in, and content of *Eichhornia crassipes* (Mart.) Solms. *Hydrobiologia* 53(1): 67–72.

Niu, J., L. Zhang, X. Jin, J. Song and Q. Shi. 2012. Research on auxin and cytokinins changes in *Eichhornia crassipes* and their correlation with it tillering at different nitrogen and phosphorus levels. *Journal of Soochow University (natural science edition)* 28(1): 76–82 (In Chinese with English Abstract).

Ornes, W. H. and D. L. Sutton. 1975. Removal of phosphorus from static sewage effluent by waterhyacinth. *Journal of Aquatic Plant Management* 13(1): 56–58.

Pan, J., Y. He, W. Deng, D. Yan and Y. Guo. 2004. Progress in the study on the function of wetland in removal phosphorus in water. *Ecology and Environment* 13(1): 102–104, 108 (In Chinese with English Abstract).

Petrucio, M. M. and F. A. Esteves. 2000. Uptake rates of nitrogen and phosphorus in the water by *Eichhornia crassipes* and *Salvinia auriculata*. *Revista brasileira de biologia* 60(2): 229–236.

Pinto-coelho, R. M., M. Karla and B. Greco. 1999. The contribution of water hyacinth (*Eichhornia crassipes*) and zooplankton to the internal cycling of phosphorus in the eutrophic Pampulha Reservoir, Brazil. *Hydrobilogia* 411: 115–127.

Pitkänen, H., J. Lehtoranta and A. Räike. 2001. Internal nutrient fluxes counteract decreases in external load: the case of the estuarial eastern Gulf of Finland, Baltic Sea. *Journal of the Human Environment* 30(4): 195–201.

Polomski, R. F., M. D. Taylor, D. G. Bielenberg, W. C. Bridges, S. J. Klaine and T. Whitwell. 2009. Nitrogen and phosphorus remediation by three floating aquatic macrophytes in greenhouse-based laboratory-scale subsurface constructed wetlands. *Water, Air, and Soil Pollution* 197(1-4): 223–232.

Qin, H. 2015. Analysis of the death causes of water hyacinth planted in large-scale enclosures in the area of Waihai in the Dianchi Lake. *Resources and Environment in the Yangtze Basin* 24(4): 594–602 (In Chinese with English Abstract).

Reddy, K. R., M. Agami and J. C. Tucker. 1989. Influence of nitrogen supply rates on growth and nutrient storage by water hyacinth (*Eichhornia crassipes*) plants. *Aquatic Botany* 36(1): 33–43.

Reddy, K. R., M. Agami and J. C. Tucker. 1990. Influence of phosphorus on growth and nutrient storage by water hyacinth (*Eichhornia crassipes* (Mart.) Solms) plants. *Aquatic Botany* 37(1): 355–365.

Reddy, K. R. and J. C. Tucker. 1983. Productivity and nutrient uptake of water hyacinth, *Eichhornia crassipes* I. effect of nitrogen source. *Economic Botany* 37(2): 237–247.

Ren, Y., S. Dong, F. Wang, Q. Gao, X. Tian and F. Liu. 2010. Sedimentation and sediment characteristics in sea cucumber *Apostichopus japonicus* (Selenka) culture ponds. *Aquaculture Research* 42(1): 14–21.

Reynolds, C. S., R. L. Oliver and A. E. Walsby. 1987. Cyanobacterial dominance: the role of buoyancy regulation in dynamic lake environments. *New Zealand Journal of Marine and Freshwater Research* 21(3): 379–90.

Rodríguez, M., J. Brisson, G. Rueda and M. S. Rodríguez. 2012. Water quality improvement of a reservoir invaded by an exotic macrophyte. *Invasive Plant Science and Management* 5(2): 290–299.

Ruttenberg, K. C. 2003. The global phosphorus cycle. In: Treatise on Geochemistry, Volume 8, ed. W. H. Schlesinger, 586–643. Amsterdam, Netherlands: Elsevier Science Ltd.

Schindler, D. W., R. E. Hecky, D. L. Findlay, M. P. Stainton, B. R. Parker, M. J. Paterson et al. 2008. Eutrophication of lakes cannot be controlled by reducing nitrogen input: results of a 37-year whole-ecosystem experiment. *Proceedings of the National Academy of Sciences* 105(32): 11254–11258.

Shiralipour, A., L. A. Garrard and W. T. Haller. 1981. Nitrogen source, biomass production, and phosphorus uptake in waterhyacinth. *Journal of Aquatic Plant Management* 19: 40–43.

Song, X. 2009. Experiment and dynamic modelling on nitrogen and phosphorus exchanging within sediment-water interface in shallow lake. MSc. Thesis, Department of Geology and Engineering. Kunming University of Science and Technology.

Wang, H. J., X. M. Liang, P. H. Jiang, J. Wang, S. K. Wu and H. Z. Wang. 2008. TN:TP ratio and planktivorous fish do not affect nutrient-chlorophyll relationships in shallow lakes. *Freshwater Biology* 53(5): 935–944.

Wang, H. and H. Wang. 2009. Mitigation of lake eutrophication: loosen nitrogen control and focus on phosphorus abatement. *Progress in Natural Science* 19(10): 1445–1451.

Wang, Z., Z. Zhang, J. Zhang, Y. Zhang, H. Liu and S. Yan. 2012. Large-scale utilization of water hyacinth for nutrient removal in Lake Dianchi in China: the effects on the water quality, macrozoobenthos and zooplankton. *Chemosphere* 89(10): 1255–1261.

Wang, Z., Z. Zhang, Y. Zhang, J. Zhang, S. Yan and J. Guo. 2013. Nitrogen removal from Lake Caohai, a typical ultra-eutrophic lake in China with large scale confined growth of *Eichhornia crassipes*. *Chemosphere* 92(2): 177–183.

Xie, L., P. Xie, S. Li, H. Tang and H. Liu. 2003. The low TN:TP ratio, a cause or a result of *Microcystis* blooms? *Water Research* 37(9): 2073–2080.

Xie, Y., M. Wen, D. Yu and Y. Li. 2004. Growth and resource allocation of water hyacinth as affected by gradually increasing nutrient concentrations. *Aquatic Botany* 79(3): 257–266.

Yuan, H., J. Song, N. Li, X. Li, Y. Zhang and S. Xu. 2009. Spatial distributions and seasonal variations of particulate phosphorus in the Jiaozhou Bay in North China. *Acta Oceanologica Sinica* 28(1): 99–108.

Zhang, F. 2004. The characteristics of phosphorus behavior at sediment-water interface and environmental risk assessment from shallow lakes. MSc. Thesis, Department of Geography. East China Normal University.

Zhang, Y., Z. Zhang, Z. Chen, H. Liu, X. Wen, H. Qin et al. 2016. Phosphorus removal pathways in water hyacinth (*Eichhornia crassipes*) ecological restoration systems and influence on phosphorus release in sediment by the macrophyte. *Journal of Nanjing Agricultural University* 39(1): 106–113 (In Chinese with English Abstract).

Zhang, Z., Z. Chang, H. Liu, J. Zheng, L. Chen and S. Yan. 2010. Effect of *Eichhornia crassipes* on removing nitrogen and phosphorus from eutrophicated water as affected by hydraulic loading. *Journal of Ecology and Rural Environment* 26(2): 148–154 (In Chinese with English Abstract).

Zhang, Z., H. Liu, S. Yan, J. Zheng, Z. Chang and L. Chen. 2009. Comparison of the removal ability of water hyacinth (*Eichhornia crassipes*) in differently eutrophic water. *Jiangsu Journal of Agricultural Sciences* 25(5): 1039–1046 (In Chinese with English Abstract).

Zhang, Z. Y., J. C. Zheng, H. Q. Liu, Z. Z. Chang, L. G. Chen and S. H. Yan. 2010. Role of *Eichhornia crassipes* uptake in the removal of nitrogen and phosphorus from eutrophic waters. *Chinese Journal of Eco-Agriculture* 18(1): 152–157 (In Chinese with English Abstract).

Zhang, Z., J. Zhang, H. Liu, L. Chen and S. Yan. 2011. Apparent removal contributions of *Eichhornia crassipes* to nitrogen and phosphorous from eutrophic water under diferent hydraulic loadings. *Jiangsu Journal of Agricultural Sciences* 27(2): 288–294 (In Chinese with English Abstract).

Zhou, Q., S. Q. Han, S. H. Yan, W. Song and J. P. Huang. 2012. The mutual effect between phytoplankton and water hyacinth planted on a large scale in the eutrophic lake. *Acta Hydrobiologica Sinica* 36(4): 873–791 (In Chinese with English Abstract).

Zhou, Q., S. Han, S. Yan, W. Song and G. Liu. 2014a. Impacts of *Eichhornia crassipes* (Mart.) Solms stress on the growth characteristics, microcystins and nutrients release of *Microcystis aeruginosa*. *Environmental Science* 35(2): 597–604 (In Chinese with English Abstract).

Zhou, Q., S. Han, S. Yan, W. Song and G. Liu. 2014b. Impacts of *Eichhornia crassipes* (Mart.) Solms stress on the physiological characteristics, microcystin production and release of *Microcystis aeruginosa*. *Biochemical Systematics and Ecology* 55: 148–155.
Zhou, Z. 2008. Effect of water quality on the growth of water hyacinth [*Eichhornia crassipes* (Mart.) Solms.]. MSc. Thesis, Department of Agronomy. Fujian Agricultural and Forestry University.
Zhu, Y. and Q. Cai. 1997. The dynamic research of the influence of wind field on the migration of algae in Taihu Lake. *Journal of Lake Sciences (China)* 9(2): 152–158 (In Chinese with English Abstract).

CHAPTER 7

Impact of Water Hyacinth on Removal of Heavy Metals and Organic Pollutants

X. Lu

7.1 Introduction

One unique and important capability of water hyacinth to remediate polluted water is the removal of heavy metals including lead, chromium, cadmium, mercury and harmful organic compounds such as detergents, pesticides, gasoline additives, hormones, pharmaceuticals and personal-care products, as well as metalloid pollutants such as arsenic. Given that the conventional sewage treatment techniques could not effectively remove them, a majority of heavy metals and organic pollutants are directly discharged into lakes, reservoirs and rivers, thus causing large scale water pollution. Although heavy metals and harmful organic pollutants are usually present at low concentrations ranging from ng L^{-1} to μg L^{-1} compared with conventional pollutants such as nitrogen and phosphorus that are usually at mg L^{-1} levels, they may significantly impact environmental safety and human health. It is notable that heavy metals could be readily concentrated in living organisms through the food chain, potentially reaching ten thousands to hundred thousand times the original concentrations and may finally enter into human body and negatively impact human health (Xu et al. 1999, Zhou et al. 2006, Xu et al. 2006).

50 Zhong Ling Street, Nanjing, China.
Email: lxdeng@126.com

7.2 Mechanism and impact of water hyacinth on heavy metal and metalloid removal

7.2.1 Presence and impact of heavy metals and metalloids on aquatic environments

Source of heavy metal and metalloid pollutants in aquatic environments

Generally, the harmful heavy metal pollutants in the aquatic ecosystem may originate from several pathways.

1) Industrial wastewater: most heavy metals that represent a threat to human health originate from wastewater discharged by industrial and mining enterprises. Wastewater produced from a variety of industries, including mining, metallurgy, chemical industry and electroplating contain heavy metals (such as chromium, cadmium, copper, mercury, nickel and zinc) could cause water pollution if they are directly discharged into natural water bodies without treatment.
2) Municipal sewage: with a rapid increase in urban population, the discharge of municipal sewage is continually increasing. Given that the environmental infrastructure construction in some countries/regions falls behind the speed of urban development, a large amount of municipal sewage is discharged without treatment, becoming an important source of environmental pollution. Municipal sewage contains abundance of Oxygen-Consuming Organisms (COD), relatively high concentration of ammonium nitrogen (NH_4^+-N) and phosphate (PO_4^{3-}), as well as trace heavy metals.
3) Inappropriate landfill and storage of industrial solid wastes and household garbage. Some solid wastes contain high concentrations of heavy metals that could enter into surface and underground waters through runoff, leaching, percolation of rainwater and other pathways, thus polluting large areas.
4) Development of intensive livestock and poultry production systems, and intensive aquaculture. These food-producing activities result in use of additives containing essential heavy metals, such as copper and zinc, as well as arsenic; such additives are frequently used in animal feed to prevent or treat diseases, promote animal growth, and improve feed utilization rate. However, improper usage and/or dosage of feed additives may result in untreated wastewater discharged from the animal-production systems to the aquatic ecosystems becoming a pollution source of heavy metals as well as organic waste.

Early in the mid-20th century, Japan had suffered from the "minamata disease" that was caused by mercury pollution and the "itai-itai disease" that was caused by cadmium pollution (Yang et al. 1999). Since the 1980s, many reports have asserted that many water bodies, such as Jinsha River, Xiang

River, Huangpu River, Dasha River and Long River in China, were polluted by heavy metals to different extents, with concentration of heavy metals in the water reaching up to several hundred $\mu g \cdot L^{-1}$ and in sediments up to a thousand mg kg^{-1} (Li et al. 1997). Another example was Swansea (a port city in South Wales, UK), with the largest grouping of the metal smelting industry in the world, as a source of pollution of aquatic environments by copper, zinc, lead, cadmium and other heavy metals (Thornton and Walsh 2001). Also in Europe, as a former naval base, the harbor of Copenhagen in Denmark was dominated by ship manufacturing and repairing as well as chemical plants, which resulted in heavy metal pollution of shallow waters and sediments along the coast (Andersen et al. 1998).

It is clear from the above examples that the heavy metal pollution of natural waters has become a global environmental problem. The common heavy metal pollutants are listed below.

1) Copper and zinc

   Copper and zinc pollution mainly comes from wastewater discharged by the mining industry, mineral processing, aquaculture, etc. As the essential nutrients needed for plant growth, copper and zinc are important components of metalloenzymes, plastocyanin and many other molecules, and zinc is important in maintaining membrane structure and function. However, excessive concentration of copper and zinc in the water body is harmful to plant growth because of accumulation of copper and zinc in plant tissues affecting the normal physiological and biochemical processes, such as photosynthesis, nutrient uptake and others (Mazen and El Maghraby 1997, Megateli et al. 2009). Humans or animals can also be harmed if eating food with a higher content of copper and zinc.

2) Cadmium

   Cadmium is the most common heavy metal presented in aquatic environments that originated mainly from wastewater discharged by mining, metal smelting and chemical industries. Cadmium is commonly used as plating material on steel; and its compounds are used as pigments and plastic stabilizers. During the production processes, cadmium can be released to environment via discharge of solid waste or wastewater and then can accumulate in food products. Excessive intake of cadmium by humans could lead to "itai-itai" disease (i.e., osteoporosis), hypertension, as well as kidney problems.

3) Lead

   Lead enters aquatic environments mainly through wastewaters discharged by mining, smelting, paint and coating industries, and combustion emission of coal and leaded petrol. Lead is a highly toxic pollutant, as well as a potential carcinogen. Excessive lead would cause damage to the human brain and kidneys resulting in dysgnosia of adolescents, and cause spasms in elderly.

4) Mercury

Mercury pollution comes from wastewater discharged by a specific industry as well as from extensive application of pesticide containing mercury. Mercury is easily assimilated and accumulated by fishes in the form of methylmercury compounds (CH_3Hg and $(CH_3)_2Hg$). People consuming fish and mussels containing methylmercury in large amounts and for a long time could suffer damage to central nervous system, causing sensory disturbance, dyskinesias, language disorder, constriction of visual field and hearing disorder, commonly known as "minamata disease".

5) Chromium

Chromium has extensive applications in electroplating, leather and pharmaceutical industries, lapping compound preservatives, pigment and synthetic catalysis, etc.; these industries could produce chromium-containing wastewater. Moreover, improper stocking of chromium-containing waste residues may result in dissolution and runoff of chromium if washed by rains, and thus pollution of aquatic environments. Chromium can induce carcinogenesis, chromosomal aberrations and mutations. In addition, chromate salts could damage the human liver and cause skin allergy.

6) Arsenic

As a metalloid element, arsenic could exist in the form of cation, anion, as well as complexing anions. Generally, arsenic contaminations in aquatic environments are mainly through wastewater (gas) discharged by mining, metal smelting and fossil fuel combustion. The extensive application of pesticides containing arsenic is another important pathway of water pollution. As one of the most poisonous substances, arsenic could cause damages to the human liver and kidney functions, disrupt the central nervous, blood and immune systems, and cause cancer.

7) Other heavy metals

There are other heavy metal pollutants in water bodies. For instance, vanadium is present in fossil fuels such as gasoline and coal; combustion of these fossil fuels is the main source of vanadium pollution, with other sources being mining and smelting of vanadium ores (Wu and Lan 2004). Moreover, other heavy metals in the waters, such as silver, cobalt and iron, are derived mainly from mining of silver, cobalt and iron ores, as well as discharging of waste gas, waste residues and wastewater of metal smelting industry.

Due to the universality and perniciousness of heavy metal pollution, it is important to combine measures (including physical, chemical and biological treatment methods) to prevent and control the pollution. Some heavy metals are carcinogenic, mutagenic, teratogenic and endocrine disruptors, whereas others cause neurological and behavioral changes especially in children. Hence, remediation of heavy metal pollution deserves high attention.

Control and remediation of heavy metal pollution

It is difficult to remove heavy metal pollutants from waters because they are relatively stable and non-biodegradable in the environment. It is known that the conventional water treatment methods could not decompose and remove heavy metals, but can change their position as well as the physical and chemical forms. The presence of heavy metals in aquatic environments has been a matter of concern for governments and environmental scientists worldwide due to their persistence and toxicity. Therefore, research has been pursued to develop cost-effective, efficient and eco-friendly remediation methods for removal of heavy metals from polluted waters.

At present, the remediation of heavy metal-polluted water is practiced by two basic pathways: (i) lower the content and bio-availability of heavy metals in the water, and (ii) completely remove heavy metals from the polluted water. Technologies for treating heavy metal-containing wastewater include traditional physical and chemical methods, such as dilution, water replacement, chemical flocculation and precipitation, physical adsorption, electro-remediation, chemical reduction, ion exchange, membrane separation and reverse osmosis processes, as well as the emerging biological remediation. Different physical and chemical methods mentioned above perform well under certain conditions, but suffer from serious limitations like high cost, intensive labor, high energy consumption, difficult implementation and easily causing secondary pollution; these problems result in limited application in treating large volumes of polluted water.

Since the 1980s, bioremediation has gradually developed as an alternative to conventional treatment methods of contaminated water; it refers to a new promising solution to remediate the contaminated environment through lowering the concentration of pollutants in water to an acceptable level by using microorganisms, plants and animals to absorb and/or transform the toxic and harmful pollutants in the soil and water environments, making bioremediation a sustainable solution with low maintenance costs and low energy consumption. According to the organisms employed in the remediation operations, bioremediation could be divided into three types: microbial remediation, phytoremediation and animal remediation, with microbial remediation and phytoremediation being frequently used in treating polluted waters.

Phytoremediation refers to the methods whereby plants and associated microbes are used to decrease the concentration or bioavailability of contaminants such as heavy metals and organic pollutants in the environments. Phytoremediation technologies use naturally-occurring or genetically-modified plants to absorb and accumulate heavy metal ions in plant tissues, thus removing heavy metals through harvesting the plants. Some aquatic plant species with high rate of heavy metal uptake can have accumulation 100 times greater than that in majority of aquatic plants, while their growth is not affected. These hyperaccumulators are usually specific for one heavy metal, but some hyperaccumulators could accumulate two or more heavy metals. Common floating and emergent plants, such as water hyacinth (*Eichhornia crassipes*), common duckweed (*Lemna minor*), cattail (*Typha* spp.), common reed (*Phragmites* spp.) and alligator weed (*Alternanthera philoxeroides*), have been used widely in phytoremediation of aquatic ecosystems polluted by heavy metals, including copper, cadmium, lead, zinc and many others. Indeed, the literature indicates that many macrophytes have a good potential for remediating heavy metal pollution in aquatic environments (Raskin et al. 1997, Miretzky et al. 2004, Ha et al. 2009, Cai et al. 2009, Zhang et al. 2010). Aquatic plants used for phytoremediation often have adaptive characteristics (such as well-developed root systems that could absorb and immobilize heavy metals) and high growth rate and yield.

Water hyacinth is a floating aquatic macrophyte with floating rhizomes, subfloat, stolons and a fibrous root system, it can produce relatively high biomass, increasing the capacity to remove cadmium, chromium, copper and other heavy metals from contaminated waters and improving water quality in the treatment systems (Chen et al. 2004, Espinoza-Quiñones et al. 2013, Lu et al. 2014). Water hyacinth was successfully used to treat various types of sewage, including sewage discharged from mining factory, that contain high concentration of heavy metals, such as cadmium, silver, nickel, copper, zinc and vanadium (Zhu et al. 1999, Agunbiade et al. 2009, Ye and Qiu 2010).

Compared with the conventional water treatment methods, phytoremediation could overcome some disadvantages, such as high cost, complex installation and maintenance, incomplete purification, secondary pollution and eco-unfriendliness. Therefore, phytoremediation has been widely considered to be a wastewater treatment method with the best development prospects, particularly in treating large water bodies (Miretzky et al. 2004, Ha et al. 2009, Cai et al. 2009, Hua et al. 2010), becoming the focus of pollution remediation research in recent years.

7.2.2 Effects of water hyacinth on heavy metal removal from aquatic environments

Research progress

The application of phytoremediation technology to remedy heavy metal-contaminated waters is an important research topic in environmental management and weed control (Espinoza-Quiñones et al. 2013, Lu et al. 2014). Water hyacinth can quickly grow to very high densities (over 60 kg m^{-2}) and easily form a dominant species in the aquatic environment (Malik 2007); hence, a negative impact on biodiversity and a positive impact on phytoremediation need to be appropriately managed. Water hyacinth has a highly developed root system, ranging in length from 5 to 200 cm (Rodríguez et al. 2012). The root biomass of water hyacinth reaches up to 50% of the weight of the whole plant, with dense root hairs contributing to a large specific root surface area (2.1–8.0 $m^2 g^{-1}$) (Kim and Kim 2000, Liu et al. 2003, Zheng 2010). Due to high tolerance to a variety of pollutants, water hyacinth is commonly regarded as an aquatic floating macrophyte for remediation of heavily polluted water and assimilation of multiple pollutants, including nitrogen and phosphorus, heavy metals, Polycyclic Aromatic Hydrocarbons (PAHs) and other organic and inorganic pollutants (Xia 2002, Malik 2007, Agunbiade et al. 2009, Hua et al. 2010, Smolyakov 2012). Although water hyacinth is quite a versatile plant regarding phytoremediation capacity to remove inorganic nutrients, toxic metals as well as persistent organic pollutants, it is often seen as a noxious weed responsible for many eco-social problems. Therefore, research concerning control (especially biological control) and utilization (especially wastewater treatment or phytoremediation) of water hyacinth has boomed in the last few decades. Several phytoremediation practical examples clearly showed that the confined growth of water hyacinth in Lake Taihu and Lake Dianchi, China, achieved the good water purification results with no significant adverse effects on the ecosystem due to the timely harvest and proper post-treatment (Zheng et al. 2008, Wang et al. 2012).

In recent years, water hyacinth has attracted considerable attention (Rezania et al. 2015) world-wide in treating different kinds of wastewaters (Tchobanoglous et al. 1989). Many studies have employed water hyacinth for removal of metals from metal-rich synthetic solutions/industrial wastewaters and showed that water hyacinth could rapidly absorb and accumulate cadmium, lead, mercury, nickel, silver, cobalt, chromium, copper, manganese, zinc and other heavy metals in large amounts (Cordes et al. 2000, Malik 2007) and even at the relatively low pollution concentration (less than 10 $mg \cdot L^{-1}$ of combined heavy metals) (Soltan and Rashed 2003).

Through assimilation and transformation by plant tissues, these heavy metals could be potentially removed from waters (Kamal et al. 2004, Ali et al. 2013). The complex root system of water hyacinth has a relatively high bio-accumulation factor for copper, reaching up to 2.5×10^3 times in the root compared to the concentration of copper in water (Zaranyika and Ndapwadza 1995). Water hyacinth could also effectively absorb and accumulate zinc to a higher extent than submerged plants (Cai et al. 2004). In general, water hyacinth shows a strong capacity to remove and accumulate heavy metals with bioaccumulation factors ranging from tens to ten thousands of times (Mishra and Tripathi 2009, Smolyakov 2012, Lu et al. 2014). The most recent reports on the removal of heavy metals using water hyacinth are listed in Table 7.2.2-1.

Table 7.2.2-1. Phytoremediation of heavy metal-rich waters by water hyacinth.

Wastewater source	**Main pollutants**	**Treatment efficiency**	**Ref.**
Wastewater-collection pond of urban waste	Mn, Cd, Fe, Zn, Cu, Cr, Hg and Pb	It provided more than 80% efficiency of removal of various heavy metal species. The treatment efficiency could be increased by increasing water hyacinth coverage.	(Chunkao et al. 2012)
Synthetic solutions	Hg	Accumulation of mercury (in mg g^{-1} dry weight) was 1.99, 1.74 and 1.39 in root, leaf and petiole tissues, respectively.	(Malar et al. 2014)
Wastewater from simulated wetland	Cr, Cu	Approximately 65% removal of heavy metals	(Lissy and Madhu 2011)
Metal-contaminated coastal water	As, Cd, Cu, Cr, Fe, Mn, Ni, Pb, V	The optimum removal was at the water pH between 5.5 and 6.5.	(Agunbiade et al. 2009)
A reservoir in the Novosibirskoye	Zn, Cu, Pb, and Cd	After 8 days the amounts of metals relative to their initial concentrations for multi-metal pollution treatments were 8 and 24% for Cu and 18 and 57% for Zn at pH 8 and pH 6.	(Smolyakov 2012)
Artificial wastewater	Cr, Zn	Metal removal efficiency for Cr was highest (84%) at 1.0 mg L^{-1}, whereas for zinc it was highest (95%) at 10 mg L^{-1}, with 63–84% Cr and 88–94% zinc removed in 11 days.	(Mishra and Tripathi 2009)
Hydroponics	Cu	The removal percentages of copper from solution varied from 41 to 73%.	(Lu et al. 2014)
Industrial wastewater	Zn, Cu, Cd, Cr	Maximum removal efficiency of metal was recorded on the 10th day, with water hyacinth leaves accumulating considerably less than roots.	(Yapoga et al. 2013)

Moreover, many studies indicated that water hyacinth could not only strongly assimilate copper and zinc from waters, but also showed an excellent capacity to remove toxic heavy metals, such as cadmium (Soltan and Rashed 2003), even growing in the wastewater from electroplating. One kilogram dry weight of water hyacinth biomass could remove 135 mg copper, 437 mg zinc, 50 mg nickel and 89 mg cadmium, which would significantly improve water quality (Li et al. 1995a). When water hyacinth was applied to cadmium-contaminated wastewater, the content of cadmium in roots, stems and leaves increased significantly, showing a positive correlation with the cadmium concentration in water. An increase in the heavy metal concentration in water resulted in an increase in heavy metal concentration in plant tissues; the concentration of cadmium in roots was 21 times higher than that in stems and leaves (Chen 2011). Zhang et al. (1989) and Agunbiade et al. (2009) reported that 1 kg dry weight biomass of water hyacinth could assimilate 3.8 g lead and 3.2 g cadmium in 7–10 days.

Different tissues of water hyacinth had different bioaccumulation factors for heavy metals, with heavy metals mainly concentrated in plant roots, particularly on the surface of roots through adsorption. Generally speaking, total concentration of heavy metals in roots was several to tens of times higher than that in stems and leaves (Li et al. 1995b, Lu et al. 2014). In other cases, the report on uptake by water hyacinth of heavy metals such as cadmium, lead and strontium in the Nile, Egypt, showed that more than 50% of heavy metals were accumulated in roots, and only 20 and 30% in stems and leaves, respectively (Mazen and El Maghraby 1997). Chen et al. (2004) also found that different tissues of water hyacinth had differential capacity to accumulate heavy metals, with unwashed roots ≥ washed roots > leaves and stems, and the concentration of heavy metals in roots being two-five times higher than that in leaves and stems. In another case, water hyacinth showed strong removal of copper in the aquatic environment. When concentration of copper in water ranged from 1 to 5 mg L^{-1}, the rate of copper removal by water hyacinth reached up to 70% after 20 days. Water hyacinth roots play an important role in removal of copper from water, with each kilogram dry weight of root biomass assimilating 200–2000 mg copper; by comparison, uptake of copper by stems and leaves was only 30–90 mg per kg dry weight (Lu et al. 2014). Tan et al. (2009) indicated that water hyacinth roots were vital for removing lead, zinc and chromium, with concentration of heavy metals in roots being tens to hundreds of times higher than that in stems and leaves. El-Gendy et al. (2006) reported that floating macrophytes, mainly water hyacinth, effectively removed copper, cadmium, lead, nickel and chromium from drainage liquid from urban wastes; the maximum accumulation (dry weight basis) in roots reaching up to 9000 mg copper kg^{-1}, 8300 mg chromium kg^{-1} and 4900 mg cadmium kg^{-1}, with concentration in stems and leaves being much lower. The concentration of heavy metals (such as copper, zinc, cadmium and lead) remaining in water after water hyacinth transplantation decreased gradually over time, with heavy metal concentration in the solution rapidly decreasing in the first day (more than 50%), with strong accumulation of heavy metals in roots. When roots reach saturation, further uptake of heavy metals was affected, even though accumulation of heavy metals in roots, stems and leaves continued with duration of cultivation (Cornwell et al. 1977, Lu et al. 2007, Zhang et al. 2010, Smolyakov 2012, Lu et al. 2014).

The accumulation capacity of water hyacinth differs for various heavy metals. Zheng (2010) found that the complex root system of water hyacinth had a high capacity to accumulate cadmium and copper. However, in the wastewater with mixed heavy metals, the existence of copper strongly reduced the absorption of cadmium because the root system had a stronger affinity for copper than cadmium.

In spite of the extensive use of water hyacinth in treating heavy metal-contaminated water, this species is also regarded as a potential biological invasion weed species. When water hyacinth is employed to treat polluted water, there is a requirement to strictly control the growth of the species, keep the population in a certain manageable quantity range, leaving a necessary space between plant patches to ensure sufficient light and oxygen for the aquatic ecosystem, with the coverage by water hyacinth generally less than 50% of total water surface area according to actual experience (Huang and Xu 2008). Moreover, because of the non-biodegradability of heavy metals, water hyacinth should be regularly harvested and properly treated after harvest to remove the pollutants from the water and ensure good remediation results.

Main factors influencing the phytoremediation effects of water hyacinth

The heavy metal removal capacity of water hyacinth has close relation to its biomass volume, heavy metal concentration in polluted water and types, water temperature and pH, and other factors. Literature reports showed that the heavy metal content accumulated in plants tissue had a significant positive correlation with the initial heavy metal concentration in the aquatic environment (Rai and Tripathi 2009). Water pH could affect the absorption and accumulation of heavy metals by water hyacinth. In case of slightly acidic waters, the root tissue of water hyacinth had the greatest metal accumulation for various heavy metals (El-Gendy et al. 2006, Zheng 2010). Dai and Zhang (1988) reported that the absorption and accumulation of heavy metals in plant tissues was correlated to water temperature (season), pH, developmental stages of plant and other factors. Cai et al. (2009) found that when pH was about 5.5, water hyacinth showed an excellent capacity to accumulate divalent copper; in contrast, when pH ranged from 7.9 to 8.5, the accumulation factor decreased significantly. Tan et al. (2009) showed that tolerance and short-term removal capacity of water hyacinth to different heavy metals were in order of lead > chromium > zinc, with pH, biomass of water hyacinth and initial concentration of heavy metal having significant effects on removal rate; the best water pH values for water hyacinth removing lead, chromium and zinc were 4.3, 5.2 and 5.2 respectively. However, in a slightly acidic environment, the effects of pH on remediation could be neglected.

An increase in water hyacinth biomass is closely related to the treatment results. When plant biomass ranged between 100–400 g L^{-1}, it would remove 55 to 99% of lead, chromium and zinc (initial concentration 20 mg L^{-1}) after 4 days. Generally, the lower the concentration of heavy metals, the better the treatment effects. The absorption and accumulation capacity of young plants is higher than that of mature plants.

In cases of pollution with various heavy metals, there may be accumulative, synergetic and antagonistic effects among different heavy metal ions (Bliss 1939). Different metals could generate variable interactions, which might cause variation in plant capacity to accumulate heavy metals. The removal rate of some heavy metals present with other heavy metals may be slightly lower than that of single heavy metal, suggesting an antagonistic effect among heavy metals. Zheng (2010) studied the absorption and accumulation of copper (II), cadmium (II) and chromium (VI) by roots of water hyacinth, and found that calcium (Ca^{2+}), magnesium (Mg^{2+}), potassium (K^+) and hydrogen (H^+) could be released during the process of heavy metal absorption, indicating an existence of ion exchange.

7.2.3 *Mechanisms governing removal of heavy metals from the aquatic environments by water hyacinth*

During phytoremediation, aquatic macrophytes could tolerate and accumulate heavy metals through several pathways, such as precipitation, adsorption, filtration, uptake and accumulation (Wang and Ma 2000), with many other complex biochemical reactions, including complexation, chelation and compartmentalization being important (Wang et al. 1994, Zhang et al. 2004). Generally, a desirable remediation plant species should possess a rapidly growing and fibrous root system, which could absorb or immobilize heavy metals in polluted water. The root system of water hyacinth played an important role in the process of phytoremediation, and the rhizosphere of its root system could contain abundant microbial communities and root exudates for pollutants removal (Liu et al. 2003, Wild et al. 2005, Mishra and Tripathi 2009, Smolyakov 2012).

Absorption and accumulation of heavy metals in water hyacinth

Water hyacinth is a common phytoremediation plant with strong capacity to accumulate heavy metals in plant tissues (with bioaccumulation factors up to ten thousands of times) resulting in effective removal of most heavy metals (i.e., zinc, copper, lead, cadmium, chromium, etc.) from aquatic environments. Water hyacinth is widely used for phytoremediation of heavy metal-polluted waters by either physical or biological purification or the combination.

Physical purification involves adsorption and precipitation of some suspended solids. A decrease in water flow through the water hyacinth area is helpful to adsorption and precipitation of suspended solids. Heavy metals and other pollutants enclosed in suspended solids could co-precipitate (Huang and Xu 2008). The fibrous root system and high biomass yield of water hyacinth could aggregate and adsorb a large number of microorganisms, protozoa, detritus, colloids and suspended particles (Zhou et al. 2005, Nawirska 2005).

Biological purification includes further uptake and transfer of heavy metals. After being taken up by water hyacinth, a large proportion of heavy metals is retained in the roots, with a relatively small fraction being transferred to stems and leaves; hence, roots of water hyacinth, like in most plants, play an important role in protecting stem and leaf tissues from toxicity of heavy metals (Cai et al. 2004). Generally, the greater the accumulation of pollutants in stems and leaves, the stronger the toxicity. Among the five heavy metal (copper, lead, zinc, cadmium and iron), the highest accumulation of copper in stems and leaves is associated with the lowest tolerance of water hyacinth to the copper-contaminated water (Cai 2005). In general, uptake of heavy metals includes accumulation in the apoplast (cell wall) and transfer across the plasma membrane. The cell wall of water hyacinth roots is the first barrier impeding uptake of pollutants, with the pectin substance such as polygalacturonic acid and cellulose molecules containing carboxyl and aldehyde groups that provide a large number of exchange sites for heavy metal pollutants. It was found through X-ray Photoelectron Spectroscopy analysis (XPS analysis) that the chelation of heavy metals with amino- and/or oxygen-containing functional groups is one of the important mechanisms for removal of heavy metal from water (Zheng 2010).

Heavy metals with large ionic radii, such as lead, and weak coordination capacity do not easily enter the cytosol through the cell wall and the plasma membrane. Uptake of lead by water hyacinth mainly relies on the non-metabolic diffusive transport into cell walls and apoplastic precipitation, resulting in relatively low toxicity. After reaching the saturation on the root surface, the excessive metal ions begin to pass through the cell wall; and the portion of non precipitation in the cell wall will be across the plasma membrane and enter into the cytosol.

Tolerance of water hyacinth to heavy metal stress

Among the tolerance mechanisms, chelation in the cytosol attracts considerable attention: heavy metal stress could induce formation of biomacromolecules that form chelates with heavy metal ions, thus lowering the activity of free heavy metal ions in plant cells and relieve the toxicity (Wang et al. 1994, Wang and Ma 2000, Sun and Zhou 2005, Hu 2007, Flores-Cáceres et al. 2015). Many aquatic macrophytes tolerate and accumulate heavy metals through chelation (Hu et al. 2007). Two main kinds of heavy metal binding peptides, namely metallothionein and phytochelatins, were found in plant cells. MT is a low-molecular-weight polypeptide rich in cysteine. It could form non-toxic or low-toxic complex through the combination of thiol (–SH) on the cysteine residue with heavy metal ions, thus lowering the toxicity of heavy metals (Margoshes and Vallee 1957). The expression of mRNA for MT in mousear cress (*Arabidopsis thaliana*) had a positive correlation with plant tolerance to heavy metals (Murphy and Taiz 1995). However, the most common heavy metal-binding peptides separated from higher plants are phytochelatins, which are low-molecular-weight (<4000 kDa) polypeptides rich in cysteine produced through enzymatic synthesis (Yadav 2010). They chelate heavy metal ions through thiolate coordination to enhance metal accumulation and detoxification in plants. The structure of heavy metal chelating peptides in plants is similar to heavy metal M-binding peptides I in *Schizosaccharomyces*, which can be represented by general formula [γ-Glu-Cys]$_n$-Gly, in which n value is variable (2–11) for different plant species and even genotypes within a species (Ding and Wang 1993). Many heavy metals could induce the synthesis of phytochelatins, such as cadmium (II), copper (II), silver (I), mercury (II), lead (II) and zinc (II) ions, and the complexes formed with phytochelatins play an important role in detoxification and accumulation of many metals (Maitani et al. 1996).

Hydrilla verticillata, a submerged plant, could relieve toxicity of lead through the synthesis of phytochelatins (Gupta et al. 1995). Similarly, *Vallisneria spiralis*, a submerged plant, could also generate phytochelatins under mercury stress (Gupta et al. 1998). Grill et al. (1985) found that water hyacinth could generate PC induced by a variety of heavy metals, thus to relieve toxicity of heavy metals and store them in plant tissues. Moreover, hexazinc cyclohexane-1,2,3,4,5,6-hexayl hexakis (phosphate), carbonate ester lead salt and silicate ester lead salt could be synthesized in the plant cells, and the formation of these chelates may play an important role in the detoxification of relevant heavy metals. X-ray microscopic analysis was employed to investigate the structure of calcium oxalate crystals in the roots, leaves and leaf petiols of water hyacinth; it was found that calcium oxalate crystals acted as a sink of cadmium, lead and other heavy metals, which was considered to be the important heavy metal tolerance mechanism of water hyacinth (Mazen and El Maghraby 1997).

Besides chelation in the cytosol, higher plants could tolerate and accumulate heavy metals by several other mechanisms, for example, immobilization of heavy metals by the cell wall and plant exudates, sequestration of heavy metals in the vacuole (compartmentalization) (Salt et al. 1998, Küpper et al. 1999), and biotransformation of heavy metals (Murphy and Taiz 1995, Wollgiehn and Neumann 1999, Raskin and Ensley 1999, Lewis et al. 2001).

7.2.4 Resource utilization of water hyacinth after treating water polluted with heavy metals and metalloids

Water hyacinth can improve water quality better than other aquatic plants during remediation and treatment of wastewater, including nutritionally-rich domestic wastewaters and industrial effluents, metal-rich synthetic solutions. After taking up pollutants, water hyacinth must be harvested periodically to completely remove pollutants from waters with appropriate post-harvest processing and treatment to ensure success of remediation. At present, the post-treatments of water hyacinth biomass via landfilling and burning are not accepted as ecologically-friendly methods. Burning could pollute air, and landfill is not only costly, but also may cause secondary pollution. In contrast, resource utilization is a good option for disposal of water hyacinth biomass after remediation, with recycling of plant nutrients and re-use in agriculture following adequate consideration of heavy metals in the tissues of water hyacinth biomass, as well as of the advantages and risks of different utilization patterns. For effective utilization the following three points are recommended:

1) *Disposal of water hyacinth after harvesting from heavy metal-rich wastewaters*

   The treatment of heavy metal-rich wastewater by water hyacinth only transfers heavy metals from water to the plant tissues. The basic properties of heavy metals remain unchanged during the transfer processes. Therefore, the treatment and utilization of water hyacinth enriched with heavy metals become an important issue. Singh et al. (2015) employed resource utilization for the disposal of water hyacinth harvested after phytoremediation of the wastewater in the Amingoan industrial district of India, and investigated the effects of composting fermentation on the form and availability of heavy metals. Their results showed that bioavailability of heavy metals (such as zinc, copper, manganese, iron, nickel, lead, cadmium and chromium) after composting of water hyacinth biomass were generally lowered. After composting for 30 days, water-soluble nickel, lead and cadmium and diethylenetriamine pentaacetic acid extractable nickel, lead and cadmium were below detection limit. The main reason was that the organic compounds formed during composting of water hyacinth biomass strongly complex nickel, lead, cadmium, making them biologically inactive and non-available (Lazzari et al. 2000). However, bioavailability of other heavy metals was relatively high, posing a pollution risk to the environment. Given that accumulation of heavy metals in the water hyacinth biomass increased with an increase in initial heavy metal concentrations in wastewater (Lu et al. 2014), it is expected that applying water hyacinth to remediate heavy metal-polluted wastewater (including mining, metal smelting, leather industry and electroplating industry wastewater) would result in plant biomass accumulating large amounts of heavy metals, particularly in roots. For this reason, in disposing of water hyacinth containing large amounts of heavy metals, the plant might be divided into root, stem and leaves. As the main accumulation tissue, roots should be burned after dehydration and the ash might need to be stored or sent to the processing plant for recycling; importantly, ash content is only 0.13% of fresh weight of water hyacinth biomass. The content of heavy metals is far lower, and the content of nutrients (nitrogen, phosphorus, potassium, etc.) far higher, in stems and leaves than roots. Hence, stems and leaves of water hyacinth could be composted to make cultivation medium for ornamental plants. Moreover, given that water hyacinth is rich in cellulose, it could also be used as the raw material in the paper industry to enhance crumpling and moisture resistance in paper for packaging, printing and posters.

2) *Disposal of water hyacinth harvested from nutrient-rich waters*

   In the phytoremediation of nutrient-rich domestic wastewaters, eutrophic lakes and urban rivers, water hyacinth can not only assimilate nitrogen, phosphorus and other nutrients, but it also takes up the coexisting harmful heavy metals and other pollutants. A field scale 1-year-long study was conducted to investigate the pollutant removal from nutrient-rich wastewaters using different kinds of macrophytes; it was found that 78–81% removal of NO_3^- and NH_4^+ and 54% of P was achieved in 3.6 days by using water hyacinth-based wastewater treatment that performed better than submerged species *Chara* sp. (Reddy and Tucker 1983). Water hyacinth effectively removed metals (20–100%) and nitrates and phosphates (>90%) and decreased BOD (97%) in the sewage treatment (Sinha and Sinha 2000).

It is necessary to investigate a potential risk of heavy metals regarding food safety, that is, whether resource utilization of water hyacinth harvested from nutrient-rich domestic wastewaters would result in exceeding the National Environmental Quality Standard for heavy metals. The study of Jiang (2011) showed that even though living in the natural water free of heavy metal pollution, water hyacinth could still accumulate certain amounts of heavy metals from the environment, with the tissue content of heavy metals varying according to water source, background concentration of heavy metals, disposal techniques and the harvest time. Because different countries and regions have different environmental quality standards for heavy metals, it is necessary to consider them individually when considering resource utilization for water hyacinth. In China, the content of heavy metals in water hyacinth growing in the natural waters is lower than that specified in National Environmental Quality Standard for Soils of the People's Republic of China (MEP-PRC 1995). The content of heavy metals in organic fertilizer made from the biomass of such water hyacinth is lower than that specified in National Industrial Standard for Organic Fertilizers of the People's Republic of China (MA-PRC 2012). Another study in China also produced organic fertilizer from water hyacinth harvested after phytoremediation of polluted water and composted under different pile height, shape and turning time, with the content of heavy metals in the organic fertilizer products (Luo et al. 2014) satisfying the standard regarding lead, mercury, arsenic, cadmium and chromium for Organic Fertilizers of the People's Republic of China (MA-PRC 2012).

3) *Strategies for disposal of harvested water hyacinth biomass*

The resource utilization of water hyacinth needs to be considered in combination with the type of water pollution remediation, environmental context for the heavy metals, as well as the local conditions and environmental management demands. Firstly, in the disposal of water hyacinth harvested from the heavy metal-contaminated water, it is essential to investigate accumulation of heavy metals in plant tissues. The biomass with heavy metal contents exceeding the heavy metals standard needs to be disposed of by incineration after dehydration. Water hyacinth juice from dehydration might be analyzed, and a re-treatment should be conducted if necessary. Secondly, for plants with relatively low heavy metal accumulation in stems and leaves but high accumulation in roots, the roots should be separated from stems and leaves. Roots should be burned after dehydration and the ash should be properly stored or sent to the processing plant for recycling, whereas stems and leaves could be composted to make cultivation medium for flowers. Thirdly, in general, water hyacinth harvested after the phytoremediation of polluted water does not create risks of heavy metals in excess of the standards. So the resource utilization of water hyacinth could be implemented via production of organic fertilizers and various cultivation media according to the environmental quality standard for heavy metals in different countries and regions. Fourthly, the water hyacinth biomass growing in natural waters with heavy metal contents below the environmental quality standards could be preferentially used as animal fodder so as to increase economic value of water hyacinth utilization.

7.3 Mechanisms underlying water hyacinth removal of organic pollutants

7.3.1 The sources and extent of organic pollution in the aquatic environment

With the rapid development of economy, especially of industry and agriculture, the organic pollutants enter aquatic environments via different pathways, causing increasingly serious organic pollution (Jiang et al. 2000, Jurado et al. 2012, Lei et al. 2014). Organic pollutants in the aquatic environment could be divided into two types: natural and synthetic organic compounds. At present, there are about 7 million known classes of organic compounds, among which synthetic compounds account for over 100 thousand classes, increasing by 2,000 classes per year. The common classes including organic pesticides, polychlorinated biphenyls, phthalate esters, alkanes and polycyclic aromatic hydrocarbons, as well as the emerging classes causing widespread concern, i.e., pharmaceuticals and personal care products. The concentration of organic pollutants is very low compared with other pollutants, but standard sewage treatment technology could not effectively remove organics. Therefore, many organic pollutants are discharged directly into surface waters such as lakes and rivers, potentially ending up in drinking water. Although concentration of organic pollutants is relatively low (generally ranging from $\mu g\ L^{-1}$ to $ng\ L^{-1}$), most of them are relatively stable, needing several years to tens of years to be degraded into harmless substances in aquatic environments (Guo et al. 2009, Loos et al. 2010, Wang et al. 2011).

1. Organic pesticides

   Synthetic organic pesticides are widely distributed in surface waters. Pesticides play an important role in promoting agricultural production. However, due to limitations of the application methods, approximately 20–70% of the applied pesticides finally enter the environment. One of the commonly found pesticide classes in the surface waters is mainly organo-chloride pesticides. They enter waters via spraying, leaching from farmland soils, surface runoffs and wastewater discharged by pesticide factories. As relatively stable substances, organo-chloride pesticides are difficult to be degraded, so they could remain in aquatic environments for a long period; in addition, they have a low K_{ow} coefficient[1] with a large proportion of them transferred into the sediments and aquatic organisms. At present, organo-chloride pesticides, such as benzene hexachloride and dichloro-diphenyl-trichloroethane, have been banned all over the world, but their residues could still be detected in some waters and sediments due to their persistence and biological accumulation capacity (Ming Zhang et al. 2010). Organo-chloride compounds were frequently detected in ground water in northern China. As reported in 2000, there were 17 classes of organic polychlorinated compounds found in the water and sediment samples in middle and lower reaches of Liaohe River in China, including 13 classes of organo-chloride pesticides and four classes of polychlorinated biphenyls, among which benzene hexachloride, dichloro-diphenyl-trichloroethane and polychlorinated biphenyls were the main detected substances (Zhang et al. 2000, Zhang and Dong 2002).

2. Polychlorinated biphenyls

   Polychlorinated biphenyls, also known as chlorinated biphenyls, are synthetic organic compounds with hydrogen atoms on the biphenyl benzene ring substituted by chlorine. In the industry, PCBs are used as heat carriers, insulating oils and lubricating oils. Because polychlorinated biphenyls have low utilization efficiency, the main pollution source is the industrial waste (waste gas, wastewater and solid waste residue) (Schecter et al. 1997, Sakurai et al. 2000). Hence, it is clear that the widespread use of polychlorinated biphenyls in the industry is causing global environment problems. It was reported that the content of polychlorinated organic pollutants in waters, suspended solids and sediments in Nanjing section of China's Yangtze River was lower than that in the main rivers of Europe in 2000 (Zhou et al. 2010). The content of polychlorinated biphenyls in the surface layer of sediments (485 ng g^{-1} dry weight) at Guangzhou

[1] K_{ow} coefficient is defined as the ratio of the solubility of a compound in octanol (a non-polar solvent) to its solubility in water (a polar solvent). The higher the K_{OW}, the more non-polar the compound. Log K_{OW} is generally used as a relative indicator of the tendency of an organic compound to adsorb to soil particles. Log K_{OW} values are generally inversely related to aqueous solubility and directly proportional to molecular weight (EPA 2004).

section of Pearl River in China (Wang et al. 2011) was 16 times higher than the maximum range limit (2 ~ 30 ng g^{-1} dry weight) of the reference background of surface sediments (Fowler 1990, Kang et al. 2000).

3. Phenolic compounds

Phenolic compounds are the main industrial raw materials to produce epoxy resins, makrolon, polysulfone resins and other polymers. Some phenolic compounds, such as phenol, nitrophenol, dichlorophenol, pentachlorophenol and octyl phenol, have the properties of estrogens, are difficult to be degraded in the environment and are easily accumulated in organisms. Even when present at relatively low concentration, phenolic compounds could cause endocrine dyscrasia in animals and humans. Moreover, due to a potential of phenolic compounds to induce carcinogenesis, chromosomal aberrations and gene mutations (Zhou 2011), they have been placed on the blacklist of priority control organic pollutants in aquatic environments by the Environmental Protection Agencies of China and the US. The industrial wastes discharged by petrochemical enterprises, paper-making industry, and pesticide and electroplating manufacturers are the main discharge source of phenolic compounds in water environments.

4. Alkanes

The alkane pollutants in the aquatic environment mainly come from industrial wastewater and domestic sewage. Petroleum pollutants (including a variety of hydrocarbon compounds) in the industrial wastewater mainly originate from oil exploration, processing and transportation, as well as industries that utilize refined petroleum compounds as raw materials. Due to low density compared with water, petroleum hydrocarbon compounds could float on water surface, which affects oxygen exchange between air and water surface. Oils dispersed in waters, absorbed on the suspended particles and existing in emulsified state in water could be oxidized and decomposed by microorganisms, resulting in consumption of dissolved oxygen and causing deterioration in water quality.

5. Polycyclic Aromatic Hydrocarbons (PAHs)

Polycyclic aromatic hydrocarbons contain more than two phenyl rings in the molecule; they include over 150 kinds of compounds, such as naphthalene, anthracene, phenanthrene, pyrene, etc. Polycyclic aromatic hydrocarbons are mainly derived from the process of incomplete combustion of fuels (such as petroleum and coal) and hydrocarbon compounds (such as wood, natural gas, gasoline and crop straw). A variety of polycyclic aromatic hydrocarbon compounds have been detected in natural waters. Seventeen kinds of polycyclic aromatic hydrocarbon compounds were detected in the water and sediments of

Yangtze River and Liaohe River in China, with 11 of those being priority control organic pollutants in the aquatic environment identified by the US Environmental Protection Agency, and the remaining six being polycyclic aromatic hydrocarbons on the blacklist of priority control organic pollutants in the aquatic environments in China (Xu et al. 2000). The content of polycyclic aromatic hydrocarbons in the urban sewage in Karak Governorate, Jordan, was 56–220 ng L^{-1}, exceeding the PAHs standard limits by World Health Organization for water (0.05 μg L^{-1}) (Jiries et al. 2000). In conclusion, the presence of organic pollutants in the global aquatic environments have attracted much attention; and at least 15 polycyclic aromatic hydrocarbons can induce cancers to experimental animals (Harmon 2015).

6. Pharmaceuticals and Personal-Care Products (PPCPs)

   Pharmaceuticals and personal-care products include various prescription drugs, over-the-counter drugs (such as antibiotics, anti-inflammatory drugs, hormones and cosmetics, etc. (Daughton and Ternes 1999). In the late 1990s, with the development of detection and analysis technology and enhancement of people's environmental protection consciousness, PPCPs were detected in a range of aquatic environments, including sewage, surface water and underground water, attracting attention because of their hazards to the environment and human health (Nassef et al. 2010, Wang et al. 2014). The pollution source of PPCPs is mainly domestic sewage, livestock and poultry production wastewater and agricultural non-point source pollution. Moreover, wastewater derived from the medical and health institutions and synthetic industry contain a high concentration of pharmaceuticals and personal-care products. Fifteen kinds of PPCPs compounds (such as ibuprofen, naproxen, diclofenac, clofibric acid, triclosan, galaxolide abbalide, tonalid abbalide, etc.) were found in the influent of urban sewage treatment plant near Ontario, Canada, with individual species reaching concentration of up to 17 μg L^{-1} (Lishman et al. 2006).

The investigation of three rivers in Northern China (including Yellow River, Haihe River and Liaohe River) detected naproxen in 36% of samples, with the maximum concentration of 41 ng L^{-1} (Wang et al. 2010); whereas 23% of samples in Pearl River contained naproxen, with the maximum concentration of 328 ng L^{-1} (Peng et al. 2008). Ibuprofen has a relatively large production in China, and its concentration as a pollutant in the surface water was relatively high in Liaohe River (up to 246 ng L^{-1}) and Pearl River (up to 1417 ng L^{-1}) (Peng et al. 2008, Wang et al. 2010). From 1999 to 2000, there were 21 kinds of antibiotics detected in 139 rivers of the 30 states in the US, including sulfonamide, tetracycline, lincomycin and tylosin, etc. with a detectable rate reaching up to 60% (Kolpin et al. 2002). Chen et al. (2010) investigated the pollution situation of veterinary antibiotics in the wastewater discharged by a large-scale livestock farm in the Tiaoxi River basin in Zhejiang Province, China, and the results showed that four kinds of tetracycline antibiotic species (tetracycline, oxytetracycline, aureomycin and doxycycline) were the most extensive organic pollutants, with a maximum pollution concentration up to 14 μg L^{-1} individually.

There are many other organic pollutants such as volatile chlorinated hydrocarbons, phthalate esters, Benzene Toluene Ethylbenzene and Xylene, chlorobenzenes, nitrobenzene, anilines and nitrosamines that are widely present in aquatic environments. Most of them are stable in natural conditions, usually taking several years to tens of years to be degraded to harmless substances in the environment. Because organic pollutants are difficult to decompose biologically, they (i) could be accumulated in living organisms to toxicity levels, inducing mutations, chromosomal aberrations and carcinogenesis, and (ii) could be transferred into the food chain. Therefore, organic pollutants represent a significant potential hazard to the environmental and human health; how to efficiently remediate the water polluted by various organic compounds has always been the focus of research projects around the world.

7.3.2 Effects of water hyacinth on the removal of organic pollutants from the aquatic environments

Traditionally, treating sewage is done by collecting sewage for centralized treatment, which is expected to have ideal effects of reducing and controlling chemical oxygen demand, biological oxygen demand, total nitrogen and total phosphorus, etc. However, given that the traditional sewage treatment systems have been designed and built especially for the treatment of conventional pollutants, there is no technique specifically designed for the removal of various organic pollutants. Moreover, although there are several physicochemical methods, such as flocculation and precipitation (Choi et al. 2008), adsorption (Rivera-Utrilla et al. 2009), membrane technology (Koyuncu et al. 2008) and chemical oxidation method (Li et al. 2011), these methods generally have some disadvantages such as difficult operation or high cost, making them suitable for treating heavy pollution only on a small-scale.

Phytoremediation technology could directly use plants to remove organic pollutants from water through transfer, degradation and fixation mechanisms. Instead of transferring the polluted water to other places for centralized and specialized treatment, phytoremediation is an *in situ* remediation technique conducted directly in the polluted water by specifically planting aquatic plants, with several advantages, including low cost, efficient, easy to accomplish without secondary pollution. Therefore, since 1980s, phytoremediation technology has become an important research focus in the field of organic pollution remediation.

Aquatic macrophytes frequently used in the remediation of organic polluted waters include emergent plants, submerged plants, floating leaved plants and floating plants. Research showed that aquatic plants, such as softstem bulrush (*Scirpus validus*), common bulrush (*Typha latifolia*) and acorus (*Acorus tatarinowii*), could accelerate dissipation of dimethoate (a pesticide) in aqueous solution, with the removal efficiency as follows: *Scirpus validus* > *Typha latifolia* > *Acorus tatarinowii*; the removal efficiency was closely related to the plant growth conditions (Fu et al. 2006). Huesemann et al. (2009) conducted an *in situ* remediation research for the removal of PAHs and PCBs by using common eelgrass (*Zostera marina*). The removal rates of PAHs and PCBs were 73 and 60% respectively, and the pollutant concentrating factor in roots exceeded that in the aerial part four-fold; the control (non-plant) treatment removed only 25% of pollutants. Chen et al. (2012) studied the phytoremediation efficiency of antibiotic-polluted water and found that the antibiotic removal rates by water lettuce (*Pistia stratiotes*) and water hyacinth (*Eichhornia crassipes*) were in excess of 80 and 90% respectively, with the remediation efficiency of water hyacinth under different concentrations of antibiotics was higher than that of water lettuce. Xia (2002) investigated the remediation effects of water hyacinth (*Eichhornia crassipes*), saka siri (*Canna indica*), weeping willow (*Salix babylonica*) and tea (*Camellia sinensis*) on wastewater polluted by the esticide malathion and found that the remediation efficiencies were: *Eichhornia crassipes* > *Canna indica* > *Salix babylonica* > *Camellia sinensis*. According to the studies mentioned above, phytoremediation has been widely applied in treating wastewaters polluted with various organic compounds, achieving relatively high remediation efficiency. Due to its superior biological characteristics compared with other plants, water hyacinth has an excellent capacity to purify water.

Water hyacinth could accelerate removal of phenol (concentrations ranging from 0.6 to 10 mg L^{-1}), with the plant purification efficiency being two–three times higher than that of natural purification processes (Wang et al. 1986). Between 17 and 37°C, water hyacinth could significantly accelerate dissipation of phenol in the water with an increase in temperature.

The performance of water hyacinth in remediation of water contaminated by pesticides (1 mg individual forms per L) was excellent, with 10–11 g water hyacinth (dry weight biomass) in 250 mL of polluted water increasing removal by 260, 80 and 357% of ethion, kelthane and cyhalothrin, respectively, compared with natural processes; water hyacinth roots accumulated more than 70% of the organic pesticide (Xia et al. 2002).

Water hyacinth could remove petroleum hydrocarbons and decrease total organic carbon in the petrochemical wastewater (de Casabianca and Laugier 1995). Water hyacinth could degrade free cyanide and remediate cyanide-containing wastewater in combination with other treatment methods (Granato 1993). Within a wide range of concentrations (0.2–5 mg L^{-1}) of antibiotic tetracycline, water hyacinth cultivation for 20 days decreased these concentrations by 96% (Lu et al. 2014). Water hyacinth could also effectively remove other pharmaceuticals and personal-care products, such as triclosan, acetaminophen and linear alkyl benzene sulfonate, with the removal rate reaching up to 98% after 5 days of cultivation (Yamamoto et al. 2014).

The removal efficiency of water hyacinth for naphthalene-contaminated water at different concentrations (4.3–13.2 mg L^{-1}) ranged from 84 to 92%, with no naphthalene detected in roots, stems and leaves of water hyacinth, suggesting naphthalene is not accumulated by water hyacinth (Yuan et al. 2004). It was hypothesized that the rhizospheric microorganism played a very important role in the treatment process. Water hyacinth removed 100% of naphthalene from sewage under non-sterilized conditions after 9 days, but only 45% when rhizosphere bacteria were removed (Nesterenko-Malkovskaya et al. 2012). The absorption of naphthalene by roots of water hyacinth could be divided into two stages: the first stage was rapid adsorption of naphthalene on the root surface (in about 2.5 hours); the second stage was that the naphthalene adsorbed on the root surface was transferred and accumulated in the root intracellular spaces at a rather slow speed in the period 2.5–225 hours.

Wang and Wen (2008) investigate the effects of six species of common aquatic macrophytes on dissipation of aniline-containing wastewater; the results showed that the removal efficiency of anilines ranged from 51 to 97% in the order of *Eichhornia crassipes* > *Pistia stratiotes* > *Canna indica* > *Alternanthera philoxeroides* > *Typha latifolia*, indicating the superiority of water hyacinth compared to other aquatic plants in the remediation of anilines. Removal of methyl naphthylamine (3 mg L^{-1}) by water hyacinth reached up to 100% after 1 day of cultivation, and no aniline (starting concentration 2 mg L^{-1}) was detected in the water after 2 days of water hyacinth cultivation, with a majority of pollutants accumulated in roots, and accumulation in stems and leaves being only about 10% of that in roots (Tao et al. 1998).

7.3.3 Mechanisms of water hyacinth underpinning remediation of organic pollutants

Water hyacinth tissues contain many kinds of enzymes that perform various biochemical processes to transfer and degrade organic compounds in joint action with rhizosphere bacteria. The developed root system of water hyacinth is crucial in removing organic pollutants from water, mainly through two mechanisms. (1) roots could absorb PCBs, PAHs as well as other organic pollutants from the aquatic environment and/or transfer pollutants to other parts of the plant, where they can be degraded into harmless molecules; and (2) the root system provides habitat for various microorganisms by releasing exudates and enzymes to stimulate microbial activity and strengthen mineralization and other forms of biotransformation by rhizosphere microorganisms (Voudrias and Assaf 1996, Xia et al. 2003). Given that the root-cell plasma membrane is semi-permeable, only some ions and molecules could be directly assimilated and further degraded. Organic pollutants, including naphthalene and phenol, that could not pass through the root-cell plasma membrane could only be adsorbed in the root cell walls and/or on the root surface (Yuan et al. 2004), and then be decomposed into small molecules with the joint action with microorganisms.

Water hyacinth had a significant effect on removal of ethion from wastewater due to direct absorption (69%) and degradation by microorganisms (12%), with pollutants accumulating mostly in roots (Xia and Ma 2006). Water hyacinth is also effective in remediation of water polluted with PCBs. After 15 days, water hyacinth decreased concentration of PCBs in water from 15 μg L^{-1} to 0.42 μg L^{-1}, whereas in cases of low PCB concentration (≤ 10 μg L^{-1}) reduction was below the detection limit, with root system acting as the main accumulation tissue for PCBs (Auma 2014).

Unlike heavy metals, the assimilation of organic pollutants by the roots of remediation plants only accounted for a small portion of the total removal (Liu et al. 2003, Zhang et al. 2011, Lu et al. 2014). Most organic pollutants, especially non-ionic ones, might enter the root system of plants via diffusion, and only some (for instance, systemic pesticides) entered plant roots via a transpiration pull of leaves (Xia 2002, Paraíba 2007, El-Queny and Abdel-Megeed 2009). There are several steps in the uptake. Firstly, the organic pollutants in the water get absorbed in the "apparent free space" in the external root tissues, which accounted for 10 ~ 20% of the volume of plant root system (Nye and Tinker 1977); then, organic pollutants could be transferred through the cell walls and intercellular spaces (apoplastically) all the way to endodermis which blocks apoplastic pathway because it contains highly-suberinized Casparian band impermeable to water; to bypass Casparian band, ions and molecules have to be taken up into root cells, and then continue symplastically (from cell interior to another cell interior via plasmodesmata) (Wild et al. 2005). Alternatively, uptake of ions and molecules into the symplast (cell interior) can occur in peripheral root tissues before endodermis (Fig. 7.3.3-1). Once in the symplast, ions and molecules can be taken up by xylem parenchyma

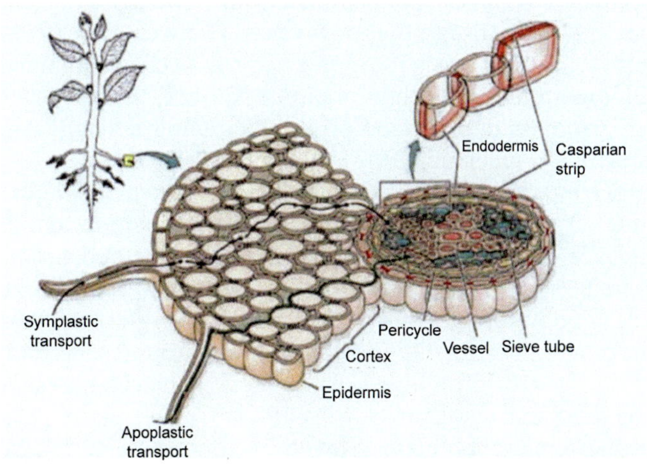

Fig. 7.3.3-1. Translocation pathways of organic contaminants among plant cells (redrawn from Wild et al. 2005).

cells and transferred into xylem vessels for transport to other plant parts (Pilon-Smits 2005).

Transport of organic pollutants in plant tissues is closely related to the physicochemical properties of compounds. There are many reports concerned with the assimilation and metabolism of systemic pesticides. In agriculture, seed treatment agents, such as orthene, demeton, carbofuran, aldicarb and imidacloprid (Westwood et al. 1998), were developed to prevent and remediate the aerial parts from diseases and pests through uptake of such pesticides by roots and transfer to the stem and leaves above the ground after germination. Once in the above-ground parts, these pesticides may be effective for some time before being metabolized.

Water hyacinth has a strong capacity to assimilate and accumulate systemic pesticides. When water hyacinth is applied to treat the water polluted by organophosphorus insecticide dimethoate, the insecticide was mainly distributed in leaves with the concentration ratio in leaves to that in roots about 5 to 1; the transport capacity by roots was positively correlated with plant transpiration (Xia 2002, El-Queny and Abdel-Megeed 2009). In addition, the roots of water hyacinth could also take in the non-systemic pesticides, including kelthane, ethion, DDT and pyrethroid pesticides, absorbed by roots and transfer them to stems and leaves to some extent (a greater proportion of these pesticides was present in water hyacinth roots than above-ground parts). The study by Xia (2002) showed that the ratio of the concentration of non-systemic pesticides, including kelthane, ethion, DDT and pyrethroid pesticides, in water hyacinth biomass to that in the culture solution (namely biological concentration factor) reached 74–3838 times in root and about 31–918 times in stems and leaves, indicating that roots were the main accumulation organ of non-systemic pesticides.

Plants could not only absorb, transfer, metabolize and assimilate most of organic pollutants through the root system (Schwab et al. 1998, Dai 2006), but also stimulate the propagation and activity of microorganisms by root secretion of low-molecular-weight organic compounds, such as amino acids, to indirectly promote degradation of organic pollutants by the rhizosphere microorganisms (Aken et al. 2010).

Water hyacinth has a fibrous root system, 40–200 cm in length, that floats in water and might secrete nutrients and other compounds, creating an ideal habitat for various rhizosphere microorganisms and promoting their bioactivity in degrading organic pollutants (Yuan et al. 2004, Kurzbaum et al. 2010). In addition, water hyacinth could transport oxygen from the leaves and stems to the root system, forming an aerobic zone around roots and providing good conditions for the growth of aerobic bacteria to keep the bioactivity and reaction efficiency at a stable level (Ma et al. 2014).

Rhizosphere microorganisms played an important role in degradation of organic pollutants (Nesterenko-Malkovskaya et al. 2012). Temperature was positively correlated with the removal effectiveness of water hyacinth and its rhizosphere microorganisms. The total abundance of bacteria in the rhizosphere was higher than that of fungi at higher temperature, indicating that temperature played a leading role during sewage purification (Přikryl and Vančura 1980, Liu et al. 2007).

Plant roots could exude a variety of enzymes that can facilitate many biochemical reactions. The enzymes released into the rhizosphere environment by plants could directly degrade a range of compounds and play a very important role in degradation of organic pollutants (Sandermann 1992, Macek et al. 2000). For example, the capacity of water hyacinth to tolerate and remove pollutants may be related to superoxide dismutase released by roots (Li et al. 1995b). The exudates of root system could decompose benzene rings of polycyclic aromatic hydrocarbons depending on the concentration of oxygenase in the exudates (Yuan et al. 2004).

Moreover, the processes of degradation of organic pollutants in plant tissues (particularly oxidative decomposition) are also dependent on enzymes. For instance, cytochrome P450 is a kind of multifunctional enzyme, which could catalyze oxidation and peroxidation reactions and greatly increase the detoxification capacity of plants (Waterston et al. 2005). The oxidation and degradation of pollutants in plant tissues is generally brought about through a multi-step reaction. Cytochrome P450 is very important for the oxidation and degradation at every step (Boutet et al. 2004). Almost all xenobiotic pollutants, such as hexachloro-cyclohexane soprocide and DDT, could induce cytochrome P450. In general, xenobiotic pollutants could induce and increase cytochrome P450 content in the water hyacinth tissues, depending on physicochemical properties of pollutants, concentration and physicochemical properties of intermediate metabolites (Zhou and Song 2001, Boutet et al. 2004).

Although water hyacinth is quite effective in purification and remediation of organic pollution in waters, most plants had a relatively weak capacity to assimilate and accumulate organic pollutants compared with heavy metals (Yuan et al. 2004, Lu et al. 2014). After uptake, the organic pollutants were further degraded in plants by various enzymes (Xia et al. 2001, Xian et al. 2010, Nesterenko-Malkovskaya et al. 2012). Moreover, water hyacinth harvested from waters polluted with organic compounds also needs to be processed for utilization. Literature indicated that, after suitable fermentation time (4–7 weeks), the water hyacinth biomass could effectively dissipate organic pollutants, such as PAHs, PCBs, hormones and antibiotics, with a degradation rate reaching up to 80%; an addition of calcium sulfate, calcium superphosphate, grass carbon, bamboo charcoal and/or microbial inocula could enhance significantly the degradation rate of organic pollutants (Zhang et al. 2006, Arikan et al. 2007, Dong et al. 2013, Ho et al. 2013). Consequently, for the water hyacinth harvested from water polluted with organic compounds, resource utilization is relatively safe. However, given that different countries and regions have different standards of environmental quality, different fermentation technologies are needed to satisfy different environmental quality standards.

References cited

Agunbiade, F. O., B. I. Olu-Owolabi and K. O. Adebowale. 2009. Phytoremediation potential of *Eichhornia crassipes* in metal-contaminated coastal water. *Bioresource Technology* 100(19): 4521–4526.

Aken, B. Van, P. A. Correa and J. L. Schnoor. 2010. Phytoremediation of polychlorinated biphenyls: new trends and promises. *Environmental Science and Technology* 44(8): 2767–2776.

Ali, H., E. Khan and M. A. Sajad. 2013. Phytoremediation of heavy metals—concepts and applications. *Chemosphere* 91(7): 869–881.

Andersen, H., J. Kjolholt, C. Poll, S. Dahl, F. Stuerlauridsen, F. Pedersen et al. 1998. Environmental risk assessment of surface water and sediments in Copenhagen harbour. *Water Science and Technology* 37(6-7): 263–272.

Arikan, O. A., L. J. Sikora, W. Mulbry, S. U. Khan and G. D. Foster. 2007. Composting rapidly reduces levels of extractable oxytetracycline in manure from therapeutically treated beef calves. *Bioresource Technology* 98(1): 169–176.

Auma, E. O. 2014. Phytoremediation of polychlorobiphenyls (PCB's) in landfill e-waste leachate with water hyacinth (*E. crassipes*), MSc Thesis, Department of Civil Engineering. University of Nairobi.

Bliss, C. I. 1939. The toxicity of poisons applied jointly. *Annals of Applied Biology* 26(3): 585–615.

Boutet, I., A. Tanguy and D. Moraga. 2004. Molecular identification and expression of two non-P450 enzymes, monoamine oxidase a and flavin-containing monooxygenase 2, involved in phase I of xenobiotic biotransformation in the Pacific oyster, *Crassostrea gigas*. *Biochimica et Biophysica Acta - Gene Structure and Expression* 1679(1): 29–36.

Cai, C. 2005. Study on removal rate of Zn^{2+}, Cd^{2+}, and Fe^{3+} by common waterhyacinthes. *Yunnan Environmental Science* 24(1): 10–12 (In Chinese with English Abstract).

Cai, C., H. Wang and Z. Zhang. 2004. Removal of Cu, Pb, Cd, Zn and Fe by water hyacinth. *Journal of Leshan Teachers College* 19(6): 69–72 (In Chinese).

Cai, Q., Z. Lei, H. Hu and Z. Chen. 2009. Study on purification for copper polluted water by *Eichhornia crassipes*. *Journal of Yangtze University (Natural Science Edition)* 6(2): 68–71, 106 (In Chinese with English Abstract).

de Casabianca, M. and T. Laugier. 1995. *Eichhornia crassipes* production on petroliferous waste waters: effects of salinity. *Bioresource Technology* 54: 39–43.

Chen, X. 2011. The physiological and biological characteristics of the water hyacinth (*Eichhornia crassipes*) in adapting to different growth conditions, MSc. Thesis, School of Life Science. Fujian Agriculture and Forestry University (In Chinese with English Abstract).

Chen, X., F. Li and Y. Hao. 2012. The preliminary exploration of remediation the antibiotic polluted water by two hydrophytes. *Subtropical Plant Science* 41(4): 1–7 (In Chinese with English Abstract).

Chen, Y., Y. Jin, X. Wang and X. Lan. 2004. Study on accumulation of different organs of *Eichhornia crassipes*. *Environmental Protection* 30: 31–37 (In Chinese with English Abstract).

Chen, Y., H. Zhang and Y. Luo. 2010. A preliminary study on the occurrence and dissipation of antibiotics in swine wastewater. *Acta Scientiae Circumstantiae* 30(11): 2205–2212 (In Chinese with English Abstract).

Choi, K. J., S. G. Kim and S. H. Kim. 2008. Removal of antibiotics by coagulation and granular activated carbon filtration. *Journal of Hazardous Materials* 151: 38–43.

Chunkao, K., C. Nimpee and K. Duangmal. 2012. The King's initiatives using water hyacinth to remove heavy metals and plant nutrients from wastewater through Bueng Makkasan in Bangkok, Thailand. *Ecological Engineering* 39: 40–52.

Cordes, K. B., A. Mehra, M. E. Farago and D. K. Banerjee. 2000. Uptake of Cd, Cu, Ni and Zn by the water hyacinth, *Eichhornia crassipes* (Mart.) Solms from pulverised fuel ash (PFA) leachates and slurries. *Environmental Geochemistry and Health* 22(4): 297–316.

Cornwell, D. A., J. Zoltek, C. D. Patrinely, T. Furman and J. I. Kim. 1977. Nutrient removal by water hyacinths. *Water Pollution Control Federation* 49(1): 57–65.

Dai, Q. and Y. Zhang. 1988. Absorbtion of heavy metals by the water hyacinth and the second accumulation in fishes after fed on the water hyacinth. *Journal of Fisheries of China* 12(2): 135–144 (In Chinese with English Abstract).

Dai, S. 2006. *Environmental Chemistry*. 2nd Edition. Beijing, China: Higher Education Publications.

Daughton, C. G. and T. A. Ternes. 1999. Pharmaceuticals and personal care products in the environment: agents of subtle change? *Environmental Health Perspectives* 107(SUPPL. 6): 907–938.

Ding, X. and W. Wang. 1993. Extraction, purification and identification of heavy metal chelating peptide from water hyacinth. *Science in China (Series B)* 23(4): 365–370 (In Chinese).

Dong, F., X. Li, A. Lin, C. Wang and L. Wei. 2013. Effect of additives on degradation of polycyclic aromatic hydrocarbons (PAHs) in composting. *Chinese Journal of Environmental Engineering* 7(5): 1951–1957 (In Chinese with English Abstract).

El-Gendy, A. S., N. Biswas and J. K. Bewtra. 2006. Municipal landfill leachate treatment for metal removal using water hyacinth in a ploating aquatic system. *Water Environment Research* 78(9): 951–964.

El-Queny, F. and A. Abdel-Megeed. 2009. Phytoremediation and detoxification of two organophosphorous pesticides residues in Riyadh Area. *World Applied Sciences Journal* 6(4): 570–578.

EPA. 2004. How to evaluate alternative cleanup technologies for underground storage tank sites: appendix—abbreviations and definitions. Cinccinnati, Ohio, USA: United States Environmental Protection Agency.

Espinoza-Quiñones, F. R., A. N. Módenes, A. P. de Oliveira and D. E. Goes Trigueros. 2013. Influence of lead-doped hydroponic medium on the adsorption/bioaccumulation processes of lead and phosphorus in roots and leaves of the aquatic macrophyte *Eicchornia crassipes*. *Journal of Environmental Management* 130: 199–206.

Flores-Cáceres, M. L., S. Hattab, S. Hattab, H. Boussetta, M. Banni and L. E. Hernández. 2015. Specific mechanisms of tolerance to copper and cadmium are compromised by a limited concentration of glutathione in alfalfa plants. *Plant Science* 233: 165–173.

Fowler, S. W. 1990. Critical review of selected heavy metal and chlorinated hydrocarbon concentrations in the marine environment. *Marine Environmental Research* 29(1): 1–64.

Fu, Y., Y. Huang, Y. Zhang and J. Zhao. 2006. Effects of three aquatic plants on the degradation of dimethonate in water. *Journal of Agro-Environment Science* 25(1): 90–94 (In Chinese with English Abstract).

Granato, M. 1993. Cyanide degradation by water hyacinths, *Eichhornia crassipes* (Mart.) Solms. *Biotechnology Letters* 15(10): 1085–1090.

Grill, E., E. -L. Winnacker and M. H. Zenk. 1985. Phytochelatins: the principal heavy-metal complexing peptides of higher plants. *Science, New Series* 230(4726): 674–676.

Guo, W., M. He, Z. Yang, C. Lin, X. Quan and B. Men. 2009. Distribution, partitioning and sources of polycyclic aromatic hydrocarbons in Daliao River water system in dry season, China. *Journal of Hazardous Materials* 164(2-3): 1379–1385.

Gupta, M., U. N. Rai, R. D. Tripathi and P. Chandra. 1995. Lead induced changes in glutathione and phytochelatin in *Hydrilla verticillata* (l. f.) Royle. *Chemosphere* 30(10): 2011–2020.

Gupta, M., R. D. Tripathi, U. N. Rai and P. Chandra. 1998. Role of glutathione and phytochelatin in *Hydrilla verticillata* (l.f.) Royle and *Vallisneria spiralis* L. under mercury stress. *Chemosphere* 37(4): 785–800.

Ha, N. T. H., M. Sakakibara and S. Sano. 2009. Phytoremediation of Sb, As, Cu, and Zn from contaminated water by the aquatic macrophyte *Eleocharis acicularis*. *Clean - Soil, Air, Water* 37(9): 720–725.

Harmon, S. M. 2015. The toxicity of pollutants to aquatic organisms. *Comprehensive Analytical Chemistry* 67: 587–613.

Ho, Y. Bin, M. P. Zakaria, P. A. Latif and N. Saari. 2013. Degradation of veterinary antibiotics and hormone during broiler manure composting. *Bioresource Technology* 131: 476–484.

Hu, C. 2007. Absorbtion and eutrophic water restoration using macrophytes and water hyacinth, Ph.D. Thesis, Department of Plant Nutrients. Hua Zhong Agricultural University (In Chinese with English Abstract).

Hu, C., L. Zhang, D. Hamilton, W. Zhou, T. Yang and D. Zhu. 2007. Physiological responses induced by copper bioaccumulation in *Eichhornia crassipes* (Mart.). *Hydrobiologia* 579(1): 211–218.

Hua, J., L. Hu, C. Zhang, Y. Yin and X. Wang. 2010. Phytoremediation of manganese-contaminated water by three aquatic macrophytes. *Ecology and Environmental Science* 19(9): 2160–2165 (In Chinese with English Abstract).

Huang, B. and H. Xu. 2008. Water hyacinth on ecological damage and phytoremediation. *Guangdong Water Resources and Hydropower* (3): 1–3, 11 (In Chinese).

Huesemann, M. H., T. S. Hausmann, T. J. Fortman, R. M. Thom and V. Cullinan. 2009. *In situ* phytoremediation of PAH- and PCB-contaminated marine sediments with eelgrass (*Zostera marina*). *Ecological Engineering* 35(10): 1395–1404.

Jiang, R. 2011. Nutritional values of water hyacinth residue silage by high moisture and evaluate the safety of animal organization on heavy metal. MSc. Thesis, Department of Animal Nutrients and Feed Science. Nanjing Agricultural University.

Jiang, X., S. Xu, D. Martens and L. Wang. 2000. Polychlorinated organic contaminants in waters, suspended solids and sediments of the Nanjing section, Yangtze River. *China Environmental Science* 20(3): 193–197 (In Chinese with English Abstract).

Jiries, A., H. Hussain and J. Lintelmann. 2000. Determination of polycyclic aromatic hydrocarbons in wastewater, sediments, sludge and plants in Karak Province, Jordan. *Water, Air, and Soil Pollution* 121: 217–228.

Jurado, A., E. Vàzquez-Suñé, J. Carrera, M. López de Alda, E. Pujades and D. Barceló. 2012. Emerging organic contaminants in groundwater in Spain: a review of sources, recent occurrence and fate in a European context. *Science of the Total Environment* 440: 82–94.

Kamal, M., A. E. Ghaly, N. Mahmoud and R. CoteCôté. 2004. Phytoaccumulation of heavy metals by aquatic plants. *Environment International* 29(8): 1029–1039.

Kang, Y., B. Mai, G. Sheng and J. Fu. 2000. The distributing characteristics of organochlorine compounds contaminants in sediments of Pearl River Delta and nearby sea area. *China Environmental Science* 20(3): 245–249 (In Chinese with English Abstract).

Kim, Y. and W. -J. Kim. 2000. Roles of water hyacinths and their roots for reducing algal concentration in the effluent from waste stabilization ponds. *Water Research* 34(13): 3285–3294.

Kolpin, D., E. Furlong, M. Meyer, E. M. Thurman, S. Zaugg, L. Barber et al. 2002. Pharmaceuticals, hormones, and other organic wastewater contaminants in U.S. streams, 1999–2000: a national reconnaissance. *Environment Science and Technology* 36: 1202–1211.

Koyuncu, I., O. A. Arikan, M. R. Wiesner and C. Rice. 2008. Removal of hormones and antibiotics by nanofiltration membranes. *Journal of Membrane Science* 309(1-2): 94–101.

Küpper, H., F. Jie Zhao and S. P. McGrath. 1999. Cellular compartmentation of zinc in leaves of the hyperaccumulator *Thlaspi caerulescens*. *Plant Physiology* 119(1): 305–312.

Kurzbaum, E., F. Kirzhner, S. Sela, Y. Zimmels and R. Armon. 2010. Efficiency of phenol biodegradation by planktonic *Pseudomonas pseudoalcaligenes* (a constructed wetland isolate) vs. root and gravel biofilm. *Water Research* 44(17): 5021–31.

Lazzari, L., L. Sperni, P. Bertin and B. Pavoni. 2000. Correlation between inorganic (heavy metals) and organic (PCBs and PAHs) micropollutant concentrations during sewage sludge composting processes. *Chemosphere* 41(3): 427–435.

Lei, S., H. Wang, Q. Yuan, Y. Wang, H. Zhao and Y. Yang. 2014. Main kinds and distribution of organics in the river growing *Leptomitus Lacteus*. *Journal of Environmental Engineering Technology* 4(5): 385–392 (In Chinese with English Abstract).

Lewis, S., M. E. Donkin and M. H. Depledge. 2001. Hsp70 expression in *Enteromorpha intestinalis* (Chlorophyta) exposed to environmental stressors. *Aquatic Toxicology* 51(3): 277–291.

Li, R., J. Li and W. Zhao. 1997. Review of the study on the heavy metal pollution in water environment of China. *Si Chuan Environment* 16(1): 18–22 (In Chinese with English Abstract).

Li, W., M. Lan and X. Peng. 2011. Removal of antibiotics from swine wastewater by UV/H_2O_2 combined oxidation. *Environmental Pollution and Control* 33(4): 25–28, 33 (In Chinese with English Abstract).

Li, W., J. Wang, L. Wen and J. Wang. 1995a. Application of water hyacinth to the removal of heavy metals from electroplate waste water. *Chinese Journal of Ecology* 14(4): 30–35 (In Chinese with English Abstract).

Li, X., Z. Wu and G. He. 1995b. Effects of low temperature and physiological age on superoxide dismutase in water hyacinth (*Eichhornia crassipes*) Solms. *Analysis* 50: 193–200.

Lishman, L., S. A. Smyth, K. Sarafin, S. Kleywegt, J. Toito, T. Peart et al. 2006. Occurrence and reductions of pharmaceuticals and personal care products and estrogens by municipal wastewater treatment plants in Ontario, Canada. *Science of the Total Environment* 367(2-3): 544–558.

Lissy, P. N. M. and G. Madhu. 2011. Removal of heavy metals from waste water using water hyacinth. *ACEE International Journal on Transportation and Urban Development* 1(1): 48–52.

Liu, J., F. Lin, Y. Wang, Z. Xu and X. Zhang. 2003. Absorption processes of macrophyte root on polycyclic aromatic hydrocarbons (naphthalene). *Environmental Science and Technology* 26(1): 32–34 (In Chinese).

Liu, L., Z. Chen and Y. Chen. 2007. Dynamic change of *Eichhornia crassipes* root-zone microbes in the process of water purification. *Journal of Agricultural Science* 35(2): 10–511 (In Chinese with English Abstract).

Loos, R., G. Locoro and S. Contini. 2010. Occurrence of polar organic contaminants in the dissolved water phase of the Danube River and its major tributaries using SPE-LC-MS2 analysis. *Water Research* 44(7): 2325–2335.

Lu, X., Y. Gao, J. Luo, S. Yan, Z. Rengel and Z. Zhang. 2014. Interaction of veterinary antibiotic tetracyclines and copper on their fates in water and water hyacinth (*Eichhornia crassipes*). *Journal of Hazardous Materials* 280: 389–398.

Lu, X., Y. Yang and W. Wang. 2007. Removal of cadmium and zinc by *Eichhornia crassipes*. *Anhui Agricultural Science Bulletin* 13(15): 16–18 (In Chinese with English Abstract).

Luo, J., L. Liu, T. Wang, H. Liu, Y. Gao, Z. Zhang et al. 2014. Study on fermentation conditions of water hyacinth and pig manure co-composting. *Jiangsu Agricultural Sciences* 42(6): 336–339 (In Chinese).

Ma, T., N. Yi, Z. Zhang, Y. Wang, Y. Gao and S. Yan. 2014. Oxygen and organic carbon releases from roots of *Eichhornia crassipes* and their influence on transformation of nitrogen in water. *Journal of Agro-Environment Science* 33(10): 2003–2013 (In Chinese with English Abstract).

Macek, T., M. Macková and J. Káš. 2000. Exploitation of plants for the removal of organics in environmental remediation. *Biotechnology Advances* 18(1): 23–34.

Maitani, T., H. Kubota, K. Sato and T. Yamada. 1996. The composition of metals bound to class III metallothionein (phytochelatin and its desglycyl peptide) induced by various metals in root cultures of *Rubia tinctorum*. *Plant Physiology* 110(4): 1145–1150.

Malar, S., S. V. Sahi, P. J. C. Favas and P. Venkatachalam. 2014. Mercury heavy-metal-induced physiochemical changes and genotoxic alterations in water hyacinths [*Eichhornia crassipes* (Mart.)]. *Environmental Science and Pollution Research* 22(6): 4597–4608.

Malik, A. 2007. Environmental challenge vis a vis opportunity: the case of water hyacinth. *Environment International* 33(1): 122–38.

MA-PRC. 2012. Standard of organic fertilizer (NY525-2012). Beijing, China: Ministry of Agriculture of People's Republic of China.

Margoshes, M. and B. Vallee. 1957. A cadmium protein from equine kidney cortex. *Journal of the American Chemical Society* 79(8): 4813–14.

Mazen, A. M. A. and O. M. O. El Maghraby. 1997. Accumulation of cadmium, lead and strontium, and a role of calcium oxalate in water hyacinth tolerance. *Biologia Plantarium* 40(3): 411–417.

Megateli, S., S. Semsari and M. Couderchet. 2009. Toxicity and removal of heavy metals (cadmium, copper, and zinc) by *Lemna gibba*. *Ecotoxicology and Environmental Safety* 72(6): 1774–1780.

MEP-PRC. 1995. Environmental quality standard for soils (GB15618-1995). Beijing, China: Ministry of Environmental Protection of The Peoples's Republic of China.

Miretzky, P., A. Saralegui and A. F. Cirelli. 2004. Aquatic macrophytes potential for the simultaneous removal of heavy metals (Buenos Aires, Argentina). *Chemosphere* 57(8): 997–1005.

Mishra, V. K. and B. D. Tripathi. 2009. Accumulation of chromium and zinc from aqueous solutions using water hyacinth (*Eichhornia crassipes*). *Journal of Hazardous Materials* 164(2-3): 1059–1063.

Murphy, A. and L. Taiz. 1995. Comparison of metallothionein gene expression and nonprotein thiols in ten arabidopsis ecotypes, correlation with copper tolerance. *Plant Physiology* 109(3): 945–954.

Nassef, M., S. Matsumoto, M. Seki, F. Khalil, I. J. Kang, Y. Shimasaki et al. 2010. Acute effects of triclosan, diclofenac and carbamazepine on feeding performance of Japanese medaka fish (*Oryzias latipes*). *Chemosphere* 80(9): 1095–1100.

Nawirska, A. 2005. Binding of heavy metals to pomace fibers. *Food Chemistry* 90(3): 395–400.
Nesterenko-Malkovskaya, A., F. Kirzhner, Y. Zimmels and R. Armon. 2012. *Eichhornia crassipes* capability to remove naphthalene from wastewater in the absence of bacteria. *Chemosphere* 87(10): 1186–1191.
Nye, P. H. and P. B. Tinker. 1977. *Solute movement in the soil-root system*. Berkeley and Los Angeles, USA: University of California Press.
Paraíba, L. C. 2007. Pesticide bioconcentration modelling for fruit trees. *Chemosphere* 66(8): 1468–1475.
Peng, X., Y. Yu, C. Tang, J. Tan, Q. Huang and Z. Wang. 2008. Occurrence of steroid estrogens, endocrine-disrupting phenols and acid pharmaceutical residues in urban riverine water of the Pearl River Delta, South China. *Science of the Total Environment* 397(1-3): 158–166.
Pilon-Smits, E. 2005. Phytoremediation. *Annual Review of Plant Biology* 56: 15–39.
Přikryl, Z. and V. Vančura. 1980. Root exudates of plants - VI. wheat root exudation as dependent on growth, concentration gradient of exudates and the presence of bacteria. *Plant and Soil* 57(1): 69–83.
Rai, P. K. and B. D. Tripathi. 2009. Comparative assessment of *Azolla pinnata* and *Vallisneria spiralis* in Hg removal from G.B. Pant Sagar of Singrauli Industrial region, India. *Environmental Monitoring and Assessment* 148(1-4): 75–84.
Raskin, I. and B. D. Ensley. 1999. *Phytoremediation of toxic metals: using plants to clean the environment*. Hoboken, USA: John Wiley & Sons, Inc.
Raskin, I., R. D. Smith and D. E. Salt. 1997. Phytoremediation of metals: using plants to remove pollutants from the environment. *Current Opinion in Biotechnology* 8(2): 221–226.
Reddy, K. R. and J. C. Tucker. 1983. Productivity and nutrient uptake of water hyacinth, *Eichhornia crassipes* I. effect of nitrogen source. *Economic Botany* 37(2): 237–247.
Rezania, S., M. Ponraj, M. F. M. Din, A. R. Songip, F. M. Sairan and S. Chelliapan. 2015. The diverse applications of water hyacinth with main focus on sustainable energy and production for new era: an overview. *Renewable and Sustainable Energy Reviews* 41: 943–954.
Rivera-Utrilla, J., G. Prados-Joya, M. Sánchez-Polo, M. A. Ferro-García and I. Bautista-Toledo. 2009. Removal of nitroimidazole antibiotics from aqueous solution by adsorption/bioadsorption on activated carbon. *Journal of Hazardous Materials* 170(1): 298–305.
Rodríguez, M., J. Brisson, G. Rueda and M. S. Rodríguez. 2012. Water quality improvement of a reservoir invaded by an exotic macrophyte. *Invasive Plant Science and Management* 5(2): 290–299.
Sakurai, T., J. G. Kim, N. Suzuki, T. Matsuo, D. Q. Li, Y. Yao et al. 2000. Polychlorinated dibenzo-p-dioxins and dibenzofurans in sediment, soil, fish, shellfish and crab samples from Tokyo Bay area, Japan. *Chemosphere* 40(6): 627–640.
Salt, D. E., R. D. Smith and I. Raskin. 1998. Phytoremediation. *Annual Reviews Plant Physiol. Plant Mol. Biol.* 643–668.
Sandermann, H. 1992. Plant metabolism of xenobiotics. *Trends in Biochemical Sciences* 17(2): 82–84.
Schecter, A., P. Cramer, K. Boggess, J. Stanley and J. R. Olson. 1997. Levels of dioxins, dibenzofurans, PCB and DDE congeners in pooled food samples collected in 1995 at supermarkets across the United States. *Chemosphere* 34(5-7): 1437–1447.
Schwab, A. P., A. A. Al-Assi and M. K. Banks. 1998. Adsorption of naphthalene onto plant roots. *Journal of Environment Quality* 27(1): 220.
Singh, W. R., S. K. Pankaj and A. S. Kalamdhad. 2015. Reduction of bioavailability and leachability of heavy metals during agitated pile composting of *Salvinia natans* weed of Loktak lake. *International Journal of Recycling of Organic Waste in Agriculture* 4(2): 143–156.
Sinha, A. K. and R. K. Sinha. 2000. Sewage management by aquatic weeds (water hyacinth and duckweed): economically viable and ecologically sustainable bio-mechanical technology. *Environmental Education and Information* 19(3): 215–226.
Smolyakov, B. S. 2012. Uptake of Zn, Cu, Pb, and Cd by water hyacinth in the initial stage of water system remediation. *Applied Geochemistry* 27(6): 1214–1219.
Soltan, M. E. and M. N. Rashed. 2003. Laboratory study on the survival of water hyacinth under several conditions of heavy metal concentrations. *Advance in Environmental Research* 7: 321–334.

Sun, R. and Q. -X. Zhou. 2005. Heavy metal tolerance and hyperaccumulation of higher plants and their molecular mechanisms: a review. *Acta Phytoecologica Sinicato* 29(3): 497–504 (In Chinese with English Abstract).
Tan, C., Y. Lin and Z. Chen. 2009. The use of *Eichhornia crassipes* for the removal of heavy metals from aqueous solutions. *Journal of Subtropical Resources and Environment* 4(1): 47–52 (In Chinese with English Abstract).
Tao, D., Y. Huang and Y. Xue. 1998. Removal of naphthylamine and aniline by water hyacinth. *Jiangsu Environment Science and Technology* 2: 4–7 (In Chinese).
Tchobanoglous, G., F. Maitski, K. Thompson and T. H. ChadwickSource. 1989. Evolution and performance of city of San Diego pilot-scale aquatic wastewater treatment system using water hyacinths. *Research Journal of the Water Pollution Control Federation* 61(11/12): 1625–1635.
Thornton, G. J. P. and R. P. D. Walsh. 2001. Heavy metals in the waters of the Nant-y-Fendrod: change in pollution levels and dynamics associated with the redevelopment of the Lower Swansea Valley, South Wales, UK. *Science of the Total Environment* 278(1-3): 45–55.
Voudrias, E. a. and K. S. Assaf. 1996. Theoretical evaluation of dissolution and biochemical reduction of TNT for phytoremediation of contaminated sediments. *Journal of Contaminant Hydrology* 23(3): 245–261.
Wang, C., S. Xu, Z. Wang, C. Tan, G. Tai and S. Yu. 1986. Studies on removal of phenol from polluted water by water hyacinths (*Eichhornia crassipes*) I. pot and oxidation pond experiments and effects of some environmental conditions on removal of phenol. *Acta Scientiae Circumstantiae* 6(2): 207–215 (In Chinese with English Abstract).
Wang, D., Q. Sui, W. Zhao, S. Lü, Z. Qiu and G. Yu. 2014. Pharmaceutical and personal care products in the surface water of China: a review. *Chinese Science Bulletin* (*Chinese Version*) 59(9): 743–751 (In Chinese with English Abstract).
Wang, J. and M. Ma. 2000. Biological mechanisms of phytoremediation. *Chinese Bulletin of Botany* 17(6): 504–510 (In Chinese with English Abstract).
Wang, L., G. G. Ying, J. L. Zhao, X. B. Yang, F. Chen, R. Tao et al. 2010. Occurrence and risk assessment of acidic pharmaceuticals in the Yellow River, Hai River and Liao River of north China. *Science of the Total Environment* 408(16): 3139–3147.
Wang, P., D. Zhao, C. Nie and Y. Chi. 2011. Present situation of pollutions of persistent organic pollutants in water environment. *Guizhou Agricultural Sciences* 39(2): 221–224 (In Chinese with English Abstract).
Wang, Y., Y. Xiong, F. Tie, L. Li and B. Ru. 1994. Preliminary study on measurement of heavy metal-induced water pollution by metallothio-peptide of water hyacinth root. *Acta Scientiae Circumstantiae* 14(4): 431–438 (In Chinese with English Abstract).
Wang, Z. and Y. Wen. 2008. The removal of aniline, nitrate and phosphate driven by 6 kinds of aquaculture plants. *Journal of Agro-Environment Science* 28(3): 570–574.
Wang, Z., Z. Zhang, J. Zhang, Y. Zhang, H. Liu and S. Yan. 2012. Large-scale utilization of water hyacinth for nutrient removal in Lake Dianchi in China: the effects on the water quality, macrozoobenthos and zooplankton. *Chemosphere* 89(10): 1255–61.
Waterston, R. H., Z. Bao and J. I. Murray. 2005. ECB12: 12th European congress on biotechnology. *Journal of Biotechnology* 118(1): 1–189.
Westwood, F., K. M. Bean, A. M. Dewar, R. H. Bromolow and K. Chamberlain. 1998. Movement and persistence of [14C] imidacloprid in sugar-beet plants following application to pelleted sugar-beet seed. *Pesticide Science* 52(2): 97–103.
Wild, E., J. Dent, G. O. Thomas and K. C. Jones. 2005. Direct observation of organic contaminant uptake, storage, and metabolism within plant roots. *Environmental Science and Technology* 39(10): 3695–3702.
Wollgiehn, R. and D. Neumann. 1999. Metal stress response and tolerance of cultured cells from *Silene vulgaris* and *Lycopersicon peruvianum*: role of heat stress proteins. *Journal of Plant Physiology* 154(4): 547–553.
Wu, T. and C. Lan. 2004. Vanadium in environment and its harm to human health. *Guangdong Weiliang Yuansu Kexue* 11(1): 11–15 (In Chinese with English Abstract).
Xia, H. 2002. Studies on the uptake and phytoremediation of pesticides by plants, Ph.D. Thesis, Department of Plant Nutrients. Zhe Jiang University.

Xia, H. and X. Ma. 2006. Phytoremediation of ethion by water hyacinth (*Eichhornia crassipes*) from water. *Bioresource Technology* 97(8): 1050–4.
Xia, H., L. Wu and Q. Tao. 2001. Water hyacinth accelerating the degradation of malathion in aqueous solution. *China Envirnmental Science* 21(6): 553–555 (In Chinese with English Abstract).
Xia, H. 2002. Phytoremediation of some pesticides by water hyacinth (*Eichhornia crassipes* Solms). *Journal of Zhejiang University* (*Agric. & Life Sci.*) 28(2): 165–168 (In Chinese with English Abstract).
Xia, H., L. Wu and Q. Tao. 2003. A review on phytoremediation of organic contaminant. *Chinese Journal of Applied Ecology* 14(3): 457–460 (In Chinese with English Abstract).
Xian, Q., L. Hu, H. Chen, Z. Chang and H. Zou. 2010. Removal of nutrients and veterinary antibiotics from swine wastewater by a constructed macrophyte floating bed system. *Journal of Environmental Management* 91(12): 2657–2661.
Xu, Q., J. Xiangcan and C. Yan. 2006. Macrophyte degradation status and countermeasures in China. *Ecology and Environment* 15(5): 1126–1130 (In Chinese with English Abstract).
Xu, S., X. Jiang, L. Wang, X. Quan and D. Martens. 2000. Polycyclic aromatic hydrocarbons (PAHs) pollutants in sediments of the Yangtse River and the Liaohe River. *China Environmental Science* 20(2): 128–131 (In Chinese with English Abstract).
Xu, X., C. Qiu, G. Den, Y. Hui and X. Zhang. 1999. Chemical-ecological effects of mercury pollution in the three gorge reservoir area. *Acta Hydrobiologica Sinica* 23(3): 197–203 (In Chinese with English Abstract).
Yadav, S. K. 2010. Heavy metals toxicity in plants: an overview on the role of glutathione and phytochelatins in heavy metal stress tolerance of plants. *South African Journal of Botany* 76(2): 167–179.
Yamamoto, H., K. Kagota, A. Hiejima and I. Tamura. 2014. Removal of selected pollutants in household effluent by solidified coal ash and water lettuce. *Journal of Water and Environment Technology* 12(4): 389–406.
Yang, J., J. Xue and H. Kuatian. 1999. Present status on original locations of pollution deseases in Japan. *Agro-environmental Protection* 18(6): 268–271 (In Chinese).
Yapoga, S., Y. B. Ossey and V. Kouame. 2013. Phytoremediation of zinc, cadmium, copper and chrome from industrial wastewater by *Eichhornia crassipes*. *International Journal of Conservation Science* 4(1): 81–86.
Ye, X. and S. Qiu. 2010. Phytoremediation of Pb-Cd combined pollution by three aquatic plants. *Chinese Journal of Environmental Engineering* 4(5): 1023–1026 (In Chinese with English Abstract).
Yuan, R., J. Liu, D. Cheng, F. Sang and T. Cao. 2004. Removal of naphthalene by *Eichhornia crassipes* Solms and study on the mechanism. *Journal of Shanghai University (Natural Science)* 10(3): 272–276 (In Chinese with English Abstract).
Zaranyika, M. F. and T. Ndapwadza. 1995. Uptake of Ni, Zn, Fe, Co, Cr, Pb, Cu and Cd by water hyacinth (*Eichhornia crassipes*) in mukuvisi and manyame rivers, Zimbabwe. *Journal of Environmental Science and Health* 30(1): 157–169.
Zhang, M., R. Hua, X. Li, T. Zhou, F. Tang, H. Cao et al. 2010. Distribution and composition of organochlorine pesticide in surface water body of Chaohu lake. *Chinese Journal of Applied Ecology* 21(1): 209–214 (In Chinese with English Abstract).
Zhang, S., F. Zhang, X. Liu, Y. Wang and J. Zhang. 2006. Degradation of antibiotics and passivation of heavy metals during thermophilic compositing process. *Scientia Agricultura Sinica* 39(2): 337–343 (In Chinese with English Abstract).
Zhang, X. and X. Dong. 2002. Organic chlorinated pesticides in middle and lower reaches of Liaohe. *Journal of Dalian Institute of Light Industry* 21(2): 102–104 (In Chinese with English Abstract).
Zhang, X., X. Quan, J. Chen, Y. Zhao, S. Chen, D. Xue et al. 2000. Investigation of polychlorinated organic compounds (PCOCs) im middle and lower reaches of Liaohe River. *China Environmental Science* 20(1): 31–35 (In Chinese with English Abstract).
Zhang, Z., C. Cai and H. Wang. 2004. Studies on short-duration purifactory mechanism of copper(II), lead(II), cadmium(II), Etc. ions by *Eichhornia Crassipes*. *Journal of Yichun University (natural science)* 26(2): 7–9 (In Chinese with English Abstract).
Zhang, Z., Z. Rengel and K. Meney. 2010. Cadmium accumulation and translocation in four emergent wetland species. *Water, Air, and Soil Pollution* 212(1-4): 239–249.

Zhang, Z., Z. Rengel, K. Meney, L. Pantelic and R. Tomanovic. 2011. Polynuclear aromatic hydrocarbons (PAHs) mediate cadmium toxicity to an emergent wetland species. *Journal of Hazardous Materials* 189(1-2): 119–126.

Zhang, Z., Z. Wang, Q. Lu, Z. Cu and F. Gang. 1989. A study on purifying capacity of water hyacinth in Pb and Cd-polluted water. *Chinese Journal of Environmental Science* 10(5): 14–17 (In Chinese with English Abstract).

Zheng, J. C. 2010. The performance and mechanism of removal of heavy metals from water by water hyacinth roots as a biosorbent material, Ph.D. Thesis, Department of Environmental Engineering. University of Science and Technology of China.

Zheng, J., Z. Chan, L. Chen, P. Zhu and J. Shen. 2008. Feasibility studies on N and P removal using water hyacinth in Taihu Lake region. *Jiangsu Agricultural Science* 3: 247–250 (In Chinese).

Zhou, C., S. An, J. Jiang, D. Yin, Z. Wang, C. Fang et al. 2006. An *in vitro* propagation protocol of two submerged macrophytes for lake revegetation in east China. *Aquatic Botany* 85(1): 44–52.

Zhou, L., E. Hu, W. Hang, Y. Mu and L. Wang. 2010. Qualitative analysis of organic pollutants in key pollution sources along Yangtse River in Nanjing. *Environmental Monitoring and Forewarning* 2(6): 39–41 (In Chinese with English Abstract).

Zhou, Q. and Y. Song. 2001. Technological implications of phytoremediation and its application in environment protection. *Journal of Safety and Environment* 1(3): 48–53 (In Chinese with English Abstract).

Zhou, W., L. Tan, D. Liu, H. Yan, M. Zhao and D. Zhu. 2005. Research advances of *Eichhornia crassipes* and it's utilization. *Journal of Huazhong Agricultural University* 24(4): 423–428 (In Chinese with English Abstract).

Zhou, Y. 2011. Progress in research on detection of phenolic compounds. *The Administration and Technique of Environmental Monitoring* 23: 70–77 (In Chinese with English Abstract).

Zhu, Y. L., A. M. Zayed, J. H. Qian, M. de Souza and N. Terry. 1999. Phytoaccumulation of trace elements by wetland plants: II. water hyacinth. *Journal of Environment Quality* 28(1): 339.

Part Three

Application of Water Hyacinth in Phytoremediation

CHAPTER 8

Ecological Engineering Using Water Hyacinth in Phytoremediation

Z. Y. Zhang and *S. H. Yan**

8.1 Introduction

Water hyacinth [*Eichhornia crassipes* (Mart) Solms] is a large floating aquatic plant naturally existing in tropical and subtropical climates, but can adapt to temperate climates very well (Wilson et al. 2005). The plant population can increase by stolon proliferation at the appropriate temperature and humidity conditions with very high biomass production rate (usually dozens of tons to hundreds of tons dry weight per hectare annually) (Zejiang and Yang 1984, Gunnarsson and Petersen 2007). It also can reproduce and spread quickly by seed. Although water hyacinth is considered the world's most noxious aquatic weed (Montoya et al. 2013), it has also been successfully applied in treating various types of wastewaters and has been accepted widely as a desired plant for eutrophic water purification and sewage treatment because its biological behavior and capacity to absorb large amounts of inorganic plant nutrients and heavy metals and to remove organic pollutants (Tchobanoglous et al. 1989, Fox et al. 2008).

Water hyacinth can tolerate water pollution stress and high nitrogen and phosphorus concentrations in eutrophic waters (Mahujchariyawong and Ikeda 2001, Hu et al. 2010) and a variety of heavy metal pollutants (Soltan and Rashed 2003, Mishra and Tripathi 2009). At very low and very high nutrient concentrations, water hyacinth showed different absorption capacities for nitrogen and phosphorus depending on their growth environment. For example, water hyacinth can absorb 1.05 to 1.51 kg N t^{-1} fresh biomass and 0.21 to 0.35 kg P t^{-1} fresh biomass at the nitrogen concentrations 2.06 to

50 Zhong Ling Street, Nanjing, China.
Email: jaaszyzhang@126.com
* Corresponding author: shyan@jaas.ac.cn

20.1 mg N L^{-1} and phosphorus concentrations 0.14 to 1.43 mg P L^{-1} (Zhang et al. 2010). At different hydraulic loading conditions of 0.14 to 1.0 m^3 m^{-2} d^{-1} in dynamic simulation test for purifying eutrophic water with initial total nitrogen concentration 4.85 mg N L^{-1} and total phosphorus concentration 0.50 mg P L^{-1}, every tonne of fresh biomass of water hyacinth can absorb 0.94 to 1.35 kg N and 0.20 to 0.31 kg P (Zhang et al. 2011a). Another example was the data from the large-scale confined growth of water hyacinth at Cao Hai in Lake Dianchi, China. The experiment showed that every tonne of fresh biomass of water hyacinth absorbed 1.70 kg nitrogen and 0.42 kg phosphorus (Zhang et al. 2011b).

Many researchers also have reported that water hyacinth has strong capacity on heavy metals (Zhou and Yang 1984, Wang and Cheng 2010). In water with high heavy metals concentration, water hyacinth may accumulate large amounts of heavy metals, but with reduced efficiency due to heavy metal stress (Malar et al. 2014).

8.1.1 Eutrophic water purification and sewage treatment using water hyacinth

Purification of domestic and industrial wastewater

Water hyacinth-based biological system for wastewater treatment and artificial wetlands have developed fast in the field of environmental science because they showed good results with lower investment and lower energy consumption compared to a typical chemical treatment system (Koottatep and Polprasert 1997). Early in 1976, Wooten and Dodd (1976) tested the application of water hyacinth in five ponds connected with pipelines (each pond was 465 m^2, 0.8 m in depth and a total volume of 370 m^3) to provide secondary treatment of effluent from municipal sewage plant. They found that the ammonium and nitrate in effluent disappeared relatively quickly during a 105-day processing period at the flow rate of 480 L min^{-1}; total phosphorus also decreased significantly. The average bioaccumulation of fresh water hyacinth biomass was 64.5 kg m^{-2}. Sinha and Sinha (2000) combined water hyacinth, duckweed and green algae to treat domestic sewage and achieved good purification, with heavy metal removal rates of 20 to 100%, BOD removal rate of 97%, and nitrate and phosphate removal rates greater than 90%. Ning et al. (2014) investigated the application of water hyacinth as a biological agent in a planted floating bed system for phytoremediation to remove plant nutrients and other pollutants in urban river water and sediment and found that, after 2 years, NH_4^+ concentration decreased by 97.0%, COD by 64%, and total phosphorus and total nitrogen by 77% each.

Water hyacinth was also used in the treatment of various types of industrial wastewater. Jayaweera and Kasturiarachchi (2004) reported the effects of water hyacinth on different concentrations of artificial industrial wastewater and discussed the mechanism of removing nitrogen and phosphorus. Under

21-day hydraulic retention time and 63-day testing period, the total nitrogen and total phosphorus (initial concentrations of 7.0–56.0 mg N L^{-1} and 1.92–15.4 mg P L^{-1}) were completely removed. They also found that plant assimilation and denitrification were the main pathways of removing nitrogen, with plant assimilation being the main pathway of removing phosphorus. Kulkarni et al. (2007) reported using water hyacinth to process textile effluent and observed a reduction of 80% in COD and about 25 to 45% decrease in metal concentrations after 18-day exposure. In addition, water hyacinth was widely used for treatment of effluents from pulp mill, food processing and slaughtering industries to remove plant nutrients and organic and heavy metal or metalloid pollutants such as Cd, Zn, Hg, Ag, Co, Cr, Cu, Ni, Pb, and As (Zaranyika and Ndapwadza 1995, Dellarossa et al. 2001, Giraldo and Garzón 2002, Li et al. 2003).

Processing livestock production effluent

Livestock farming has become one of the main sources of pollution. The effluents from livestock farming have typically contained feed residuals and high loading of plant nutrients, antibiotics and heavy metals. Water hyacinth was often applied either in artificial wetland or oxidization pond treatment system to remove these pollutants (Koottatep and Polprasert 1997, Malik 2007). Lu et al. (2008) investigated the performance of water hyacinth by constructing 688 m^2 of wetland with water hyacinth on the edge of an intensive duck farm and reported that COD removal rate was 64%, total nitrogen was 22%, total phosphorus removal rate was 23%, and the dissolved oxygen and transparency in the water were significantly improved. Díaz et al. (2014) used water hyacinth to construct a surface-flow wetland for treating effluents from fish farming of tilapia (*Oreochromis* sp.) and red-bellied pacu (*Piaractus brachypomus*) fingerlings and reported good removal rate of nitrogen compounds (17% for NO_3^-, 27% for NO_2^-, 68% for NH_4^+), total phosphorus (23%), biological oxygen demand (32%) at hydraulic retention time of 1.6 days.

Yu and Deng (2000) studied secondary treatment of effluents from central swine waste processing facilities with water hyacinth (Wanfeng Farm with 100,000-pig output annually) at Shenzhen, Guangdong Province, China. The system consisted of seven steps: initial anaerobic fermentation, solid depositing pond with natural oxidation, enforced oxidation pond, water hyacinth pond, second natural oxidation pond, water hyacinth pond and finally sandy filter pond before water discharge to environment. The effluent was discharged at a daily flow rate of 1500 m^3 with a Chemical Oxygen Demand (COD) 14 g L^{-1} and total nitrogen (TN) > 600 mg N L^{-1} initially. After the third step, the COD was reduced to 0.6–0.8 g L^{-1} and TN to 500–600 mg N L^{-1}. When the effluent passed through the fourth step and up to final sandy filtration, the removals were 51 to 71% for COD and 55 to 72% for TN. Although good results were obtained, the system design was not clear on the size of each pond, the methods of anaerobic fermentation, etc.

Purification of surface water

Non-point source pollution from runoffs and drainage generated more than 70% of nitrogen loading in all river systems in the USA (USEPA 2000). A one-year experiment using water hyacinth pond to treat agricultural drainage revealed NO_3^- and NH_4^+ removal rates up to 78 to 81% with hydraulic retention time of 3.6 days, whereas the removal rate for total phosphorus was 54% (Reddy et al. 1982). Polomski et al. (2009) investigated the effects of three wetland macrophytes (*Eichhornia crassipes*, *Pistia stratiotes* and *Myriophyllum aquaticum*) on remediation of nursery runoff in indoor simulation experiment. During the 8-week period, the three wetlands received surface runoff containing total nitrogen 0.4–37 mg N L^{-1} and total phosphorus 0.07–6.8 mg P L^{-1}; the nitrogen and phosphorus concentrations in plant tissues were positively correlated with the concentrations of nitrogen and phosphorus in water. All three types of aquatic plants can be used for remediating nursery runoff, with water hyacinth and water lettuce removing 100% TN and nearly 100% TP.

Xia (2012) constructed a water hyacinth pond to recover plant nutrients from paddy field runoffs and reported water purification efficiency of 62–88% for total nitrogen and 79–94% for total phosphorus depending on the nutrient concentrations in runoffs. The nutrient recoveries were positively correlated with the nitrogen and phosphorus concentrations in runoffs. During the whole rice season, water hyacinth recovered 8.2 kg N ha^{-1} and 0.57 kg P ha^{-1} in runoffs from paddy fields.

Remediation of eutrophic rivers, lakes and reservoirs

Eutrophication has become a challenge on a global scale and an urgent problem to resolve (Heisler et al. 2008, Ansari et al. 2011). Use of water hyacinth to purify eutrophic waters shows good effects at relatively low investment (Qi and Gao 1999). Application of water hyacinth as a biological agent for phytoremediation has been practiced not only in ponds, harbors, small bays and lagoons, but also in large lakes and reservoirs to purify water in ecological engineering projects (Dou et al. 1995). Wang et al. (1998) grew different types of macrophytes in enclosures (5 m by 40 m) made from bamboo piles and waterproof industrial fabric at Wuli Bay in Lake Taihu, Jiangsu Province, China. The project employed combinations of species: alligator weed (*Alternanthera philoxeroides*) followed by water chestnut (*Trapa natans*) then European frogbit (*Hydrocharis* sp.), water hyacinth (*E. crassipes*), water chestnut (*Trapa natans*) again and finally water hyacinth. During a 1-year water quality monitoring program, there was a significant reduction of 58% in algal populations, 66% in NH_4^+, 60% in total nitrogen, 72% in total phosphorus, and 90% in orthophosphate.

Rodríguez et al. (2012) investigated the effects of water hyacinth in Tominé reservoir in Columbia and reported water quality improvement after the natural growth of water hyacinth. Water hyacinth mat covering about 300 hectares decreased NO_3^- concentration from 0.6 mg L^{-1} to 0.1 mg L^{-1}, NH_4^+

from 0.5 mg L^{-1} to 0.3 mg L^{-1}, total nitrogen from 2.0 mg L^{-1} to 1.2 mg L^{-1} and increased transparency from 0.3 m to 2.0 m.

Mahujchariyawong and Ikeda (2001) studied the use of water hyacinth to control eutrophication in Tha-chin River, Thailand. They used continuous harvests to control the water hyacinth population at maximum sustainable yield in the river and monitored water quality at the same time. The carrying capacity for water hyacinth was fresh biomass of 53 kg m^{-2} or 530 tonnes per hectare in the river under tropical climate. The nutrient removal at the maximum water hyacinth carrying capacity was 0.42 g m^{-2} d^{-1} for nitrogen and 0.09 g m^{-2} d^{-1} for phosphorus.

The literature clearly suggests the potential for using water hyacinth to control eutrophication in either small or large water bodies, including rivers, lakes and reservoirs. However, the indication of potential or possibility is not enough. Most water bodies are public properties and using water hyacinth to remediate eutrophic water bodies, regardless of size, is full of challenges and risks in planning, designing, preparing and executing remediation as well as in processing and disposal (utilization) of huge amounts of harvested fresh biomass.

8.2 Challenges in phytoremediation using water hyacinth

Water hyacinth has desired capabilities for remediation of water bodies, but only a few large scale applications have been put to practice due to ecological risks to biodiversity in aquatic ecosystems and high costs of harvest and disposal (utilization) of huge amounts of fresh biomass.

8.2.1 Ecological risks to biodiversity

Under optimum growth conditions water hyacinth can yield as much as 140 tonnes of dry matter per hectare (Abdelhamid and Gabr 1991) and can be potentially a noxious weed that may damage aquatic environments by reducing biodiversity, blocking water transportation, reducing fish yield and impeding hydropower stations (Navarro and Phiri 2000, Osmond and Petroeschevsky 2013). Water hyacinth is a macrophyte that not only assimilates nutrients from water, but also enhances the function of microorganisms and cleans the water and intercepts suspended solids. This implies that if the water hyacinth is not harvested or removed, the huge biomass itself becomes a pollutant and the assimilated nutrients may return to water (Greenfield et al. 2007). During the growth period of water hyacinth, the population forms a dense mat that decreases dissolved oxygen, potentially endangering most aquatic animals (Perna and Burrows 2005). Also, the mat may block incident sunlight for submerged aquatic vegetation, reducing its growth and potentially causing even the extinction of the submerged species (Schultz and Dibble 2011). Therefore, an ecological project for remediation of eutrophic waters using water hyacinth must take into account the ecosystem safety issues such as location and size of water hyacinth mat, how to avoid a decrease in dissolved oxygen, and planning of harvests and fresh biomass disposal (or utilization) (Bicudo et al. 2007).

Water hyacinth can yield as high as 600–900 tonnes of fresh biomass (35–54 tonnes of dry biomass) per hectare in eutrophic water under temperate climate zone (Zheng et al. 2008) and may contain potentially high concentrations of heavy metals in the tissues. Processing (disposal) of such biomass requires special skills to reduce the inherent risks (Li et al. 2004). Before these problems are understood and solved, using water hyacinth as a biological agent for phytoremediation is questionable.

8.2.2 Challenges of mechanical control of water hyacinth

Most literature on the macrophyte removal usually refers to mechanical equipment, biological agents, or herbicides. Unsurprisingly, hand-harvesting is not suitable for large scale application (Jiang and Zhang 2003), whereas biological agents (e.g., beetles) or herbicides are not really harvesting methods. This leaves mechanical harvesting as the only management method that controls water hyacinth in water bodies and recovers its biomass on a large scale; however, the method in the past was costly and has relatively low efficiency in terms of (i) the amount harvested per vessel per day and (ii) energy cost per ton of fresh biomass harvested and distance transported (Gettys et al. 2009).

The use of heavy-duty machines to control water hyacinth started in 1937 by the U.S. Army Corps of Engineers (Little 1979). By the 1950s, the design and application of aquatic weed harvesters (including chopper and excavator) were so popular that almost every North America mechanical engineer knew how to build a mechanical harvester for removing water hyacinths (Gettys et al. 2009).

Using water hyacinth for large scale phytoremediation to control eutrophication implies that the population of water hyacinth (regardless of whether naturally present or introduced) needs to be controlled at a given location. However, literature on water hyacinth management and control suggests that water hyacinth is difficult and sometimes even impossible to control (Villamagna and Murphy 2010), partly because of high cost of mechanical harvest and disposal versus manual removal on a large scale being almost impossible (Lindsey and Hirt 2000). The harvest costs may range from US$600 to US$1,200 per hectare plus additional cost of post-harvest processing and disposal (including transportation). Data from the European Environment Agency showed that, during 2005 to 2008, the cost of mechanically harvesting 200,000 tonnes of water hyacinth fresh biomass at the Portugal-Spain border along 75 km of the Guadiana River Basin was 14.7 million EUR (UNEP and GEAS 2013). In Mali, West Africa, annual costs of mechanically harvesting water hyacinth were US$80,000–100,000 before 2006 (Dagno et al. 2012).

In analyzing the costs of mechanically harvesting water hyacinth, one should take into account that the fresh biomass of water hyacinth contains 93–95% water (Cifuentes and Bagnall 1976). Furthermore, water hyacinth is typically harvested for disposal, whereas field crops are harvested for their value on the commercial market. To deal with these challenges, two interesting ideas have emerged: (1) reduce the volume and water content of water hyacinth biomass during or after harvesting; and (2) find commercial uses for water hyacinth biomass to generate value and thus partially compensate for the high cost of harvesting (Babourina and Rengel 2011). However, in order to utilize the fresh biomass of water hyacinth (e.g., for methane production or for silage), the effective, low-cost dehydration techniques are needed (Cifuentes and Bagnall 1976).

Cifuentes and Bagnall (1976) designed and tested a screw press for dewatering water hyacinth and concluded that the optimal pressure limit was 600 kPa (i.e., higher pressure did not remove much more water from the plant), and water content of the residual was 84%, which was compatible with other published literature (Du, Chan et al. 2010). However, at water content of 84% water hyacinth biomass cannot meet the minimum requirements for making organic fertilizer or silage or for solid state fermentation if not dried further (Jiang et al. 2011, Shi et al. 2012). Further drying may increase the costs and reduce productivity.

Gettys et al. (2009) summarized different methods of mechanical control of aquatic weeds, such as shredding boats and cut-and-remove harvest systems. These harvesters were powered by an engine and equipped with a cutter at the water surface to cut floating plants such as water hyacinth. However, using a shredding boat could create a large mat of plant fragments that are difficult to remove, thus serving a purpose of controlling the water hyacinth population but not remediation because shredded plant fragments will decay and the plant nutrients will be returned to the water.

Water hyacinth contains spongy tissue with a density of 0.3 g cm^{-3} and a water content of about 95%. This means that every cubic meter of fresh biomass occupying the available space of a harvester (a boat) weighs only 300 kg out of which about 285 kg is the weight of water. Mathur and Singh (2004) designed an on-shore chopper-cum-crusher that can reduce the volume of fresh water hyacinth by 64%, which can be interpreted as an increase in biomass density to 830 kg m^{-3}. The machine could process fresh water hyacinth at a capacity of 1.0 tonne per hour. However, in large lakes, harvesting of water hyacinth requires a capacity to process multi-million tonnes of fresh biomass a year, i.e., equivalent to a natural standing biomass of 1000–2000 hectares of water hyacinth, such as in Lake Victory in East Africa or Lagos Lagoon in West Africa. The simple calculation shows the importance of the high volume harvesting and processing capacity. For example, a standing population of one million tonnes of water hyacinth could increase 70,000 tonnes of additional water hyacinth every day, which would require 9000 choppers-cum-crushers operating for 8 hours a day. However, these

9000 choppers-cum-crushers would only be removing the newly growing biomass of the standing population, not reducing the population. This calculation highlights the need for new technology or more efficient equipment before the idea of large scale phytoremediation involving water hyacinth can be practiced. The recent advancements in mechanical harvesting have focused on (i) cost reduction of the harvesting processes and (ii) improving the efficiency during and after harvest (transportation and dehydration of fresh biomass could cost approximately US$40 per tonne of dry mass produced) (Hronich et al. 2008). However, further developments of efficient equipment for harvest, transportation and dehydration are still needed.

The biology of water hyacinth also implied that completely harvesting water hyacinth at a multi-million ton capacity only solves half the problem in large scale phytoremediation for control of eutrophication and water resources management or noxious weed management because dehydration is needed in order to avoid unaffordable high cost either during utilization or disposal of the biomass.

8.2.3 Challenges regarding disposal (utilization) of water hyacinth fresh biomass

Fresh water hyacinth biomass has a water content of about 95%, which makes the fresh biomass of water hyacinth very costly to transport after harvest and requires a huge amount of space for its disposal. A calculation on the removal of one million tonnes of fresh biomass shows the need for 75 full-size (40 m^3 each) trucks every day to transport the biomass all year around (365 days). Also, it would require multi-million cubic meters of space at the waste dump site. The calculation also indicates that one million tonnes of fresh biomass would need about 20 standard full-size football fields (about 7400 m^2 each) with biomass piled at a height of 2 meters if the drying and rotting turnover rate is 100 days. Furthermore, the result of fresh biomass deposited at a waste dump site could include high socio-economic and ecological costs due to possible high emissions of carbon dioxide and methane as well as nitrous oxide. This example highlights that reductions in the harvest and dehydration costs must be achieved before the idea of phytoremediation using water hyacinth becomes feasible.

In the past few decades, a large number of studies has been conducted globally (Ndimele et al. 2011) concerning the disposal of water hyacinth and related technologies. However, a fundamental problem facing the use of water hyacinth (i.e., high water content of fresh biomass) is difficult to solve (Malik 2007), in particular when the harvest and disposal processes are scaled up. Therefore, solving the problem of high water content is the key to practice of phytoremediation using water hyacinth.

The methods for dehydration of fresh biomass of water hyacinth are sun drying, baking and mechanical dehydration. The first two are either low efficiency (long time to dry) or have high cost of energy. Therefore, mechanical dehydration appears to be a potential solution to the water content problem.

The above discussion pointed out the essential criteria to start a large-scale application of water hyacinth in phytoremediation to control eutrophication in water bodies.

1) Phytoremediation does not only involve control of eutrophication, but also solving the problems of noxious weed spread, reclamation of water resources, production of bio-energy and recycling of plant nutrients. All these targets must be considered together;
2) Ensuring ecological safety and appropriate risk management are prerequisites;
3) Research and development of low-cost high-efficiency equipment for water hyacinth harvest and post-harvest processing are crucial;
4) Water hyacinth should not be harvested for disposal, but for commercial value of its biomass; and
5) Evaluation of phytoremediation must be based on an integrative system approach that would consider socio-economic and ecological aspects.

This chapter focuses on two case studies on phytoremediation conducted in Lake Dianchi and Lake Taihu. The detailed planning, layout of the location, enclosure manufacturing, maintenance, equipment for harvest and dehydration, and biomass utilization are all specified.

8.3 Phytoremediation in large lakes and reservoirs

8.3.1 Facilities for confined growth of water hyacinth

Water hyacinth grows fast and drifts easily along with water currents and wind in non-confined situations. To confine water hyacinth at desired location in rivers, lakes and reservoirs and to prevent the interference of waves and strong winds, even during a strong storm, enclosures were used in the phytoremediation project. The management procedures were implemented to ensure no escaping of plants and to limit damages during a storm. In both Lake Dianchi and Lake Taihu, anti-storm and anti-wave enclosures were designed for water depth of either more than 2 meters (deep water) or less than 2 meters (shallow water) and were manufactured with steel posts, anchors and polyethylene nets. In deep water, an anchor-based steel frame floating structure was used (Fig. 8.3.1-1). In shallow water with less powerful wave and wind actions, a post-fixed structure was used (Figs. 8.3.1-2 and 8.3.1-3). The initial idea was to use the anchor-based steel frame floating structure to prevent the macrophyte from escaping and to resist wind speeds of up to 18 meters s^{-1} (frequently occurring at open water locations in both lakes).

The Anchor-based Steel Frame Floating Structure (ASFFS) is made of anchors, galvanized steel frames (steel pipe size ϕ33.7 mm internal diameter,

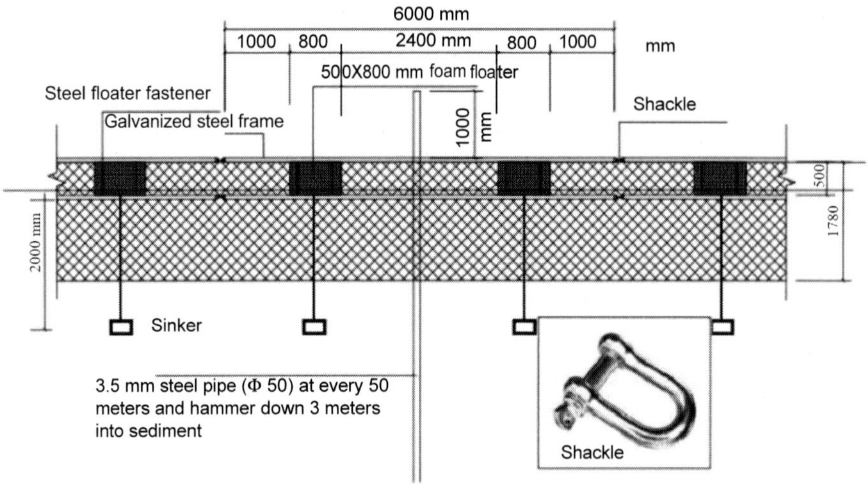

Fig. 8.3.1-1. Anchor-based steel frame floating structure illustration.

3.2 mm thickness, length 6000 mm), polyethylene net (mesh size 25 mm, width 2000–3000 mm), foam floats (ϕ500 mm wide, 800 mm long) and nylon rope (ϕ14 mm).

The post fixed structure was made of either bamboo posts or steel posts, bamboo frame, polyethylene net (mesh size 25 mm, width 2000–3000 mm) and nylon rope (ϕ14 mm). Basically, in areas with less strong wind and wave actions, bamboo posts ($>\phi$50 mm) were used, otherwise steel posts (ϕ50 mm) were used.

To limit the drift of ASFFS, a steel post (ϕ50 mm) was applied at every 50 meters, a 3-kg sinker was applied on every foam floater; and a 50-kg steel anchor was applied at every 25 meters. This setting enabled the ASFFS to have limited drift and auto-adjustment up and down water levels. The post-fixed structure was relatively easy to set up. The steel posts were fixed every 5 meters when no bamboo posts were used (Fig. 8.3.1-2) or every 10 meters when bamboo posts were used (Fig. 8.3.1-3). Although the disadvantage of PFS was that the net cannot auto-adjust up and down according to the water level, it was cheaper than ASFFS.

Each unit was constructed to cover about 7 to 8 hectares (500 meters by 150 meters). In order to prevent the depression of dissolved oxygen, each enclosure had adjacent open-water gaps at least 50 meters. However, the ecological engineering project did not test for the optimal unit shape and size and the effects of different open-water gaps surrounding the unit regarding the effects on aquatic fauna and flora. These factors may result in different costs of the enclosures and different effects on the dissolved oxygen concentration and aquatic life.

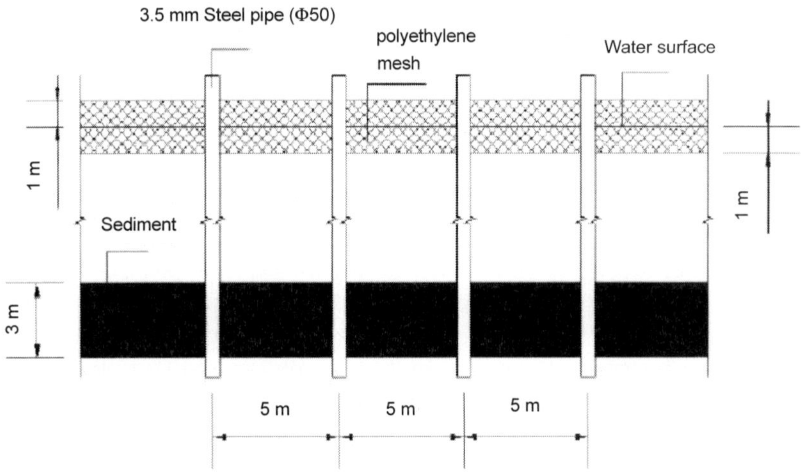

Fig. 8.3.1-2. Steel pipe fixed structure illustration.

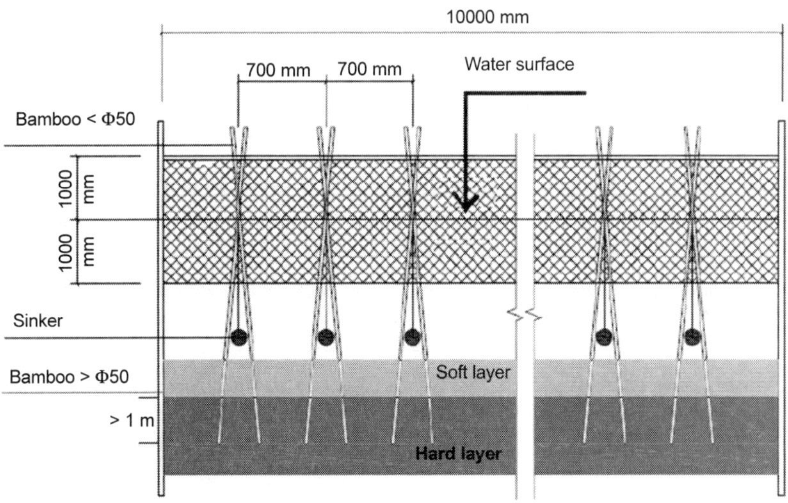

Fig. 8.3.1-3. Bamboo pipe fixed structure illustration.

After enclosures are in place, management procedures must be enforced to reduce the water hyacinth being damaged by wave forces during a storm. The damage has two aspects: (1) the wave force may increase a chance of water hyacinth escaping from enclosures; and (2) wave force may smash water hyacinth against the net. In the process of phytoremediation, such damages need to be kept to a minimum. The literature reported that when the width of a water hyacinth mat was four times longer than the wave length, the wave force was reduced to 1/3 of the wave power (Zhang et al. 2013). This means that the width of a water hyacinth mat must be longer than 93.2 meters at the wind speed of 20 m s^{-1} (gale-force winds at Beaufort number 8) and the wave length of 23.3 meters. Also, the vertically stacked height of water hyacinth biomass was 1.34 meters (0.94 m under water) at the gale-force winds. The management procedures defined 100 meters as the initial width of a water hyacinth mat and the underwater net depth must be greater than 1 meter.

8.3.2 Harvesting water hyacinth

Mechanical harvest of water hyacinth is the only option for phytoremediation given a huge amount of fresh biomass produced during the remediation process; such harvesting needs to be timely during and complete after the remediation. Otherwise, the phytoremediation may become polluted after macrophyte rotting. The second criterion is high efficiency and low cost (of both equipment and operating) because most water bodies are public properties and the costs must be affordable.

The phytoremediation projects in Lake Dianchi and Lake Taihu redesigned the harvest processes and equipment to reduce equipment cost by compacting and integrating the functions to facilitate a multi-million-tonne capacity of controlling water hyacinth.

Integrated multi-functioning harvest vessel and shredder

Spongy tissue of the fresh water hyacinth biomass contains about 70% air volume (a density 0.3 t m^{-3}). An integrated multi-functional harvest vessel was designed to have a high-efficiency shredder (Zhiyong Zhang et al. 2010) and a juice-retrieval device on board to achieve the harvesting capacity of 44 tonnes of fresh biomass per hour, with a total loading capacity of 17 tonnes of fresh biomass per vessel (Yan et al. 2010) (Fig. 8.3.2-1).

After the biomass was shredded (Fig. 8.3.2-2A), the density of fresh biomass increased to 0.9 t m^{-3}. The vessel was designed to work together with an on-board shredder (Fig. 8.3.2-2B) and a transportation boat to further increase harvesting efficiency and reduce costs of both equipment and operating.

With automatic loading and unloading capacity, the vessel can work at full speed of 3.5 km h^{-1} continuously with only three operators on board (Table 8.3.2-1).

Fixed harvest dock

A fixed harvest dock was used when the locations of water hyacinth were only a few kilometers apart, and the speed of the harvest process was not the first priority, so that further reduction in the cost of harvest was possible. The design of fixed harvest dock can take advantage of local climate, especially wind direction given that the water hyacinth floating characteristics can be

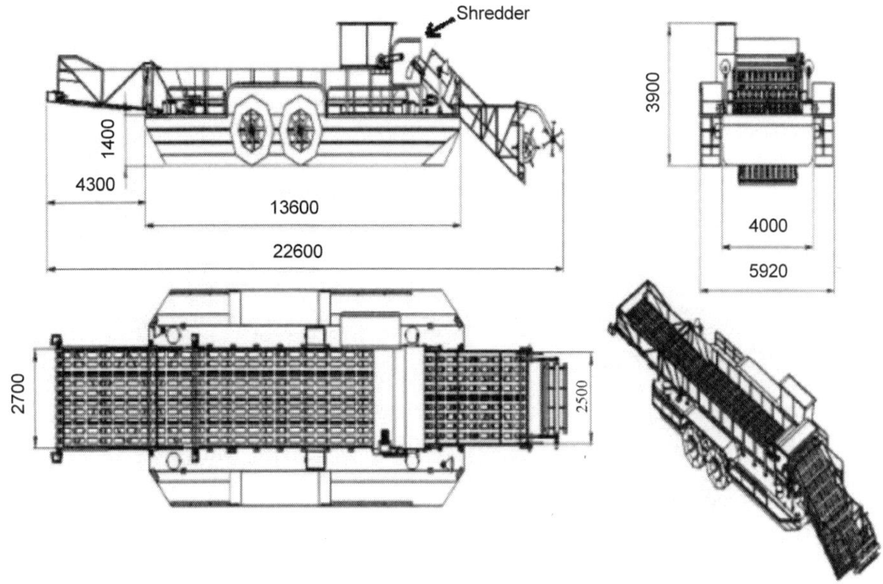

Fig. 8.3.2-1. Integrated multi-functioning harvest vessel design and size.

Fig. 8.3.2-2A. High speed shredder to work with harvest vessel for fresh biomass volume reduction.

used to speed up the transportation and to reduce the energy required. The fixed harvest dock was often used together with whole-patch haul of water hyacinth (Fig. 8.3.2-3).

The on-shore facilities (Yan et al. 2011) consisted of a collecting device, a conveyor, a shredder (Fig. 8.3.2-4) and a dehydrating device, which greatly

Fig. 8.3.2-2B. Shredder on board to work with harvest vessel at full speed.

Table 8.3.2-1. Water hyacinth harvest vessel performance parameters.

Item	Parameter	Item	Parameter
Vessel body length	13.8 m	Conveyor line speed	15 m min^{-1}
Conveyor total length	22.60 m	Loading draft	1.2 m
Vessel depth of hold	1.6 m	Engine power	120 kW
Total height	3.90 m	Working speed	3.5 km h^{-1}
Overall width	5.92 m	Full speed	7 km h^{-1}
Light draught	0.86 m	Sailing limitation	Fresh water only
Cutting width	2.50 m	Loading capacity	17 t
Full depth	1.5 m	Electricity generator capacity	2.8 kW

improved the efficiency of harvesting water hyacinth, further reducing cost. A single fixed harvest dock can harvest 600 tonnes fresh biomass of water hyacinth per day at reduced costs (only one third of equipment investment and half of the harvesting operational costs needed) compared to the integrated multi-functioning harvest vessel method.

8.3.3 Dehydration of fresh biomass of water hyacinth after harvest

There are two choices after harvesting water hyacinth: (1) disposal as waste; or (2) using it as a bio-resource. The ecological engineering project selected the second choice with a simple reason that both choices required dehydrating the biomass, and then the second choice offered a potential for juice and bagasse to create some commercial value and partially compensate for the phytoremediation costs. Dehydration of fresh biomass of water hyacinth

Fig. 8.3.2-3. Whole patch haul of water hyacinth by floating nets.

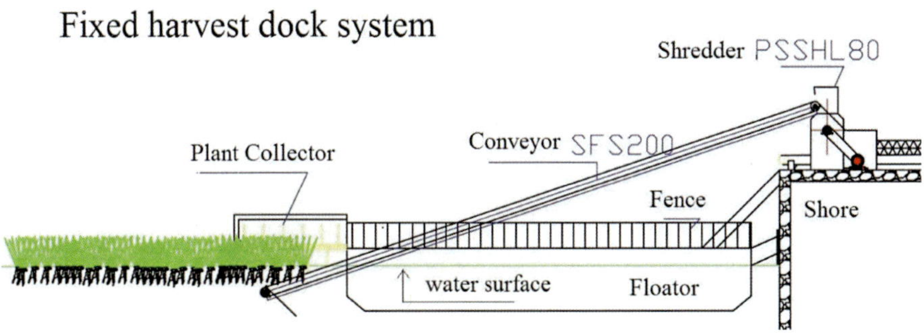

Fig. 8.3.2-4. On-shore facilities of a fixed harvest dock system.

was critical for the success of the ecological engineering project because the successful dehydration allowed production of commercially valuable goods.

To solve the problem of high water content in the water hyacinth fresh biomass, a roller press dehydration equipment (Figs. 8.3.3-1A and 8.3.3-1B) was designed to effectively reduce the water content to around 65% (Wang et al. 2013b), which was suitable for (i) solid state fermentation to produce biogas (Hu et al. 2008, Ye et al. 2011), or (ii) organic fertilizer fermentation (Shi et al. 2012) to recycle plant nutrients or (iii) silage fermentation for animal feed (Jiang 2011, Jiang et al. 2011).

However, dehydration of fresh biomass of water hyacinth requires a solution to another problem: a huge amount of juice (80–90% of fresh biomass) from the dehydration processes in the large-scale projects; juice

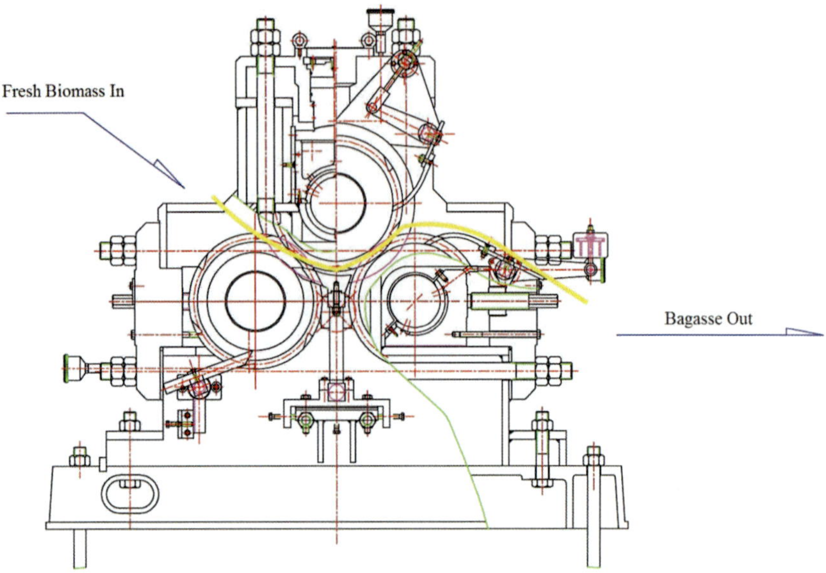

Fig. 8.3.3-1A. Roller press dehydrator design and assemblage.

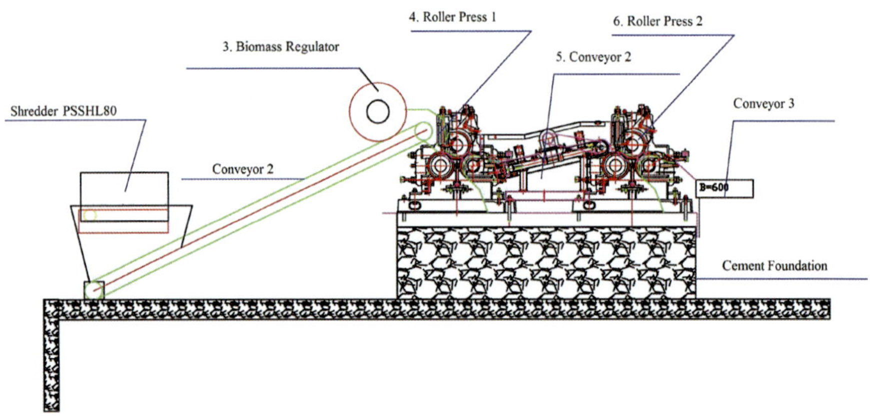

Fig. 8.3.3-1B. Design and assemblage of water hyacinth working line on-shore.

contains 1.4–2.1% of organic matter, 800–1200 mg N L^{-1} and 110–170 mg P L^{-1} (Du, Chang et al. 2010), which become pollutants if directly discharged to the surface waters. To solve the problems of large quantities of water hyacinth juice and bagasse, the ecological engineering project designed equipment for water hyacinth juice fermentation, solid state (bagasse) fermentation and organic fertilizer production; a possibility of using bagasse for silage to produce animal feed was also tested.

8.3.4 Utilization of water hyacinth

The ecological engineering project was a pilot project to test the integrated technology regarding the system design and practice in order to understand (i) how every single step impacts on aquatic ecosystem, (ii) the effects on water quality and (iii) possible ways to utilize the water hyacinth biomass. Although the water content of water hyacinth biomass did not impede the production of methane in the project, higher water content would mean lower content of total solid materials, meaning that the efficiency of fermentation equipment would be reduced, and the cost of biogas production would be increased. The lower the water content, the better the quality of the biomass for fermentation. For this reason, fermentation of water hyacinth biomass was split into liquid (juice) and solid state (bagasse) fermentation. During the execution of the project, the utilization of water hyacinth juice and bagasse focused on water hyacinth juice fermentation, solid state (bagasse) fermentation for organic fertilizers and silage, and utilization of fermentation.

Water hyacinth juice and bagasse fermentation

This process was mainly to produce methane. Research on the fermentation of water hyacinth juice revealed that a continuously stirred tank reactor could reduce hydraulic retention time to 2.5 days, which increased efficiency 10-fold compared to a batch fermentation reactor using non-dehydrated biomass (Hu et al. 2008), achieving a yield of 1.1 m^3 gas per cubic meter of the reactor per day, with a potential capacity of 219 liters of gas production per kilogram of total oxygen demand (Ye et al. 2010).

The bagasse fermentation process had a potential capacity to produce 398 liters of biogas per kilogram of total solids, with 59% of gas being methane (Ye et al. 2011).

Production of organic fertilizer

Dry water hyacinth biomass contains 3–4% of nitrogen and 0.4–0.5% of phosphorus in addition to other plant nutrients (Poddar et al. 1991). After dehydration to reduce the water content of fresh biomass to less than 75%, a faster rise in fermentation temperature, higher emission of carbon dioxide and lower emission of nitrous oxide were achieved (Shi et al. 2012). Good quality of organic fertilizer was also achieved.

The project examined the effects of (i) variable water content of water hyacinth biomass on fermentation, (ii) agents to prevent nutrient losses, and (iii) various fermentation methods such as mixing bagasse with crop stalks and other solid waste (Wang et al. 2011). The technology was optimized for water hyacinth composting, especially a rapid increase in heap temperature and shortened fermentation time to reach compost maturity.

Production of silage for animal feed

Water hyacinth biomass contains about 15.9 to 23.6% of crude protein (Abdelhamid and Gabr 1991). For green feed or silage, water content of 70% in the water hyacinth biomass was suitable (Jiang et al. 2011), but less than 60% was optimal. The project tested special feed formula for herbivores and omnivores such as lambs (Bai et al. 2010) and fattening geese (Bai, Zhu et al. 2011). In another study, an average goat could intake 2.2 kg of formulated water hyacinth silage per day and could gain up to 122 g or more in weight per day (Bai, Zhou et al. 2011). An average goose could intake 744 grams per day and could gain up to 26 g or more in weight per day (Bai, Zhu et al. 2011). It is clear that dehydrated fresh biomass of water hyacinth has great potential for being used as animal feed for herbivores and omnivores.

Dehydration technology enables various commercial uses of the water hyacinth fresh biomass. However, lowering water content of the biomass may be costly. Ideally, the cost of dehydration and the quality of bagasse should be balanced according to its intended utilization.

8.4 Phytoremediation demonstration project using water hyacinth

Two demonstration locations were selected: one at Lake Taihu and another at Lake Dianchi.

8.4.1 Demonstration project in Lake Taihu

Background of ecological engineering project in Lake Taihu

Lake Taihu (30°55'–31°32' N, 119°52'–120°26' E) is located in the Yangtze Delta plain with a water surface area of 2,250 square kilometers and an average depth of 2 meters; it is the third largest freshwater lake in China (Fig. 8.4.1-1).

Lake Taihu is linked to the renowned Grand Canal and is the origin of a number of watercourses, including Suzhou Creek. The Lake Taihu catchment is home to a population of 44 million and had a gross domestic product output of US$336.5 billion in 2005 (Zheng et al. 2008). Lake Taihu has been plagued by pollution as a result of rapid economic development and a population increase in recent years. According to the China National Water Quality Standard (MEP-PRC 2002), eutrophic water was covering about 73% the total water area, with total nitrogen exceeding 2.8 mg L^{-1}, total phosphorus exceeding 0.058 mg L^{-1} in 2011 (Kang et al. 2012), blue-green algae blooms occurring along 1000 meters of shoreline and alga biomass on water surface up to 100 mm thick at some lee sides (Han et al. 2009).

The outbreak of cyanobacteria algal bloom in Taihu Lake in 2007 threatened the domestic water supply for 4.6 million people in Wuxi City and triggered the establishment of a demonstration ecological engineering project using

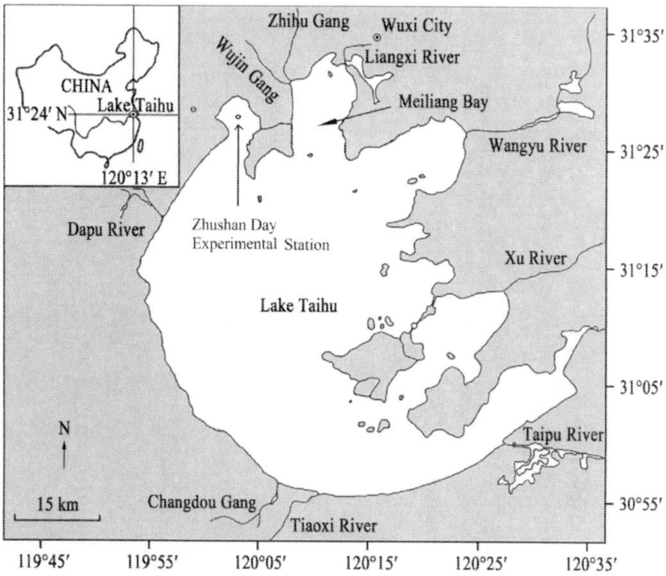

Fig. 8.4.1-1. Lake Taihu in Jiangsu Province, China has 73% surface area covered with eutrophic water.

water hyacinth for remediation in 2010. The demonstration project was established at Zhushan Bay in Lake Taihu and was designed to confine the growth of water hyacinth to 200 hectares and to completely harvest and dehydrate water hyacinth fresh biomass using specially designed heavy-duty machines together with (i) a methane fermentation tank with capacity of 2000 m^3 d^{-1} for using water hyacinth juice from dehydration and (ii) facility for organic fertilizer production of 10,000 t yr^{-1} using the dehydrated biomass. This fertilizer was to be applied on 137 hectares of peach tree orchards and 77 hectares of vegetable farmland. The confined growth of water hyacinth was aimed at removing nitrogen and phosphorus and preventing the ecological risks at the same time. The project was designed to harvest and dehydrate 600 t of fresh water hyacinth biomass per day, with the integrated technology and management system. Juice and bagasse from dehydration was used for production of methane and organic fertilizer. During the project, the water hyacinth population was monitored by a satellite system together with regular GPS surface water checking; in addition, water quality, macrozoobenthos and zooplankton at the site were regularly sampled and monitored.

Confined growth of water hyacinth at Zhushan Bay

The local climate of Lake Taihu is unsuitable for the natural growth of water hyacinth, but winter survival of water hyacinth was observed. It was necessary to protect the plantlets over winter on a large scale to allow the normal growth in early May. The facility used at Zhushan Bay was an anchor-based steel frame floating structure. According to the design, the initial population of the water hyacinth was 6000 tonnes scattered in 30 enclosures (Fig. 8.4.1-2).

Harvest took place from the end of September to the end of December each year; the designed yield was 540 tonnes ha^{-1} yr^{-1}. The harvest and dehydration capacities of 220,000 tonnes were designed as double the designed yield to ensure timely (90 days) and complete harvests and post-harvest processing of fresh biomass.

Fig. 8.4.1-2. Enclosure set up in demonstration project at Zhushan Bay in Lake Taihu.

Mechanical harvesting of water hyacinth at Zhushan Bay

The local climate makes the water hyacinth grow very slowly by the end of September and begin to die by the end of December; hence the harvest of fresh biomass started at the end of September and finished by the end of December. According to the project design, fresh biomass was harvested at 1000 tonnes every day by three integrated multi-functioning harvest vessels (Model HF226B-GP: FS) and six transportation boats (200-tonne capacity each). From 2010 to 2014, 258 t of nitrogen and 38 t of phosphorus were removed from the Lake Taihu each year.

Water hyacinth dehydration

Dehydration of fresh biomass was integrated into the system management. According to the design, three heavy duty dehydrators were set up on site with two dehydrators working and one dehydrator in a stand-by or maintenance mode. From 2009 to 2014, the dehydrators achieved the designed target of 100,000 tonnes fresh biomass each year.

Utilization of water hyacinth at Zhushan Bay

Dehydration of water hyacinth fresh biomass made the water hyacinth juice and bagasse suitable for production of methane, organic fertilizer and silage. During the project, 60,000 tonnes of organic fertilizer were produced and applied to the vegetable fields. Although the data for methane production were incomplete, all the residual liquid from methane production was applied to the vegetable fields as liquid fertilizer, and solid sludge as well as water hyacinth organic fertilizers were applied on various crops such as rice, vegetables and fruit trees (Chen et al. 2011, Liu et al. 2011, Feng et al. 2011, Wang et al. 2013a). Compared with the chemical fertilizer and chicken manure treatment, the organic fertilizer application resulted in quality improvements such as an increase in content of vitamin C by 52 to 59% in Chinese cabbage (*Brassica rapa*) and a decrease in nitrate content in green leafy vegetables (Xue et al. 2011).

8.4.2 Demonstration project in Lake Dianchi

Background of ecological engineering project in Lake Dianchi

Lake Dianchi (24°40'–25°01' N, 102°35'–102°46' E) is a rift-plateau freshwater lake at an elevation of 1,886 meters and adjacent to Kunming City, Yunnan Province, China (Fig. 8.4.2-1). It has a catchment area of 2940 km^2, water surface area of 311 km^2, about 200 kilometers of shoreline, with an average depth of 4.7 meters, the highest depth of 8 meters, and storage capacity of 1.57 billion cubic meters.

The water temperature of the lake ranges from 9.8°C to 26.5°C, with an annual average of about 16.0°C; the mean annual precipitation is about 1070 mm. There were 29 streams flowing into the lake with drinking water quality in 1975 (Institute of Environmental Science 1992, MEP-PRC 2002) that got degraded to eutrophic waters in 2009 because of a rapid increase in local population to 6.8 million in Kunming City resulting in massive discharging of treated or even untreated sewage. Lake Dianchi is in a very severe state of ultra-eutrophication that has been exacerbating over the past 30 years (Wang et al. 2009). Once the primary source of water for domestic and industry utilization and a fully-functioning aquatic ecosystem, the lake has become associated not only with frequent occurrences of toxic cyanobacteria (blue-green algae) blooms, decreased biodiversity and degraded ecosystem services, but also with diminishing capacity to supply drinking water and with water use now being limited to tree nursery purposes only (Huang et al. 2010).

In order to remediate the lake, several projects have been executed, including improving the treatments of municipal and industrial sewage as to

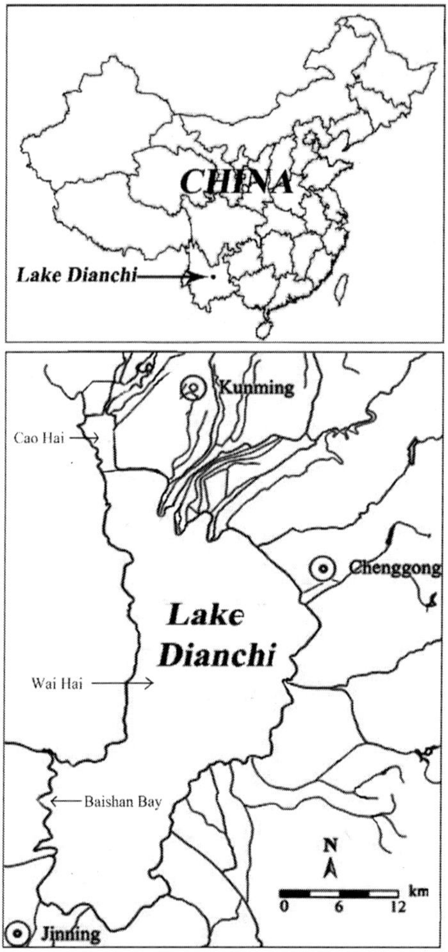

Fig. 8.4.2-1. Lake Dianchi project areas.

control external nutrient loads, sediment dredging and curing, and restoration of submerged and emerging macrophytes to reduce internal nutrient loads during the past decades. However, the nutrients concentrations in the lake remain unacceptably high, e.g., 13 mg total nitrogen L^{-1}, 1.3 mg total phosphorus L^{-1}, COD_{Mn} 8.4 mg L^{-1} and transparency 0.6 m in 2009 (Zhang et al. 2014). To understand the eutrophication management in Lake Dianchi, a pilot ecological engineering project using water hyacinth for phytoremediation was implemented in 2010. The project was designed to confine growth of water hyacinth and to completely harvest and dehydrate water hyacinth fresh biomass using specially designed heavy-duty machines together with a facility to process 50,000 tonnes of bagasse for organic fertilizer per year (Wang et al. 2012, Wang et al. 2013c).

Basic information on demonstration project in Lake Dianchi

The objective of the demonstration project in Lake Dianchi was different from that in Lake Taihu. The project was designed to remove both endogenous and exogenous pollution at Cao Hai and endogenous pollution at heavily eutrophic bay in Wai Hai. Cao Hai received about 30% of entire pollution from the catchment with only 3% of water surface area (streams brought in 100 million cubic meters of polluted water, whereas the water volume of Cao Hai is about 25 million cubic meters), while Wai Hai (although heavily eutrophic) was relatively protected.

The strategies of phytoremediation and management of eutrophication were based on the same principles as in Lake Taihu, but were intensified in Cao Hai with a coverage area up to about 50% because the calculation indicated that such extensive coverage was required for removing excessive pollution. From 2011 to 2014, confined growth of water hyacinth totaled 29.2 km^2 in Lake Dianchi. Harvested fresh biomass of water hyacinth amounted to a total of 1.3 million tonnes, with 1446 tonnes of nitrogen and 130 tonnes phosphorus being removed. The organic fertilizer production from bagasse was 50,000 tonnes. During the project, the water quality in Cao Hai was significantly improved.

Confined growth of water hyacinth

Growth of water hyacinth was confined using a post-fixed structure at Cao Hai and an anchor-based steel frame floating structure at Wai Hai due to different water depths and wave actions. The locations were selected based on possible accumulation of algae, stream mouths, endogenous pollutant sources and minimal wind influence.

During the project, the cumulative confined growth area was 12.2 km^2 at Cao Hai and 17.0 km^2 at Wai Hai. The water hyacinth population was monitored by a satellite system and GPS water surface checking (Fig. 8.4.2-2). The initial plant biomass was 30 tonnes per hectare (225 tonnes per enclosure), and the annual yield was 550 tonnes fresh biomass at Cao Hai and 375 tonnes fresh biomass at Wai Hai. During the project, a total of 671,000 tonnes of fresh biomass were harvested from Cao Hai and 595,000 tonnes from Wai Hai.

Fixed dock harvest engineering

Water hyacinth was harvested using a fixed-dock harvest system at Cao Hai and an integrated multi-functional harvest vessel at Wai Hai. One fixed-dock harvest system processed 600 tonnes of fresh biomass per day. The project comprised five fixed-dock harvest systems to ensure timely and complete harvests of all fresh biomass.

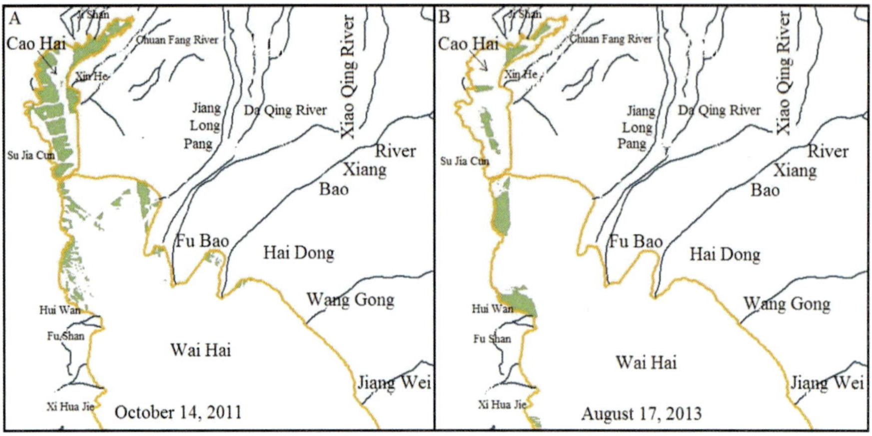

Fig. 8.4.2-2. Water hyacinth distribution at Cao Hai and Wai Hai in Lake Dianchi as assessed by satellite monitoring.

Dehydration of water hyacinth fresh biomass

For large quantities of harvested water hyacinth fresh biomass (3000 t d^{-1}), the project built five dehydration systems to ensure complete dewatering of daily harvested biomass.

Utilization of water hyacinth

The objective of the demonstration project was not only using water hyacinth for phytoremediation, but establishing an integrated technology and management system. After dewatering of harvested fresh biomass, the bagasse only contained less than 70% water and was suitable for production of organic fertilizer. From 2010 to 2014, a total of 50,000 tonnes of organic fertilizer were produced and applied to 670 hectares of horticultural fields.

Water quality as influenced by the demonstration project

From 2011 to 2014, growth of water hyacinth was confined to 12.2 km^2, 752,000 tonnes of water hyacinth fresh biomass was harvested, removing 738 tonnes of nitrogen and 72 tonnes of phosphorus. At the end of the project, averaged total nitrogen concentration in water was 7.2 mg L^{-1} (decreased by 47%), ammonium 9.1 mg L^{-1} (decreased by 77%) and phosphorus 1.0 mg L^{-1} (decreased by 70%).

The detailed assessment of the effects of water hyacinth on the macrozoobenthos and zooplankton in the two above-mentioned demonstrations projects is reported in the next chapter.

References cited

Abdelhamid, A. M. and A. A. Gabr. 1991. Evaluation of water hyacinth as a feed for ruminants. *Archiv für Tierernaehrung* 41(7-8): 745–756.

Ansari, A. A., S. S. Gill and F. A. Khan. 2011. Eutrophication: threat to aquatic ecosystems. In: Eutrophication: causes, consequences and control, ed. A. A. Ansari, S. Singh Gill, G. R. Lanza, and W. Rast, 143–170. Dordrecht, Netherlands: Springer Netherlands.

Babourina, O. and Z. Rengel. 2011. Nitrogen removal from eutrophicated water by aquatic plants. In: Eutrophication: causes, consequences and control, ed. A. A. Ansari, S. S. Gill, G. R. Lanza, and W. Rast, 355–372. Dordrecht, Netherlands: Springer Netherlands.

Bai, Y., W. Zhou, S. Yan, J. Liu, H. Zhang and L. Jiang. 2010. Effects of feeding water hyacinth silage on lambs fattening. *Jiangsu Journal of Agricultural Sciences* 26(5): 1108–1110 (In Chinese with English Abstract).

Bai, Y., W. Zhou, S. Yan, J. Liu, H. Zhang and L. Jiang. 2011. Ensilaging water hyacinth: effects of water hyacinth compound silage on the performance of goat. *Chinese Journal of Animal Nutrition* 23(2): 330–335 (In Chinese with English Abstract).

Bai, Y., J. Zhu, S. Yan, W. Wang, J. Liu, Y. Tu et al. 2011. Effects of feeding water hyacinth silage on fattening geese in large scale. *China Poultry* 33(7): 49–52 (In Chinese).

Bicudo, D. D. C., B. M. Fonseca, L. M. Bini, L. O. Crossetti, C. E. Bicudo and T. Araújo-Jesus. 2007. Undesirable side-effects of water hyacinth control in a shallow tropical reservoir. *Freshwater Biology* 52(6): 1120–1133.

Chen, L., H. Liu, X. Sun, J. Sheng and Y. Lu. 2011. Effect of hyacinth mulching on kalium uptake and utilization of rice (*Oryza sativa* L.). *Chinese Agricultural Science Bulletin* 27(9): 65–71 (In Chinese with English Abstract).

Cifuentes, J. and L. O. Bagnall. 1976. Pressing characteristics of water hyacinth. *Journal of Aquatic Plant Management* 14(0): 71–75.

Dagno, K., R. Lahlali, M. Diourté and M. H. Jijakli. 2012. Present status of the development of mycoherbicides against water hyacinth: successes and challenges. a review. *Biotechnologie Agronomie Societe Environment* 16(3): 360–368.

Dellarossa, V., J. Céspedes and C. Zaror. 2001. *Eichhornia crassipes*-based tertiary treatment of Kraft pulp mill effluents in Chilean Central Region. *Hydrobiologia* 443: 187–191.

Díaz, A. C., V. G. Atencio and S. C. Pardo. 2014. Assessment of an artificial free-flow wetland system with water hyacinth (*Eichhornia crassipes*) for treating fish farming effluents. *Revista Colombiana de Ciencias Pecuarias* 27: 202–210.

Dou, H. S., P. M. Pu, S. Z. Zhang, W. P. Hu and Y. Pang. 1995. An experimental study on culture of *Eichhornia crassipes* (Mart.) Solms on open area of Taihu Lake. *Journal of Plant Resources and Environment* 4(1): 54–60 (In Chinese with English Abstract).

Du, J., Z. Chan, H. Huang, X. Ye, Y. Ma, Y. Xu et al. 2010. Optimization of dewatering parameters for water hyacinth (*E. crassipes*). *Jiangsu Agricultural Sciences* 2(0): 267–269 (In Chinese).

Du, J., Z. Z. Chang, X. Ye and H. Huang. 2010. Losses in nitrogen, phosphorus and potassium of water hyacinth dehydrated by mechanical press. *Fujian Journal of Agricultural Science* 25(1): 104–107 (In Chinese with English Abstract).

Feng, H., Y. Xue, Z. Shi and S. Yan. 2011. Effects of water hyacinth fermentation biogas fluid on ascorbate-glutathione cycle In Chinese cabbage (*Brassica chinensis* L.). *Jiangsu Journal of Agricultural Sciences* 27(2): 301–306 (In Chinese with English Abstract).

Fox, L. J., P. C. Struik, B. L. Appleton and J. H. Rule. 2008. Nitrogen phytoremediation by water hyacinth (*Eichhornia crassipes* (Mart.) Solms). *Water, Air, and Soil Pollution* 194: 199–207.

Gettys, L. A., W. T. Haller and M. Bellaud. 2009. *Biology and control of aquatic plants: a best management practices handbook*. 2nd ed. Marietta GA, USA: Aquatic Ecosystem Restoration Foundation.

Giraldo, E. and A. Garzón. 2002. The potential for water hyacinth to improve the quality of Bogota River water in the Muña Reservoir: comparison with the performance of waste stabilization ponds. *Water Science & Technology* 45(1): 103–110.

Greenfield, B. K., G. S. Siemering, J. C. Andrews, M. Rajan, S. P. Andrews Jr. and D. F. Spencer. 2007. Mechanical shredding of water hyacinth (*Eichhornia crassipes*): effects on water quality in the Sacramento-San Joaquin River Delta, California. *Estuaries and Coasts* 30(4): 627–640.

Gunnarsson, C. C. and C. M. Petersen. 2007. Water hyacinths as a resource in agriculture and energy production: a literature review. *Waste Management* 27(1): 117–29.
Han, S., S. Yan, Z. Wang, W. Song, H. Liu, J. Zhang et al. 2009. Harmless disposal and resources utilizations of Taihu Lake blue algae. *Journal of Natural Resources* 24(3): 431–438 (In Chinese with English Abstract).
Heisler, J., P. M. Glibert, J. M. Burkholder, D. M. Anderson, W. Cochlan, W. C. Dennison et al. 2008. Eutrophication and harmful algal blooms: a scientific consensus. *Harmful Algae* 8(1): 3–13.
Hronich, J. E., L. Martin, J. Plawsky and H. R. Bungay. 2008. Potential of *Eichhornia crassipes* for biomass refining. *Journal of Industrial Microbiology and Biotechnology* 35(5): 393–402.
Hu, L., W. Hu, J. Deng, Q. Li, F. Gao, J. Zhu et al. 2010. Nutrient removal in wetlands with different macrophyte structures in eastern Lake Taihu, China. *Ecological Engineering* 36(12): 1725–1732.
Hu, X., G. Zha, W. Zhang, F. Yi and R. Xu. 2008. Experimental study on mesophilic biogas fermentation with juice of *Eichhornia crassipes*. *Energy Engineering* 2: 36–38 (In Chinese with English Abstract).
Huang, Y., H. Wen, J. Cai, M. Cai and J. Sun. 2010. Key aquatic environmental factors affecting ecosystem health of streams in the Dianchi Lake watershed, China. *Procedia Environmental Sciences* 2: 868–880.
Institute of Environmental Science. 1992. *Survey on eutrophication of Lake Dianchi*. 1st ed. Kunming, China: Kunming Science and Technology Press.
Jayaweera, M. W. and J. C. Kasturiarachchi. 2004. Removal of nitrogen and phosphorus from industrial wastewaters by phytoremediation using water hyacinth (*Eichhornia crassipes* (Mart.) Solms). *Water Science and Technology* 50(6): 217–225.
Jiang, H. and H. Zhang. 2003. A review of controlling common waterhyacinth at home and abroad. *Journal of Agricultural Science and Technology* 5(3): 72–74 (In Chinese with English Abstract).
Jiang, L., Y. Bai, S. Yan, H. Zhang, J. Liu and L. Tu. 2011. Effects of additives on dehydrated water hyacinth for silage. *Jiangsu Agricultural Sciences* 39(6): 337–340 (In Chinese).
Jiang, R. 2011. Nutritional values of water hyacinth residue silage by high moisture and evaluate the safety of animal organization on heavy metal. MSc. Thesis, Department of Animal Nutrients and Feed Science. Nanjing Agricultural University.
Kang, Z., J. Tang and H. Liu. 2012. Eco-compensation based on cost-benefit analysis of nitrogen and phosphorus removal in Taihu Lake region. In: Advances in China farming system research 2012, ed. Farming System Committee, Chinese Agricultural Association, 182–187 p. (In Chinese). Beijing, China: China Agricultural Science and Technology Press.
Koottatep, T. and C. Polprasert. 1997. Role of plant uptake on nitrogen removal in constructed wetlands located in the tropics. *Water Science and Technology* 36(12): 1–8.
Kulkarni, B. V., S. V. Ranade and A. I. Wasif. 2007. Phytoremediation of textile process effluent by using water hyacinth—a polishing treatment. *Journal of Industrial Pollution Control* 23(1): 97–101.
Li, B., C. Liao, L. Gao, Y. Luo and Z. Ma. 2004. Strategic management of water hyacinth (*Eichhornia crassipes*), an invasive alien plant. *Journal of Fudan University (Natural Science)* 43(2): 267–274 (In Chinese with English Abstract).
Li, J., Y. Zhang, Y. Wang, M. Liu, D. Pan, J. Li et al. 2003. A preliminary study on decontaminating waste water of slaughter industry by planting hyacinth in northern China. *Journal of Shengyang Agricultural University* 34(2): 103–105 (In Chinese with English Abstract).
Lindsey, K. and H. M. Hirt. 2000. *Use water hyacinth—a practical handbook of uses for water hyacinth from across the world*. Winnenden, Germany: Anamed International.
Little, E. C. S. 1979. *Handbook of utilization of aquatic plants*. Rome, Italy: Food and Agriculture Organization of The United Nations.
Liu, H., L. Chen, P. Zhu, J. Sheng, Y. Zhang and J. Zheng. 2011. Effect of hyacinth mulching on rice (*Oryza sativa* L.) uptake and utilization of nitrogen. *Environmental Science* 32(5): 1292–1298 (In Chinese).
Lu, J., Z. Fu and Z. Yin. 2008. Performance of a water hyacinth (*Eichhornia crassipes*) system in the treatment of wastewater from a duck farm and the effects of using water hyacinth as duck feed. *Journal of Environmental Sciences* 20(5): 513–9.

Mahujchariyawong, J. and S. Ikeda. 2001. Modelling of environmental phytoremediation in eutrophic river—the case of water hyacinth harvest in Tha-chin River, Thailand. *Ecological Modelling* 142(1-2): 121–134.

Malar, S., S. V. Sahi, P. J. C. Favas and P. Venkatachalam. 2014. Mercury heavy-metal-induced physiochemical changes and genotoxic alterations in water hyacinths [*Eichhornia crassipes* (Mart.)]. *Environmental Science and Pollution Research* 22(6): 4597–4608.

Malik, A. 2007. Environmental challenge vis a vis opportunity: the case of water hyacinth. *Environment International* 33(1): 122–38.

Mathur, S. M. and P. Singh. 2004. Development and performance evaluation of a water hyacinth chopper cum crusher. *Biosystems Engineering* 88(4): 411–418.

MEP-PRC. 2002. Environmental quality standards for surface water (GB3838-2002). Baijing, China: Ministry of Environmental Protection of The Peoples's Republic of China.

Mishra, V. K. and B. D. Tripathi. 2009. Accumulation of chromium and zinc from aqueous solutions using water hyacinth (*Eichhornia crassipes*). *Journal of Hazardous Materials* 164(2-3): 1059–63.

Montoya, J. E., T. M. Waliczek and M. L. Abbott. 2013. Large scale composting as a means of managing water hyacinth (*Eichhornia crassipes*). *Invasive Plant Science and Management* 6(2): 243–249.

Navarro, L. and G. Phiri. 2000. *Water hyacinth in Africa and the Middle East: a survey of problems and solutions*. 1st ed. Ottawa, Canada: International Development Research Centre.

Ndimele, P. E., C. A. Kumolu-Johnson and M. A. Anetekhai. 2011. The invasive aquatic macrophyte, water hyacinth (*Eichhornia crassipes*) (Mart.) Solm-Laubach: Pontedericeae: problems and prospects. *Research Journal of Environmental Sciences* 5(6): 509–520.

Ning, D., Y. Huang, R. Pan, F. Wang and H. Wang. 2014. Effect of eco-remediation using planted floating bed system on nutrients and heavy metals in urban river water and sediment: a field study in China. *The Science of the Total Environment* 485-486: 596–603.

Osmond, R. and A. Petroeschevsky. 2013. *Water hyacinth - control modules*. Final V2. ORANGE NSW 2800, Australia: NSW Department of Primary Industries.

Perna, C. and D. Burrows. 2005. Improved dissolved oxygen status following removal of exotic weed mats in important fish habitat lagoons of the tropical Burdekin River floodplain, Australia. *Marine Pollution Bulletin* 51(1-4): 138–148.

Poddar, K., L. Mandal and G. C. Banerjee. 1991. Studies on water hyacinth (*Eichhornia crassipes*) – chemical composition of the plant and water from different habitats. *Indian Veterinary Journal* 68: 833–837.

Polomski, R. F., M. D. Taylor, D. G. Bielenberg, W. C. Bridges, S. J. Klaine and T. Whitwell. 2009. Nitrogen and phosphorus remediation by three floating aquatic macrophytes in greenhouse-based laboratory-scale subsurface constructed wetlands. *Water, Air, and Soil Pollution* 197(1-4): 223–232.

Qi, Y. and W. Gao. 1999. About the purifying water quality by *Eichhornia crassipes* and the after treatment technology for *Eichhornia crassipes*. *Advances in Environmental Science* 7(2): 136–140 (In Chinese with English Abstract).

Reddy, K. R., K. L. Campbell, D. A. Graetz and K. M. Portier. 1982. Use of biological filters for treating agricultural drainage effluents. *Journal of Environmental Quality* 11(4): 591–595.

Rodríguez, M., J. Brisson, G. Rueda and M. S. Rodríguez. 2012. Water quality improvement of a reservoir invaded by an exotic macrophyte. *Invasive Plant Science and Management* 5(2): 290–299.

Schultz, R. and E. Dibble. 2011. Effects of invasive macrophytes on freshwater fish and macroinvertebrate communities: the role of invasive plant traits. *Hydrobiologia* 684(1): 1–14.

Shi, L. L., M. X. Shen, Z. Z. Chang, H. H. Wang, C. Y. Lu, F. S. Chen et al. 2012. Effect of water content on composition of water hyacinth *Eichhornia crassipes* (Mart.) Solms residue and greenhouse gas emission. *Chinese Journal of Eco-Agriculture* 20(3): 337–342 (In Chinese with English Abstract).

Sinha, A. K. and R. K. Sinha. 2000. Sewage management by aquatic weeds (water hyacinth and duckweed): economically viable and ecologically sustainable bio-mechanical technology. *Environmental Education and Information* 19(3): 215–226.

Soltan, M. E. and M. N. Rashed. 2003. Laboratory study on the survival of water hyacinth under several conditions of heavy metal concentrations. *Advance in Environmental Research* 7: 321–334.

Tchobanoglous, G., F. Maitski, K. Thompson and T. H. ChadwickSource. 1989. Evolution and performance of city of San Diego pilot-scale aquatic wastewater treatment system using water hyacinths. *Research Journal of the Water Pollution Control Federation* 61(11/12): 1625–1635.

UNEP and GEAS. 2013. Water hyacinth – can its aggressive invasion be controlled? *Environmental Development* 7: 139–154.

USEPA. 2000. Agricultural nonpoint source pollution. In: National management measures for the control nonpoint pollution from agriculture, 9–30. Washington, D.C. USA: EPA-841-B-03-00. Office of Water (4503T), U.S. Environmental Protection Agency.

Villamagna, A. M. and B. R. Murphy. 2010. Ecological and socio-economic impacts of invasive water hyacinth (*Eichhornia crassipes*): a review. *Freshwater Biology* 55(2): 282–298.

Wang, F. S., C. Q. Liu, M. H. Wu, Y. X. Yu, F. W. Wu, S. L. Lu et al. 2009. Stable isotopes in sedimentary organic matter from Lake Dianchi and their indication of eutrophication history. *Water, Air, & Soil Pollution* (199): 159–170.

Wang, G., P. Pu, S. Zhang, W. Li, W. Hu and C. Hu. 1998. The purification of artificial complex ecosystem for local water in Taihu Lake. *China Environmental Science* 18(5): 410–414 (In Chinese with English Abstract).

Wang, H., M. Shen, Z. Chang, C. Lu, F. Chen, L. Shi et al. 2011. Nitrogen loss and technique for nitrogen conservation in high temperature composting of hyacinth. *Journal of Agro-Environment Science* 30(6): 1214–1220 (In Chinese with English Abstract).

Wang, J., Y. Cao, Zh. Chang, Y. Zhang and H. Ma. 2013a. Effects of combined application of biogas slurry with chemical fertilizers on fruit qualities of *Prunus persica* L. and soil nitrogen accumulation risk. *Plant Nutrition and Fertilizer Science* 19(2): 379–386 (In Chinese with English Abstract).

Wang, Q. and S. Cheng. 2010. Review on phytoremediation of heavy metal polluted water by macrophytes. *Environmental Science and Technology* 33(5): 96–102 (In Chinese with English Abstract).

Wang, Y., Z. Zhang, Y. Zhang, X. Weng, X. Wang and S. Yan. 2013b. A new method of dehydration of water hyacinth. *Jiangsu Agricultural Sciences* 41(10): 286–288 (In Chinese).

Wang, Z., Z. Zhang, Y. Han, Y. Zhang, Y. Wang and S. Yan. 2012. Effects of large-area planting water hyacinth (*Eichhornia crassipes*) on water quality in the bay of Lake Dianchi. *Chinese Journal of Environmental Engineering* 6(11): 3827–3832 (In Chinese with English Abstract).

Wang, Z., Z. Zhang, Y. Zhang, J. Zhang, S. Yan and J. Guo. 2013c. Nitrogen removal from Lake Caohai, a typical ultra-eutrophic lake in China with large scale confined growth of *Eichhornia crassipes*. *Chemosphere* 92(2): 177–183.

Wilson, J. R., N. Holst and M. Rees. 2005. Determinants and patterns of population growth in water hyacinth. *Aquatic Botany* 81(1): 51–67.

Wooten, J. W. and J. D. Dodd. 1976. Growth of water hyacinths in treated sewage effluent. *Economic Botany* 30(1): 29–37.

Xia, X. 2012. Study on nitrogen and phosphorus runoff losses from paddy field and control technology in Tai Lake area. Ph.D. Thesis, Nanjing Agricultural University.

Xue, Y., H. Feng, Z. Shi, S. Yan and J. Zheng. 2011. Effect of biogas slurry of *Eichhornia crassipes* on the seedling quality of Chinese cabbage. *Pratacultural Science* 28(4): 687–692.

Yan, S., H. Liu, Z. Zhang, Y. Zhang and G. Liu. 2010. Integrated water hyacinth harvest vessel. China: Patent # ZL201020100681.6.

Yan, S., Z. Zhang, Y. Zhang, X. Yang and H. Liu. 2011. Floating plant fixed harvest system. China: Patent # ZL201120373563.7.

Ye, X., J. Du, Z. Chang, Y. Qian, Y. Xu and J. Zhang. 2011. Anaerobic digestion of solid residue of water hyacinth. *Jiangsu Journal of Agricultural Sciences* 27(6): 1261–1266 (In Chinese with English Abstract).

Ye, X., L. Zhou, S. Yan, Z. Chang and J. Du. 2010. Anaerobic digestion of water hyacinth juice in CSTR reactor. *Fujian Journal of Agricultural Science* 25(1): 100–103 (In Chinese with English Abstract).

Yu, Y. and R. Deng. 2000. Application of *Eichhornia crassipes* aquatic plant system for treatment of sewage from large-scale pig farm. *Agro-environmental Protection* 19(5): 301–303 (In Chinese with English Abstract).

Zaranyika, M. F. and T. Ndapwadza. 1995. Uptake of Ni, Zn, Fe, Co, Cr, Pb, Cu and Cd by water hyacinth (*Eichhornia crassipes*) in mukuvisi and manyame rivers, Zimbabwe. *Journal of Environmental Science and Health* 30(1): 157–169.

Zhang, L., P. Zhu, Y. Gao, Z. Zhang and S. Yan. 2013. Design of anti-stormy wave enclosures for confined growth of water hyacinth in lakes. *Jiangsu Journal of Agricultural Sciences* 29(6): 1360–1364 (In Chinese with English Abstract).

Zhang, Y., Z. Zhang, Y. Wang, H. Liu, Z. Wang, S. Yan et al. 2011b. Research on the growth characteristics and accumulation ability to N and P of *Eichhornia crassipes* in different water areas of Dianchi Lake. *Journal of Ecology and Rural Environment* 27(6): 73–77 (In Chinese with English Abstract).

Zhang, Z., Y. Gao, J. Guo and S. Yan. 2014. Practice and reflections of remediation of eutrophicated waters: a case study of haptophyte remediation of the ecology of Dianchi. *Journal of Ecology and Rural Environment* 30(1): 15–21 (In Chinese with English Abstract).

Zhang, Z., H. Liu, G. Liu and Y. Zhang. 2010. Floating plant shredder. China: Patent # ZL201020100683.5.

Zhang, Z. Y., J. C. Zheng, H. Q. Liu, Z. Z. Chang, L. G. Chen and S. H. Yan. 2010. Role of *Eichhornia crassipes* uptake in the removal of nitrogen and phosphorus from eutrophic waters. *Chinese Journal of Eco-Agriculture* 18(1): 152–157 (In Chinese with English Abstract).

Zhang, Z., J. Zhang, H. Liu, L. Chen and S. Yan. 2011a. Apparent removal contributions of *Eichhornia crassipes* to nitrogen and phosphorous from eutrophic water under diferent hydraulic loadings. *Jiangsu Journal of Agricultural Sciences* 27(2): 288–294 (In Chinese with English Abstract).

Zheng, J., Z. Chan, L. Chen, P. Zhu and J. Shen. 2008. Feasibility studies on N and P removal using water hyacinth in Taihu Lake region. *Jiangsu Agricultural Science* 3: 247–250 (In Chinese).

Zhou, Z. and J. Yang. 1984. The role of waterhyacinth in sewage ecological treatment system and the approaches to their utilization—part 1: the biological characteristics of water hyacinth and the effects of environmental factors on their growth. *Chinese Journal of Ecology* (5): 36–40 (In Chinese with English Abstract).

CHAPTER 9

Impact of Water Hyacinth on Aquatic Environment in Phytoremediation of Eutrophic Lakes

Z. Wang[1] and *S. H. Yan*[2,*]

9.1 Introduction

Water hyacinth has a strong capacity to absorb nutrients, heavy metals and organic pollutants and is an excellent candidate for the water pollution control and eutrophic water restoration. In large lakes, there were few successful examples of submerged or emerging macrophyte restoration. This might be due to: (1) growth of submerged or emerging macrophytes is affected by the depth and transparency of the water; (2) submerged or emerging macrophytes might assimilate nutrients from sediment instead of from water; and (3) macrophyte biomass, if not harvested, decomposes in lakes, and as a result, the assimilated nutrients are returned back to the aquatic ecosystem.

It is mostly nitrogen and phosphorus that are responsible for eutrophication in the aquatic environments (Xu et al. 2010, Smith 2003). Both of these elements are considered to be the control targets for the restoration of aquatic ecosystems (Elser et al. 2007) especially using water hyacinth due to low-cost and easy management (Babourina and Rengel 2011). However, these conclusions were obtained mostly under the experimental conditions or from small scale studies.

[1] 340 Xudong dajie Road, Wuhan, China.
Email: Wazh519@hotmail.com
[2] 50 Zhong Ling Street, Nanjing, China.
* Corresponding author: shyan@jaas.ac.cn

Before using water hyacinth for large scale application on sewage treatment and eutrophic water management, there are some challenges: (1) how to confine water hyacinth in a targeted area in eutrophic lakes or in an open system in sewage treatments, (2) dynamics of water hyacinth growth and propagation; (3) when and how much to harvest and how to dehydrate the fresh biomass; (4) disposal and/or utilization of water hyacinth juice and bagasse; (5) the ecological risk to species diversity and abundance in lakes and other large water bodies on the application of this invasive species; (6) the effects on ecosystem health regarding biogeochemical changes in the environment and water chemistry; and (7) the effects on targeted water quality.

This chapter presents the results on water quality and ecological risk assessment in two ecological engineering projects on phytoremediation of eutrophic lakes: Lake Dianchi (24°40'–25°01' N, 102°35'–102°46' E) and Lake Taihu (30°55'–31°32' N, 119°52'–120°26' E) in China. The challenges, effects and problems in applying water hyacinth for large-scale phytoremediation in these two case studies are discussed.

9.1.1 Background: ecological engineering using water hyacinth for phytoremediation

After two years of preliminary survey in Lake Taihu, an ecological engineering project was established at Zhushan Bay in Lake Taihu in 2010 to investigate the above-mentioned challenges. Lake Taihu is in the early stage heavy eutrophication development due to large number of people in the catchment and rapid development of agriculture and industries (Zheng et al. 2008). According to the China National Water Quality Standard (MEP-PRC 2002), eutrophication is characterized by total nitrogen exceeding 2.8 mg L^{-1} and phosphorus exceeding 0.058 mg L^{-1} (Kang et al. 2012) and blue-green algal blooms occurring frequently (Han et al. 2009).

The survey of eutrophication status of Lake Dianchi started in 1983 and many remediation projects were executed since then (Huang et al. 2010); however, the eutrophication status has not significantly changed (Institute of Environmental Science 1992, MEP-PRC 2002) because a rapid increase in local population and massive amounts of treated or untreated sewage being discharged (Wang et al. 2009, Zhang et al. 2014). A demonstration ecological engineering project using water hyacinth for phytoremediation was implemented in 2010 including areas of Cao Hai and Wai Hain. The project was designed to confine growth of water hyacinth and to completely harvest and dehydrate water hyacinth fresh biomass. This ecological engineering application of the floating macrophyte has been confirmed to improve the water quality in the lake (Wang et al. 2012, Wang et al. 2013).

9.2 Impact of water hyacinth on water quality in Lake Dianchi

9.2.1 Impact of water hyacinth on water quality at Baishan Bay

Lake Dianchi consists of two sections: Cao Hai and Wai Hai. Wai Hai has an area of 286.5 km^2. There is an embankment to separate the waters between the two sections and with a sluice to control water flow. Baishan Bay is located in Wai Hai (Fig. 9.2.1-1) and has a water depth of 2.5 meters.

About a 70-hectare area of water hyacinth was planted in enclosures at Baishan Bay in 2010. The bay had a relatively large population of cyanobacteria due to nutrient loading, local climate characteristics and hydraulic forces.

The project monitored water quality, macrozoobenthos and zooplankton at sites indicated in Fig. 9.2.1-1; the sites were divided in three groups: area covered with water hyacinth (sites 5, 7, 9, 10, 12, 13), near that area (sites 4, 6, 8, 11) and far from it (sites 1–3), with the latter two groups representing controls. Water hyacinth was planted in June 2010 in enclosures and harvested after 20 October 2010. Samples of water, macrozoobenthos and zooplankton were collected at 2-week intervals during the water hyacinth growth season. Water properties were presented in Table 9.2.1-1.

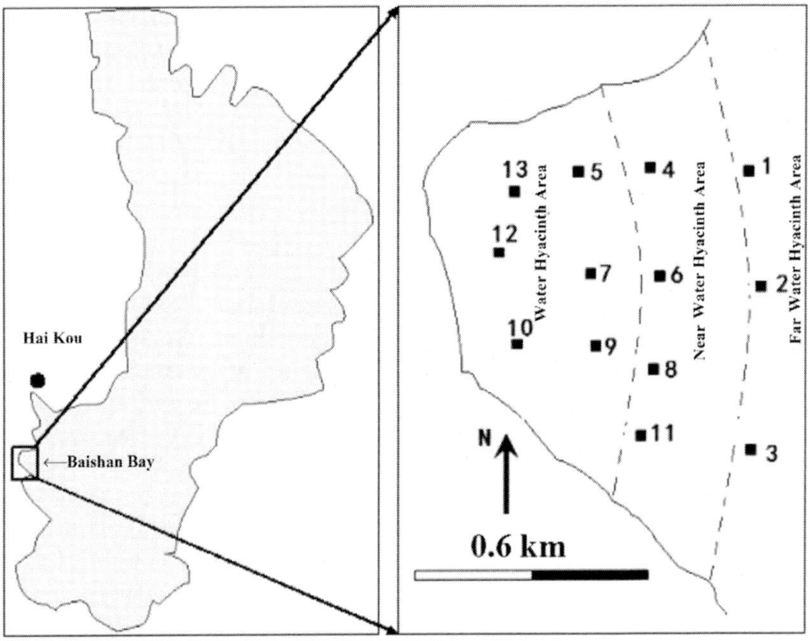

Fig. 9.2.1-1. Sampling sites in the testing area of Baishan Bay (24°45' N, 102°36' E), Lake Dianchi (redrawn with permission).

Table 9.2.1-1. Water quality summary before and after planting water hyacinth.

Sampling	TN mg L^{-1}	NH_4^+ mg L^{-1}	TP mg L^{-1}	COD_{Mn} mg L^{-1}	DO mg L^{-1}	Secchi[a] M	Chl-*a* mg L^{-1}	Ref.
Before Planting 4 May 2010	2.60	0.44	0.26	10.60	8.57	0.34	0.10	Monitoring Data From KCEP[b]
After Planting 20 October 2010	2.40	0.50	0.13	17.50	6.00	0.42	0.07	(Wang et al. 2012)

[a] Transparency measured by Secchi Disk; [b] KCEP refers to Environmental Protection Department of Kunming City.

Water temperature

The water temperature in the area before and after planting water hyacinth showed no significant difference among the three sampling areas.

Dissolved oxygen content, pH and transparency

The basic water chemistry of the Baishan Bay shows slight alkalinity (pH 8–10) due to calcium–magnesium–sodium carbonates (Wang et al. 2010). The results were presented in Fig. 9.2.1-2.

Dissolved Oxygen (DO) in water near and far the water hyacinth area declined in August and then gradually increased. Among the three areas, DO was significantly lower in the water hyacinth area than both near and far away areas ($p < 0.05$), with no significant difference between two areas away

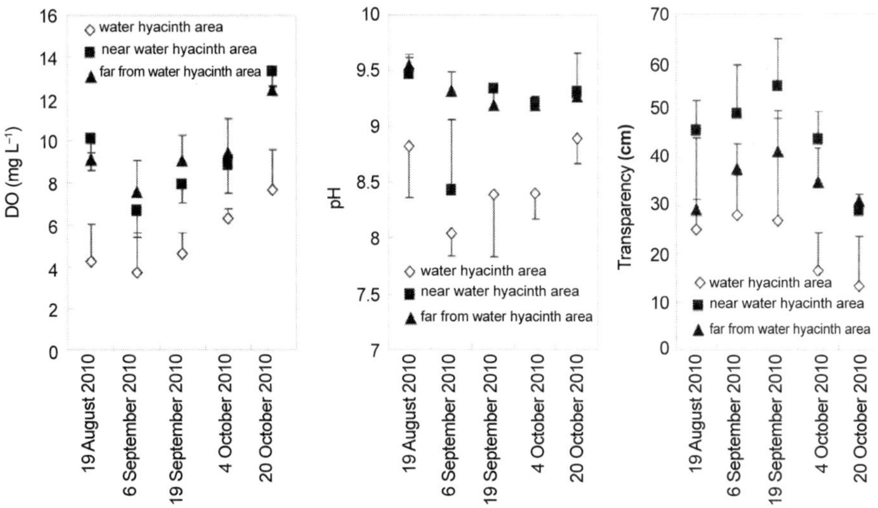

Fig. 9.2.1-2. Changes of dissolved oxygen, pH and transparency in water column of the testing area, vertical bars representing standard error (redrawn with permission).

from water hyacinth. Due to (i) water hyacinth mat being relatively small compared with the open area in the bay and (ii) water exchange caused by frequent winds, the dissolved oxygen content stayed above 4 mg L^{-1} during the water hyacinth growing season (Fig. 9.2.1-2).

In the water hyacinth area and the nearby area, water pH showed a decrease between the August and September sampling followed by an upward trend and stayed stable around 9.3 in the far water hyacinth area. Statistical analysis showed that the pH of the water body was significantly lower in the water hyacinth area than in the areas near and far from water hyacinth ($p < 0.05$). Lowered pH in the water hyacinth mat area was expected and consistent with literature (Dai and Che 1987, Giraldo and Garzón 2002).

Water transparency showed a slow increase in August and September, and then a gradual decline in October; generally, transparency was significantly lower in the water hyacinth area than in the areas near and far ($p < 0.05$). Literature often showed that water hyacinth can increase water transparency because its roots can intercept detritus and blue-green algae (Kim and Kim 2000). The results from Baishan Bay presented a slightly different pattern that might have been caused by the interactions among the size of water hyacinth mat, wind speed (6 m s^{-1}) and quantities of blue-green algae. The hypothesis was that wind brought large amounts of blue-green algae and, while water hyacinth roots intercepted the algae, they were not decomposed quickly enough and thus water transparency decreased. However, this hypothesis needs to be further tested.

Total phosphorus and orthophosphate

The concentrations of Total Phosphorus (TP) and orthophosphate (PO_4^{3-}) showed the same pattern among the three sampling areas: declined in August then increased afterwards, especially in the water hyacinth area (Fig. 9.2.1-3). This phenomenon indicated that water hyacinth assimilation capacity exceeded a contribution of detritus and algae to increasing water phosphorus, but the trend changed due to a rapid rise in the amount of detritus and blue-green algae brought in by wind. Among the three areas, TP was lowest in the area near water hyacinth ($p < 0.05$), showing that water hyacinth has a potential to improve water quality around its mat.

Total nitrogen, ammonium and nitrate

Total nitrogen showed a decline in August and early September, then increased in late September in the water hyacinth area. It was stable during August and September in the areas near and far from water hyacinth, but with the same trends as in the water hyacinth area (Fig. 9.2.1-4).

Among the three areas, the concentration of total nitrogen was highest in the water hyacinth area, suggesting a rapid accumulation of detritus and blue-green algae during the testing period. In aquatic environments, total nitrogen,

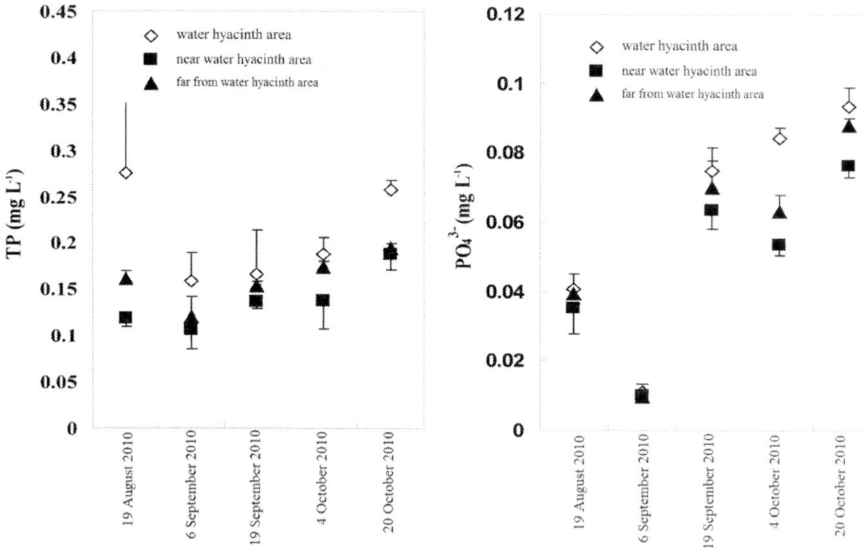

Fig. 9.2.1-3. Changes of total phosphorus and orthophosphate in water column of the testing area (redrawn with permission).

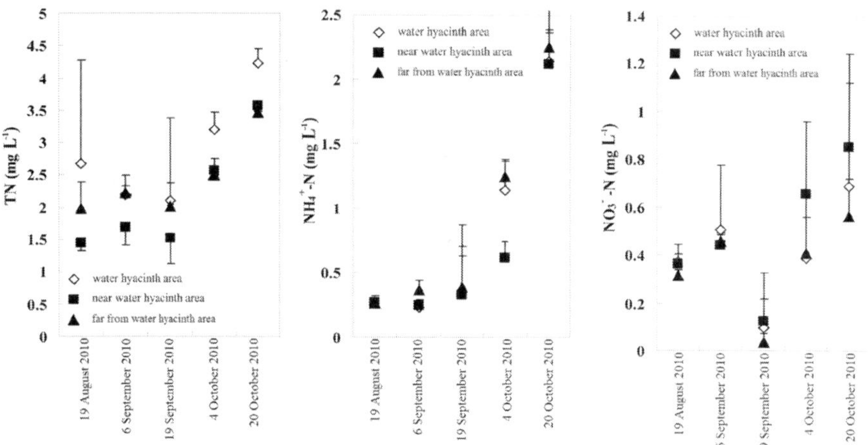

Fig. 9.2.1-4. Changes of total nitrogen, ammonium nitrogen and nitrate nitrogen in water column of the testing area (redrawn with permission).

ammonium and nitrate nitrogen are closely related because mineralization of organic matter (including decomposition of blue-green algae) increases ammonium concentration, which is easily converted to nitrate in aerobic conditions such as during the testing period discussed here. The testing results may suggest that the growth of water hyacinth trapped detritus and blue-green algae to enhance decomposition and removal from endogenous nitrogen pool. Concentration of ammonium nitrogen showed the same pattern in all three areas: stable during August and early September, and increased rapidly afterwards. Nitrate nitrogen increased on 6 September 2010, then declined and increased afterward. The results may suggest a time lag in the total nitrogen -> ammonium -> nitrate transformation, implying that in using phytoremediation to control nitrogen pollution, consideration needs to be taken of the water volume, nitrogen loading, local climate (such as wind strength and direction, and water current) and growth rate of macrophyte in order to achieve desired targets.

Chlorophyll-a and COD_{Mn} index

The results of the changes of chlorophyll-α and COD_{Mn} index were presented in Fig. 9.2.1-5.

In the areas near and far from water hyacinth, chlorophyll-*a* significantly increased in the period from 19 August to 6 September, and then stabilized; in contrast, in the water hyacinth area, it showed a continuous rising trend. This phenomenon further confirmed the above discussion on total nitrogen, ammonium and nitrate nitrogen changes during the testing period. Analysis of variance showed that chlorophyll-*a* concentration in the area near water hyacinth was significantly lower than that in the areas of water hyacinth and far away ($p < 0.05$) (Fig. 9.2.1-5).

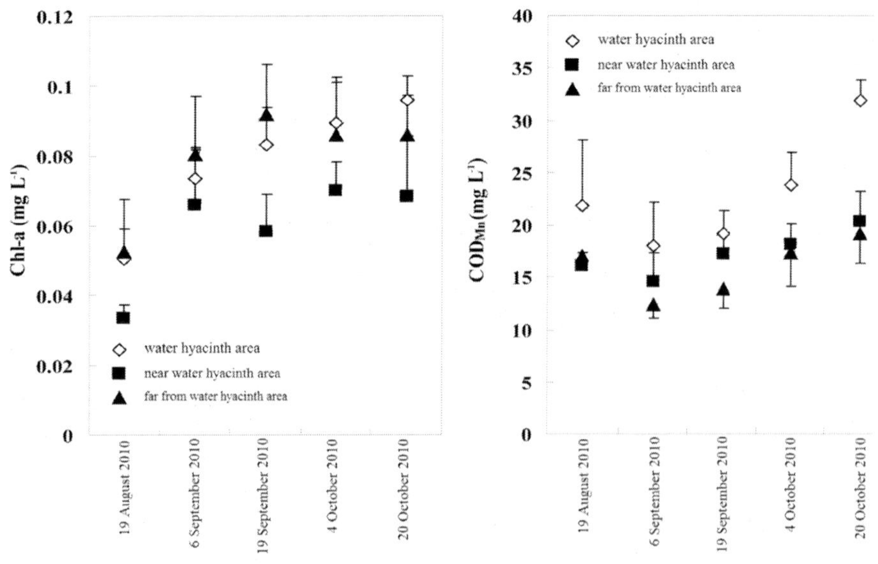

Fig. 9.2.1-5. Changes of chlorophyll-*a* and COD_{Mn} in water column of the testing area (redrawn with permission).

COD_{Mn} had a downward trend in the initial stage and rose later, especially in the water hyacinth area. Among the three regions, the concentration of COD_{Mn} was highest in the water hyacinth area, medium in the near area and lowest in the area far from water hyacinth ($p < 0.05$). Chlorophyll-*a* mainly came from algae, and COD_{Mn} was mainly contributed by small organic molecules and reducing agents. When algae and detritus were decomposed, large amounts of small organic molecules and reducing agents were produced, and nitrogen and other nutrients were released.

9.2.2 Impact of water hyacinth on water quality at Wai Hai

Water quality at Wai Hai

The experiment at Baishan Bay in 2010 had an area of water hyacinth cover of about 70 hectares representing only 0.24% of the total area in Wai Hai, meaning it was difficult to expect significant effects on the physical and chemical properties of the lake. In 2012, the use of water hyacinth for phytoremediation of eutrophic water was extended at the northern part of Wai Hai and the whole of Cao Hai in Lake Dianchi (Fig. 9.2.2-1).

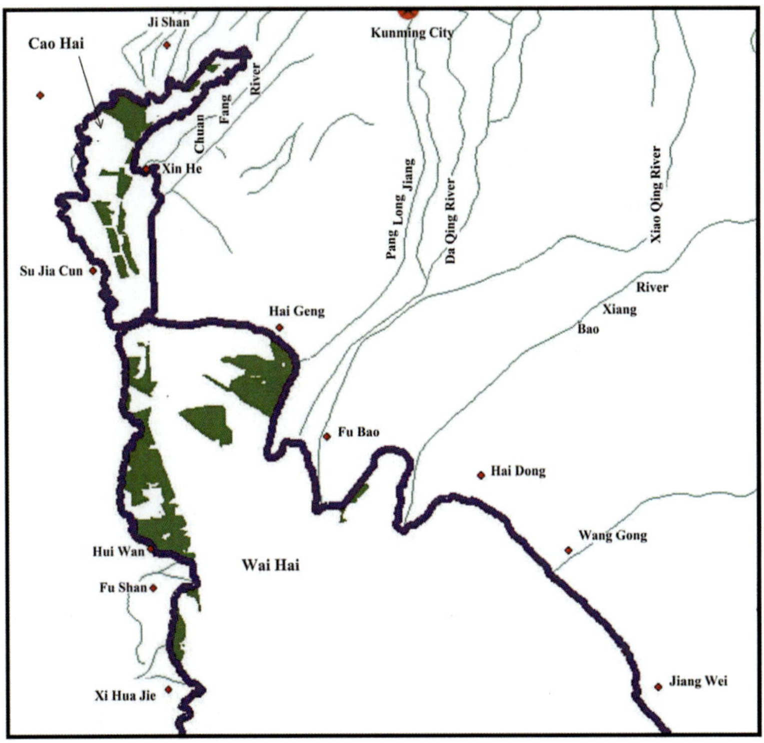

Fig. 9.2.2-1. Water hyacinth distribution at Cao Hai and Wai Hai in Lake Dianchi on 17 August 2012 (redrawn with permission).

The sites were in ultra-eutrophic state with heavily accumulation of blue-green algae. The project planted 621 hectares, or 1.5% of the total area, of water hyacinth in Wai Hai in 2012. The concentrations of nitrogen, phosphorus and ammonium were monitored at eight sites (Fig. 9.2.2-2) by sampling once a month in 2011 (before planting water hyacinth) and 2012 (after planting water hyacinth).

In 2012, the total amount of fresh water hyacinth harvested was 0.32 million tonnes, removing about 567 tonnes of nitrogen. Total nitrogen concentration was reduced from average of 2.8 mg L^{-1} in 2011 down to 2.09 mg L^{-1} in 2012 (about 25% reduction, Table 9.2.2-1).

An increase in ammonium concentration in 2012 was likely due to trapping and decomposition of blue-green algae. The data showed a significant reduction in total nitrogen, non-significant reduction in total phosphorus, and a non-significant increase in ammonium concentration (Table 9.2.2-1).

Fig. 9.2.2-2. Sampling sites in Wai Hai, Lake Dianchi in 2012.

Table 9.2.2-1. Changes in nutrient concentration (mg L^{-1}) from 2006 to 2012; the data were pooled over 12 monthly measurements each year.

Nutrients	Average from 2006 to 2010[a]	Average in 2011[b]	Average in 2012[b]
Total nitrogen	2.62	2.80	2.09
Total phosphorus	0.21	0.16	0.17
Ammonium	0.26	0.25	0.33

[a] Data from Kunming Environmental Monitoring Station; [b] Data from project monitoring.

9.2.3 Impact of water hyacinth on water quality at Cao Hai

The Cao Hai section of Lake Dianchi has an area of 10.5 km^2 with an average depth of 2.5 meters and a water holding capacity of 25 million cubic meters with hydraulic retention time of 3.8–4.2 months. Around Cao Hai, there are six rivers (R1–R6), and two large municipal waste treatment plants (marked R1 and R5), discharging effluent (about 30 tonnes d^{-1}) into the lake (Fig. 9.2.3-1).

The green patches in Fig. 9.2.3-1 represent the enclosures for water hyacinth growth, with about 500 meters in length and 150 meters in width separated by 50-meter open gaps. Xi Yuan channel (C1) is the only water outflow site of Cao Hai (discharged 94,098,100 m^3 water in 2011). Due to municipal effluent and other industrial wastewater discharge, Cao Hai is the most polluted area in Lake Dianchi. Before the ecological engineering project using water hyacinth, concentrations of total nitrogen and total phosphorus were 12–20 mg N L^{-1} and 1.2–1.6 mg P L^{-1} from 2007 to 2011; nutrient loading in 2011 was about 2,000 tonnes of nitrogen and 200 tonnes of phosphorus.

The ecological engineering project was implemented during 2011 to 2013 to investigate possibility of phytoremediation in Cao Hai. The project was designed to confine the growth of water hyacinth to 5.3 km^2 from May to October and to harvest water hyacinth starting from November (planned to finish by the end of December each year). A special heavy duty machine was designed for harvesting and dehydration. Water hyacinth juice was processed for methane fermentation, and bagasse was used for production of organic fertilizer.

Water sampling sites were set at each river near the Cao Hai (labels R1 to R6) and along the water flowing from northern Cao Hai to discharge outflow at Xi Yuan channel (C5 to C1). Water samples were collected in the period from May to November 2011 (from the initial water hyacinth planting to harvest) at a frequency of one to three times a month for 11 sites (R1–R6 and C5–C1). Three mixed water samples were collected at each site at three layers (from the water surface 0–0.5 m and 1.0–1.5 m as well as 0.5 m above the sediment surface). One liter of water was collected for each sample and immediately transported to the laboratory for chemical analysis on total nitrogen, total phosphorus, ammonium, nitrate and orthophosphate. Meanwhile, water hyacinth plants were sampled each month for the determination of nitrogen,

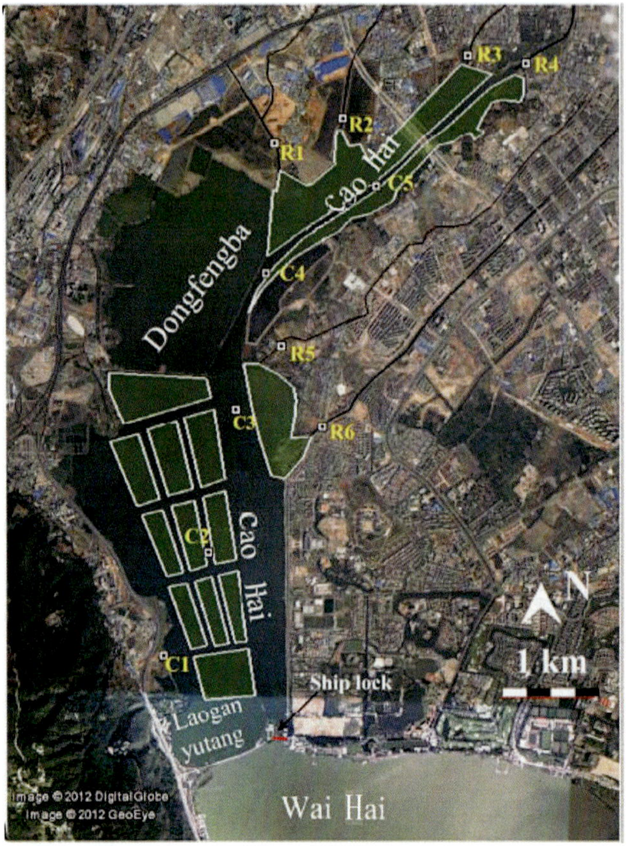

Fig. 9.2.3-1. Experiment layout and sampling sites at Cao Hai in Lake Dianchi.

phosphorus and water contents. Total water hyacinth biomass was measured by the combination of a GPS and on-site weighing using quadrants.

For the period before May 2011, the physical and chemical properties of the water on each site were provided by the Kunming Environmental Monitoring Station and the data on water discharge through Xi Yuan outflow site C1 were provided by Xi Yuan Channel Discharge Administration, Kunming City.

Dissolved oxygen before and during the experiment

Dissolved oxygen (DO) at R1 to R6 increased gradually from 1.8 to 4.5 mg L^{-1} from 2007 to 2011 due to management strategies for pollution control at the source. After growth of water hyacinth, DO stabilized at 4.1–5.2 mg L^{-1} at sites C5 and C4 and was 5.5–7.9 mg L^{-1} at C3 to C1 sites. Compared with the DO concentration in 2010, DO increase by 33% from May to November in 2011 (Fig. 9.2.3-2A).

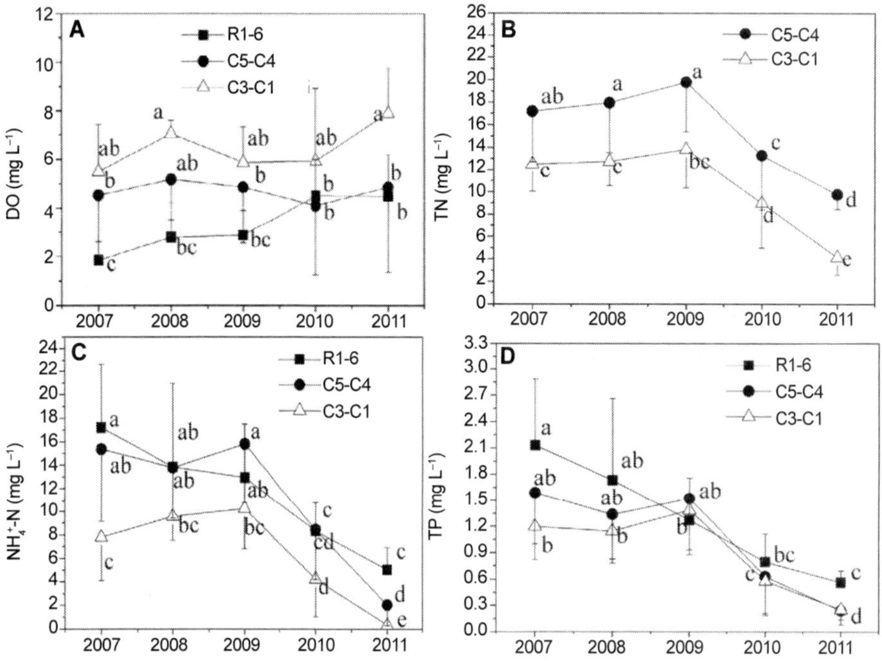

Fig. 9.2.3-2. Changes of dissolved oxygen, nitrogen and phosphorus before and during the growth of water hyacinth at Cao Hai, Lake Dianchi (redrawn with permission).

During the experiment in 2011, water hyacinth coverage was maximized by the end of October (50% total Cao Hai surface). However, it did not cause a significant adverse impact on DO in Cao Hai. It is true that a large number of reports have shown that water hyacinth can reduce DO significantly (Rommens et al. 2003, Villamagna and Murphy 2010). In the present study, DO concentration slightly increased at sampling sites C5 to C1 after growth of water hyacinth. This phenomenon was unexpected and the reasons might have been as follows: (1) in Cao Hai, there were serious organic pollutants from 2007 to 2009 that could lower the background DO to 4–6 mg L^{-1}. After planting water hyacinth, organic pollutants were filtered by water hyacinth roots or removed by enhanced bacterial activities, reducing the consumption of oxygen in water; and (2) water hyacinth improved transparency especially in the open areas, which can enhance photosynthesis of algae and submerged plants to increase the release of oxygen.

Total nitrogen before and during the experiment

Due to a lack of data on total nitrogen at the river mouth (R1 to R6) from 2006–2010, only the total nitrogen concentration before and during the experiment in Cao Hai (C5 to C1) can be discussed. In 2010, there were only

200 hectares of natural growth of water hyacinth in Cao Hai, with the total nitrogen concentrations of 13.3 mg N L^{-1} at C5 and C4 but 9.0 mg N L^{-1} at C3–C1; the average total nitrogen concentrations from 2007 to 2009 were 18.3 mg N L^{-1} at C5–C4 and 13.0 mg N L^{-1} at C3–C1, significantly decreasing by 27 and 31% at the two sites ($p < 0.05$) (Fig. 9.2.3-2B). In 2011, the implementation of the project increased water hyacinth coverage to 420 hectares; and the total nitrogen concentrations further dropped to 9.8 at C5–C4 and 4.1 mg N L^{-1} at C3–C1 ($p < 0.05$) (decreasing by 26 and 54% compared to those in 2010).

Ammonium nitrogen before and during the experiment

During 2006 to 2011, the concentrations of ammonium correlated with those of total nitrogen (Fig. 9.2.3-2C). The concentrations of ammonium decreased gradually from 2007 to 2010 due to the same reason as mentioned in the paragraph above on dissolved oxygen. In 2011, the ammonium concentrations further dropped along the water pass from river mouth to Xi Yuan outflow, being 40% lower at R1–R6, 67% lower at C5–C4 and 89% lower at C3–C1 than the values in 2010.

Total phosphorus before and during the experiment

From 2007 to 2008, due to the management and technology improvements at the municipal wastewater treatment plants, Total Phosphorus (TP) in rivers (site R1 to R6) showed a slight downward trend. Starting from 2009, TP concentration in the rivers further decreased due to water hyacinth growth (Fig. 9.2.3-2D). In Cao Hai, TP concentration was stable during 2007 to 2009 and significantly decreased after 2010 due to the growth of water hyacinth. Compared to 2009, TP concentration decreased 59% at sites C5 and C4 and 58% at sites C3–C1 in 2010, and further decreased 60% at sites C5 and C4 and 55% at sites C3–C1 in 2011 compared to 2010.

Spatial variation in nitrogen and phosphorus concentrations

Total Nitrogen (TN) concentration decreased along the way from the entering sites (R1–R6) to Cao Hai, then exiting at Xi Yuan (C1) channel during the water hyacinth growing period (June to November). The average TN concentration of 13.8 mg N L^{-1} gradually decreased to 3.3 mg N L^{-1} at Xi Yuan (C1) (down by 76%) (Fig. 9.2.3-3A).

Statistical analysis showed that total nitrogen concentrations at R1–R6 (river into the lake), at C5 and C4 (in Cao Hai) and at C3, C2 and C1 (Xi Yuan Exit) were significantly different (Fig. 9.2.3-3A). The decreases were due to (i) denitrification enhanced by the presence of water hyacinth and (ii) nitrogen removed by water hyacinth. The natural denitrification effect can be estimated using the data from the period 2007 to 2009, during which the average TN

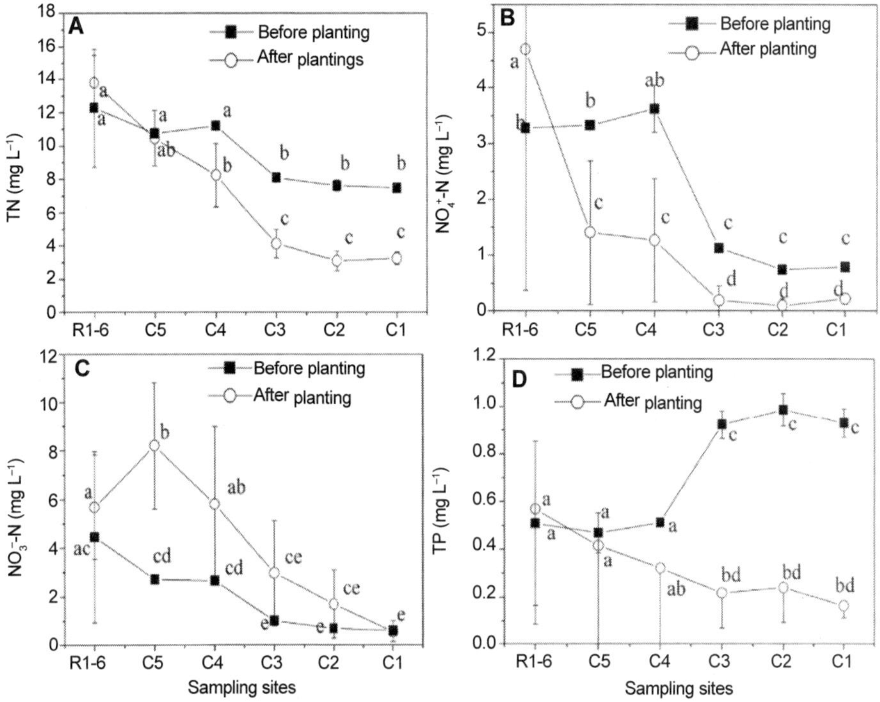

Fig. 9.2.3-3. Spatial changes of nitrogen and phosphorus before and after growth of water hyacinth (redrawn with permission).

concentration was 12.3 mg N L^{-1} at R1–R6 and 7.5 mg N L^{-1} at C1 (down by 39%). The presence of water hyacinth was estimated to contribute a net decrease of 5.1 mg N L^{-1} (37% decrease).

Ammonium (NH_4^+) and nitrate concentrations (Figs. 9.2.3-3B and 9.2.3-3C) showed a similar relationship as total nitrogen: (1) before the growth of water hyacinth, decreases of 79% in ammonium (3.3 to 0.8 mg N L^{-1}) and 89% in nitrate concentration (4.5 to 0.5 mg N L^{-1}) from sites R1–R6; (2) after planting water hyacinth, decreases of 96% in ammonium (4.7 to 0.2 mg N L^{-1}) and 94% in nitrate concentration from R1–R6 (5.5 to 0.5 mg N L^{-1}). Statistical analysis showed that and assimilation by water hyacinth were all significantly different from R1–R6 to C5–C1. Similar to total nitrogen concentration changes, the water hyacinth contribution was estimated to be a net decrease in ammonium of 0.8 mg N L^{-1} (17% decrease). One interesting phenomena was an increase in nitrate concentration at site C5 (Fig. 9.2.3-3C) due to the growth of water hyacinth.

Total phosphorus (TP) concentration indicated a more interesting impact of water hyacinth on the nutrient concentration in eutrophic water. Before growth of water hyacinth, the TP concentration increased significantly from site C4 to site C3 (Fig. 9.2.3-3D). However, after growth of water hyacinth, the TP concentration decreased from 0.55 mg P L^{-1} at sites R1–R6 to 0.15 mg P L^{-1}

at C1 site. If compared to the concentration before planting water hyacinth (0.9 mg P L^{-1}), a decrease of 83% occurred from sites of river outflows to lake water exit site.

Nitrogen concentration changes during the growth of water hyacinth in Cao Hai

From May to November 2011, total nitrogen concentrations were 11.7–16.3 mg N L^{-1} at river outflow sites, with no significant difference among the months. During the water hyacinth growing season, after the effluent passed through 147 hectares of water hyacinth area, the total nitrogen concentration dropped to 7.7 mg N L^{-1} at the middle of water ways in May to 3.5 mg N L^{-1} near lake water exit site in June (to 2.6 mg N L^{-1} in August), and then fluctuated during September to November (Fig. 9.2.3-4A).

Ammonium (NH_4^+) and nitrate concentrations fluctuated from May to November (3.3–6.7 mg N L^{-1}), which reflected the effluent discharging characteristics. The growth of water hyacinth assimilated ammonium nitrogen and kept the ammonium concentration below 1 mg N L^{-1} and

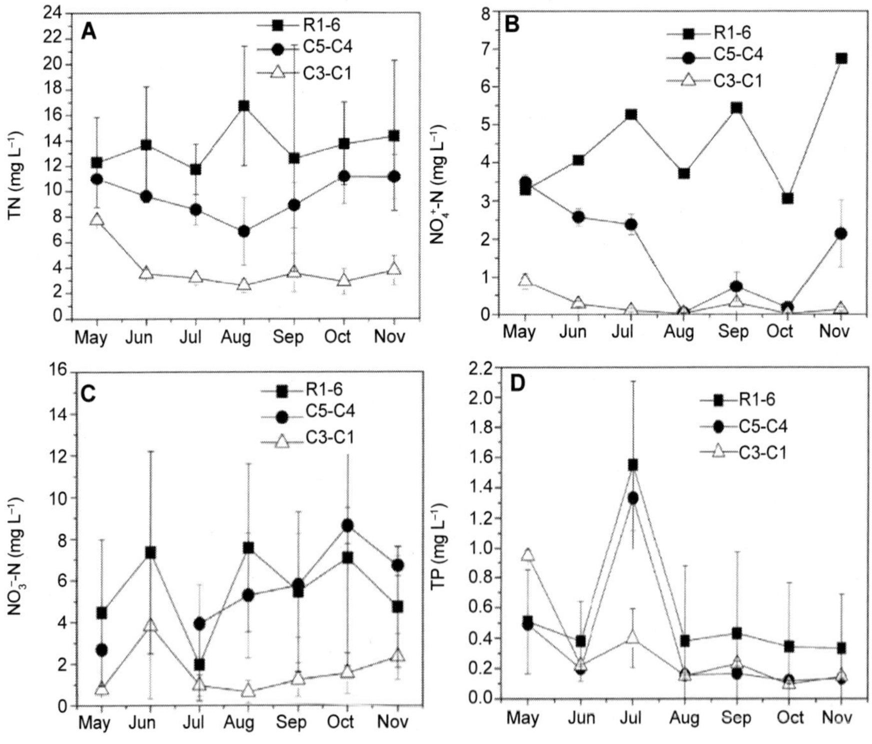

Fig. 9.2.3-4. Seasonal changes in nitrogen and phosphorus at the sampling areas after planting water hyacinth in Cao Hai (redrawn with permission).

nitrate concentration below 3.5 mg N L^{-1} at sites C5–C1 in May and almost 0 mg N L^{-1} for ammonium and 0.5 mg N L^{-1} for nitrate in August. At sites R1–R6, there were fluctuations in concentrations during September to November, decreasing significantly to 0.02 mg N L^{-1} for ammonium and 2.1 mg N L^{-1} for nitrate at C3–C1 in November (Figs. 9.2.3-4B and 4C).

The TP concentration was 1.55 mg P L^{-1} at R1–R6, reduced to 1.35 mg P L^{-1} and then to 0.4 mg P L^{-1} along the water flowing direction in July; in other months, it was stable at 0.15 mg P L^{-1} passing through the testing sites. Based on decreases in TP concentration, there was removal efficiency of 74% in July and 57% in October and November (Fig. 9.2.3-4D).

Fate of nitrogen and phosphorus in Cao Hai in 2011

Cao Hai is an open ecosystem with ~100 million cubic meters of water flowing from six rivers and leaving at Xi Yuan channel each year. The water in rivers comes mainly from the catchment and the discharge of municipal wastewater, industrial wastewater and non-point agricultural leachate and drainage. The removal of the nutrient can be estimated by the formula:

$$R_i = (IN_i - OU_i) + (E5_i - E11_i) \qquad [9.2.3\text{-}1]$$

Where R_i is nutrient removed; i = nitrogen or phosphorus; IN_i is the amount of nutrient flow into Cao Hai; OU is the amount of nutrient outflow from Cao Hai; E5 is the amount of nutrient remained in water in May; E11 is the amount of nutrient remained in water in November.

IN_i is calculated as the volume of water flowing into Cao Hai multiplied by average concentration of the nutrient in water. OU is calculated as the volume of water drained from Cao Hai multiplied by average concentration of the nutrient in water. Nitrogen and phosphorus assimilated by water hyacinth are calculated by multiplying harvested biomass (211 kt) by average nitrogen and phosphorus concentration in dry matter (41 g N kg^{-1} and 2.8 g P kg^{-1}) (Table 9.2.3-1).

Water hyacinth can significantly enhance the activity of microorganisms to promote nitrification and denitrification processes and increase nitrogen removal via nitrous oxide and molecular nitrogen (Gao et al. 2012). In natural aquatic ecosystem, excessive nitrogen is removed mainly through mineralization, nitrification and denitrification to nitrous oxide (N_2O) and molecular nitrogen (N_2) back to atmosphere. The processes are driven by microorganisms in water or sediment (Risgaard-Petersen and Jensen 1997, Zhao et al. 1999). In phytoremediation, nitrogen is removed by both plant uptake and the microorganism-driven processes. The experiment showed that 761 tonnes of total N were removed, among which 63.8% (or 486 tonnes) was removed via uptake by water hyacinth (Table 9.2.3-1). Literature suggested that water hyacinth may contribute 42–83% of total removed nitrogen (Zhang et al. 2010). The higher the nitrogen concentration in effluent, the lower the removal proportion water hyacinth contributed.

Table 9.2.3-1. Nitrogen uptake by water hyacinth and nitrogen removed from Cao Hai in Lake Dianchi from May to November 2011.

Month	Water outflow million t	Outflow TN conc. mg N L^{-1}	Outflow TN tonne	Inflow water million t	TN conc. inflow mg N L^{-1}	Inflow TN t	TN removed t	Uptake by water hyacinth t	Removed by water hyacinth %
5	12	3.48	42.1	12	12.3	153			
6	14	5.2	72.9	14	13.7	196			
7	11	3.3	36.7	11	11.7	134			
8	6	2.94	17.6	6	16.7	105			
9	14	1.99	28.2	14	12.6	182			
10	6	2.22	13.5	6	13.7	88			
11	5	2.6	13.5	6	14.3	79			
Total	68		224.5			937	761[a]	486	63.8

[a] Including the nitrogen removed from the pool originally in lake water.

Excessive phosphorus in eutrophic waters is mainly removed by sedimentation and biological assimilation. The experiment at Cao Hai showed that water hyacinth removed 139% of phosphorus from the water (Table 9.2.3-2). This result implied the release of P from the sediment. The activity of phosphorus on the interface of sediment-water is dominated by the concentrations in sediment and water (Surridge et al. 2007). When the concentration of phosphorus is lower in surface water than sediment, the phosphorus can be released from sediment to surface water, and vice versa. In the Cao Hai case, when water hyacinth removed most of phosphorus from surface water, sediment may release phosphorus to the surface water. This result is consistent with the literature (Fan and Aizaki 1997).

9.3 Impact of water hyacinth on biodiversity and structure of biological communities in Lake Dianchi

9.3.1 Macrozoobenthos at Baishan Bay in Lake Dianchi

Sample collection and analysis

The samples were collected from Baishan Bay experiment area (Fig. 9.2.1-1) from August to October 2010. Macrozoobenthos was collected monthly at 13 sampling sites of 0.0625 m^2 each with a Peterson sampler. All bottom samples were sieved through a 420-μm sieve. Specimens were manually sorted out from sediment on a white porcelain plate and preserved in 10% v/v formaldehyde solution. The macrozoobenthos was identified to be at the lowest feasible taxonomic level and counted under a microscope in the laboratory and weighed.

Table 9.2.3-2. Phosphorus uptake by water hyacinth and phosphorus removed from Cao Hai in Lake Dianchi from May to November 2011.

Month	Water outflow million t	Outflow TP concn mg P L^{-1}	Outflow TP t	Inflow water million t	TP concn inflow mg P L^{-1}	Inflow TP t	TP removed t	Uptake by water hyacinth t	Removed by water hyacinth %
5	12	0.335	4.05	12	0.58	7.19			
6	14	0.718	10.07	14	0.48	6.95			
7	11	0.362	4.03	11	0.51	5.85			
8	6	0.214	1.28	6	0.53	3.34			
9	14	0.122	1.73	14	0.58	8.46			
10	6	0.128	0.78	6	0.53	3.37			
11	5	0.082	0.43	6	0.89	4.91			
Total	68		22.36			40.05	23.8	33.1	139

To estimate the biodiversity, Shannon-Wiener, Margalef, Gini-Simpson and Pielou indices were calculated using the formulae as defined below.

Shannon-Wiener index:

$$H' = -\sum_{i=1}^{s} P_i \ln P_i = -\sum_{i=1}^{s} (N_i/N)\ln(N_i/N) \qquad [9.3.1\text{-}1]$$

Margalef index:

$$d = (S-1)/\ln(N) \qquad [9.3.1\text{-}2]$$

Gini-Simpson index:

$$D = 1-\sum_{i=1}^{s} P_i^2 \qquad [9.3.1\text{-}3]$$

Pielou index:

$$J = H'/\ln(S) \qquad [9.3.1\text{-}4]$$

Where P_i is the proportion of the *i*th species in a specific group and is calculated as the number of individuals (N_i) belonging to the *i*th species in the total number (N) of individuals; S is the total number of species in the samples; H' is the Shannon-Wiener index.

Statistical method of one-way ANOVA was employed to identify significant differences among three groups (water hyacinth area, near water hyacinth area and area far away from water hyacinth) on the biodiversity indices as well as environmental parameters in surface water and sediment. When the data did not follow normal distribution and homogeneity, the Mann–Whitney U test was used to identify differences among the three areas. All statistical analyses were performed using software SPSS v16.0 for Windows. The significant level was set at $p < 0.05$.

Benthos community and density

The sample data showed a total of 18 species present in all the samples of three areas, among which, eight species of Oligochaeta (44.4% of total species), five species of Insecta (27.8% of total), Crustacea three species (16.7% of total) and Gastropoda and Nematoda one species each (5.6% of total species each). In the water hyacinth area, total of 14 species were collected including Oligochaeta seven species, Gastropoda one species, Insecta two species, Crustacea three species and Nematoda one species (unidentified class). In the area near water hyacinth, total of 10 species were collected including Oligochaeta six species, Insecta three species and Nematoda one species. Far from water hyacinth area, only six species were collected including Oligochaeta four species and Insecta two species. The common species presented in all three areas were *Limnodrilus hoffmeisteri*, *Limnodrilus grandisetosus* and *Tubifex tubifex*; whereas Gastropoda (*Radix swinhoe*) and Crustacea (Decapoda, *Caridina* sp. and *Gammaridae* spp.) were only present in the water hyacinth area (Table 9.3.1-1).

Table 9.3.1-1. Species compositions of benthic macrozoobenthos at the sampling areas from August to October 2010.

Taxon	Water hyacinth area	Area near water hyacinth	Area far from water hyacinth
Nematoda			
1. *unidentified* sp.	+	+	
Oligochaeta			
Naididae			
2. *Dero digitata*	+	+	
3. *Stephensoniana trivandrana*	+		
Tubificidae			
4. *Limnodrilus hoffmeisteri*	+	+	+
5. *Limnodrilus grandisetosus*	+	+	+
6. *Limnodrilus* sp.	+		
7. *Tubifex tubifex*	+	+	+
8. *Branchiura sowerbyi*	+	+	
9. *unidentified* sp.		+	+
Gastropoda			
Mesogastropoda			
Lymnaeidae			
10. *Radix swinhoe*	+		
Insecta			
Diptera			
Chironomidae			
11. *Chironomus Plumosus*	+		+
12. *Dicrotendipes* sp.		+	
13. *Orthocladius* sp.	+	+	
Phemeroptera			
Baetidae			
14. *Baetis* sp.			+
Odonata			
15. *unidentified* sp.		+	
Crustacea			
Decapoda			
16. *unidentified* sp.	+		
Atyidae			
17. *Caridina* sp.	+		
Amphipoda			
Gammaridae			
18. *unidentified* sp.	+		

The densities (expressed as individuals per square meter) of macrozoobenthos were 295 in the water hyacinth area, 159 near water hyacinth and 261 far from the water hyacinth area. In all three sampled areas, the dominant species of macrozoobenthos were mainly Oligochaeta (*Limnodrilus hoffmeisteri* and *Limnodrilus grandisetosus*), which accounted for 264 individuals in the water hyacinth area, 151 individuals near water hyacinth and 250 individuals far from the water hyacinth area, representing 90, 95 and 96% of the total densities in three areas, respectively (Table 9.3.1-2).

Due to different weight of individual benthic fauna, the biomass distribution and density were different in the three experiment areas. In the water hyacinth area, the biomass of macrozoobenthos was mainly of Oligochaeta (*Limnodrilus hoffmeisteri* and *Limnodrilus grandisetosus*) and Insects (Chironomidae) representing 56 and 30% of the total, respectively. Similar numbers were recorded in the area far from water hyacinth (Oligochaeta 61% and Insects 39% of the total). However, in the area near water hyacinth, the biomass was mainly Oligochaeta, representing 99.3% of the total (Table 9.3.1-2).

Functional feeding groups analysis showed that the species composition was collector-gatherer (density: 97%, biomass: 93%), parasite (density: 1.3%, biomass: 0.6%), scraper (density: 1.3%, biomass: 1.8%) and shredders (density: 0.6% biomass: 3.0%) in the water hyacinth area; in the area near water hyacinth: collector-gatherer (density: 97%, biomass: 96%), parasites (density: 1.8%, biomass: 2.9%) and predator (density: 1.0%, biomass: 0.7%); and only collector-gatherers in the area far from water hyacinth.

Dominant species

In the all sampling areas, Oligochaeta were the dominant taxa with the average density of 259 individuals m^{-2} representing 93% of the total number, among which *Limnodrilus hoffmeisteri* was the main species with an average density of 201 individuals m^{-2} or 72% of the total. The average densities of *Limnodrilus hoffmeisteri* were 219, 104 and 281 individuals m^{-2} in the areas of water hyacinth, and near and far from water hyacinth, respectively, representing 68, 60 and 86% of the total. From August to October, density of *Limnodrilus hoffmeisteri* increased first then dropped in the water hyacinth area and far from water hyacinth, while it showed a continuous drop near the water hyacinth area (Fig. 9.3.1-1A). However, biomass in all three sampling areas showed a similar pattern (Fig. 9.3.1-1B). This is because the biomass of the second dominant family Chironomidae was larger than the biomass of Oligochaeta so that the ratio of the total biomass was similar among the three areas (Fig. 9.3.1-1B).

Characteristics of community indices

Variance analysis showed that in the water hyacinth area and the nearby area, Margalef, Shannon-Wiener, Gini-Simpson and Peilou indices decreased slightly (non-significantly) from August to September, and then increased

Table 9.3.1-2. Density (individuals m^{-2}) and biomass ($g\ m^{-2}$) of macrozoobenthos at the sampling areas in August to October 2010.

Taxon	Water hyacinth area				Area near water hyacinth				Area far from water hyacinth			
	Density	%	**Biomass**	%	**Density**	%	**Biomass**	%	**Density**	%	**Biomass**	%
Oligochaeta	264	89.6	0.36	55.8	151	95.0	0.37	99.3	250	95.8	0.25	61.1
Gastropoda	1	0.3	0.03	3.9	0	0	0	0	0	0	0	0
Insecta	11	3.7	0.19	29.7	5.3	3.3	0	0.7	11	4.2	0.16	38.9
Crustacea	15	5.1	0.07	10.5	0	0	0	0	0	0	0	0
Nematoda	4	1.2	0	0.2	2.7	1.7	0	0	0	0	0	0
Total	295	100	0.65	100	159	100	0.38	100	261	100	0.40	100

significantly in October ($p < 0.05$). In the area far from water hyacinth, diversity indices of Margalef, Gini-Simpson and Shannon-Wiener gradually reduced from August to October, and were significantly lower in October than August. However, Peilou index was similar in August, September and October in that area (Table 9.3.1-3).

Comparing different areas, Margalef diversity, Gini-Simpson and Shannon-Wiener indices were significantly higher in the water hyacinth area than areas near and far from water hyacinth. However, Peilou index at the three areas was not significantly different (Table 9.3.1-3).

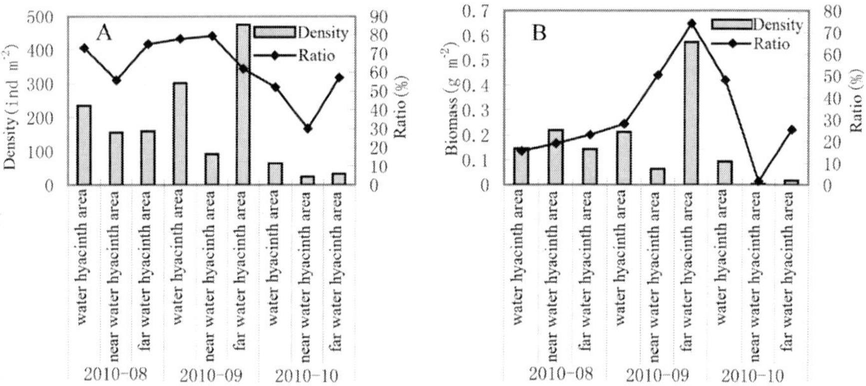

Fig. 9.3.1-1. Density (A) and biomass (B) changes of *Limnodrilus hoffmeisteri* in the sampling areas.

Table 9.3.1-3. Characteristics of community indices of macrozoobenthos in various sampling areas in August to October 2010 (mean ± SD).

Area	Month	Margalef	Gini-Simpson	Shannon-Wiener	Peilou
Water hyacinth area	8	0.38±0.22a	0.40±0.24a	0.71±0.43a	0.60±0.35ab
	9	0.42±0.29a	0.36±0.21a	0.68±0.42a	0.54±0.34a
	10	0.56±0.12b	0.60±0.14b	1.10±0.27b	0.84±0.12b
	8–10	0.43±0.23A	0.42±0.22 A	0.77±0.41A	0.62±0.31A
Area near water hyacinth	8	0.27±0.20ab	0.31±0.28a	0.54±0.47a	0.52±0.42a
	9	0.18±0.25a	0.18±0.21a	0.38±0.44a	0.39±0.48a
	10	0.46±0.03b	0.62±0.01b	1.03±0.02b	0.93±0.02b
	8–10	0.27±0.21B	0.32±0.26 B	0.57±0.45B	0.55±0.43A
Area far from water hyacinth	8	0.38±0.20a	0.40±0.23a	0.72±0.44a	0.65±0.20a
	9	0.18±0.26ab	0.29±0.41ab	0.49±0.70ab	0.45±0.63a
	10	0.11±0.15b	0.22±0.31b	0.32±0.45b	0.46±0.65a
	8–10	0.25±0.21B	0.32±0.26 B	0.54±0.46B	0.54±0.40A

Notes: Different lowercase letters represent significant ($p < 0.05$) differences in the same area but different months; different uppercase letters represent significant ($p < 0.05$) differences among different areas.

9.3.2 Zooplankton at Baishan Bay in Lake Dianchi

Sample collection and analysis

The location and sampling periods were the same as mentioned earlier, but focused on planktonic crustaceans Cladocera, Copepoda and Rotifera.

Composition of zooplankton as influenced by water hyacinth

At all three sampling areas, 36 species of zooplankton were obtained: 15 species in six genera belonged to Cladocera, nine species in four genera were Copepoda, and 11 species in seven genera belonged to Rotifera. In the water hyacinth area, 24 species of zooplankton included 11 species of Cladocera, nine species of Copepoda and four species of Rotifera. In the area near water hyacinth, 28 species comprised 11 species of Cladocera, eight species of Copepoda and nine species of Rotifera. In the area far from water hyacinth, 24 species included 12 species of Cladocera, seven species of Copepoda and five species of Rotifera (Table 9.3.2-1). Species co-occurring in all three areas were *Bosmina longirostris, Bosmina coregoni, Bosmina fatali, Ceriodaphnia cornuta, Alona rectangular, Daphnia hyalina, Daphnia cucullata, Mesocyclops leuckarti, Mesocyclops pehpeiesis, Cyclops strenuus, Cyclops vicinus, Microcyclops varicans, Microcyclops robustus, Microcyclops intermedius, and Keratella quadrata. Limnoithona sinensis* and *Lepadella ovalis* were only found in the water hyacinth area. *Brachionus forficula, Filinia maior* and *Lecane luna* were only found in the area near water hyacinth, whereas *Ceriodaphnia cornigera, Ceriodaphnia quadrangula* and *Trichocera gracilis* were only found in the area far from water hyacinth (Table 9.3.2-1).

Spatial and temporal distribution of zooplankton as influenced by water hyacinth

During the sampling period (August to October 2010), the number of Cladocera gradually declined in the water hyacinth area, gradually increased in the nearby area, and increased in September but decreased in October in the area far from water hyacinth (Fig. 9.3.2-1). Analysis of variance showed that the density of Copepoda was not significantly different in the water hyacinth area and the area far from water hyacinth during the sampling period, but was significantly different between the water hyacinth and nearby areas in the period August to October ($p < 0.05$, Fig. 9.3.2-1). The density of Rotifera was also not significantly different in the water hyacinth and nearby areas from August to October. The Copepoda density did not differ in the water hyacinth and far away areas in the August-October period. In October, Rotifera were not present in the area far from water hyacinth.

Table 9.3.2-1. Species composition of zooplankton in the sampling areas from August to October 2010.

Taxon	WHA	ANWH	AFWH
Cladocera			
Bosmina			
1. *Bosmina longirostris*	+	+	+
2. *Bosmina coregoni*	+	+	+
3. *Bosmina fatali*	+	+	+
Ceriodaphnia			
4. *Ceriodaphnia cornuta*	+	+	+
5. *Ceriodaphnia pulchella*	+		+
6. *Ceriodaphnia cornigera*			+
7. *Ceriodaphnia megalops*			+
8. *Ceriodaphnia quadrangula*			+
Moina			
9. *Moina macrocopa*		+	+
Alona			
10. *Alona rectangular*	+	+	+
Daphnia			
11. *Daphnia hyalina*	+	+	+
12. *Daphnia cucullata*	+	+	+
13. *Daphnia longispina*	+	+	
14. *Daphnia pulex*	+	+	
Diaphanosoma			
15. *Diaphanosoma brachyurum*	+	+	
Copepoda			
Mesocyclops			
16. *Mesocyclops leuckarti*	+	+	+
17. *Mesocyclops pehpeiesis*	+	+	+
Cyclops			
18. *Cyclops strenuus*	+	+	+
19. *Cyclops vicinus*	+	+	+
Microcyclops			
20. *Microcyclops varicans*	+	+	+
21. *Microcyclops robustus*	+	+	+
22. *Microcyclops intermedius*	+	+	+
23. *Microcyclops longiramus*	+	+	
Limnoithona			
24. *Limnoithona sinensis*	+		

Table 9.3.2-1. contd....

Table 9.3.2-1. contd.

Taxon	WHA	ANWH	AFWH
Rotifera			
Brachionus			
25. *Brachionus angularis*	+	+	
26. *Brachionus calyciflorus*		+	+
27. *Brachionus forficula*		+	
28. *Brachionus falcatus*		+	+
Keratella			
29. *Keratella quadrata*	+	+	+
30. *Keratella valga*	+	+	
Monostyla			
31. *Monostyla closterocerca*		+	+
Filinia			
32. *Filinia maior*		+	
Lepadella			
33. *Lepadella ovalis*	+		
Lecane			
34. *Lecane luna*		+	
Trichocera			
35. *Trichocera gracilis*			+

Notes: WHA = water hyacinth area; ANWH = area near water hyacinth; AFWH = area far from water hyacinth; + refers to species represent at site.

The total densities of Cladocera were 26, 28 and 54 individuals L^{-1} in the areas of water hyacinth, nearby and far away, respectively, and were not significantly different among the areas. The total densities of Copepoda were 33, 32 and 25 individuals L^{-1} in the areas of water hyacinth, nearby and far away, respectively, with no significant difference among the areas. The densities of Rotifera were 7, 38 and 34 individuals L^{-1} in the areas of water hyacinth, nearby and far away, respectively, with significant differences between the water hyacinth area and the nearby area ($p < 0.05$).

Dominant species

The dominant species in Cladocera was *Bosmina longirostris* with density of 10 individuals L^{-1} (37% of the total Cladocera density) in the water hyacinth area, 13 individuals L^{-1} (47% of the total Cladocera density) in the area near water hyacinth and 32 individuals L^{-1} (60% of the total Cladocera density) in the area far from water hyacinth. Due to large variations in standard deviation, the densities of the dominant species in Cladocera were not significantly different among the three areas.

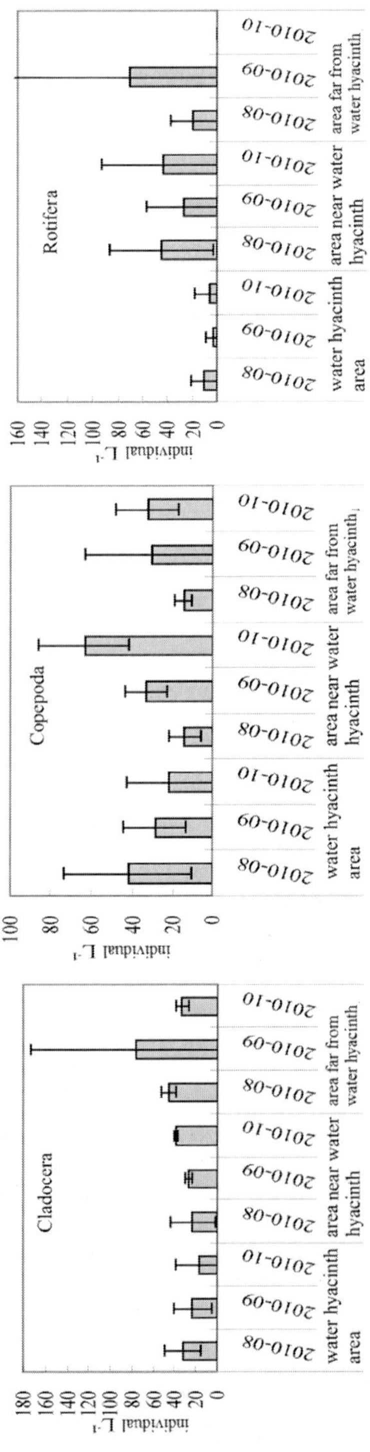

Fig. 9.3.2-1. Variations of zooplankton in the sampling areas.

The dominant species of Copepoda was *Microcyclops varicans* with density of 25 individuals L^{-1} (75% of the total Copepoda density) in the water hyacinth area, 23 individuals L^{-1} (74% of the total Copepoda density) in the area near water hyacinth and 21 individuals L^{-1} (86% of the total Copepoda density) in the area far from water hyacinth, with no significant difference among the three areas.

The dominant species of Rotifera were different among the three areas: *Keratella quadrata* in the water hyacinth area (4.4 individuals L^{-1}, or 68% of the total Rotifera density), *Keratella valga* in the area near water hyacinth (11 individuals L^{-1}, or 28% of the total Rotifera density) and *Monostyla closterocerca* in the area far from water hyacinth (15 L^{-1}, or 39% of the total Rotifera density).

Biodiversity indices

Shannon-Wiener (H'), Gini-Simpson (D) and Margalef (d) indices in Cladocera decreased gradually during August to October in all three sampled areas, but without statistically significant difference (Table 9.3.2-2). All the indices in Copepoda showed a similar pattern as in Cladocera, although they were lower in the area far from water hyacinth compared to the other two areas. However, these indices in Rotifera were higher in the area near water hyacinth than those in the water hyacinth and far away areas ($p < 0.05$, Table 9.3.2-2).

Correlation analysis

There was no significant correlation among main physical and chemical indicators of water, the total densities of Cladocera and Rotifera species and the dominant species of Rotifera. The total density and the dominant species (*Microcyclops varicans*) of Copepoda were significantly and positively correlated only with total nitrogen and total phosphorus in water. The dominant species (*Bosmina longirostris*) of Cladocera was significantly and negatively correlated with chlorophyll-*a* concentration in water. Shannon-Wiener index in Cladocera showed a significant and positive correlation with the water temperature and negative correlation with NH_4^+ and PO_4^{3-}. Shannon-Wiener index in Copepoda showed a significant and positive correlation with the water temperature and negative correlation with NH_4^+. Shannon-Wiener index in Rotifera showed a significant and positive correlation with the dissolved oxygen and water pH (Table 9.3.2-3).

9.4 Impact of water hyacinth on biodiversity and structure of biological community in Lake Taihu

The total water hyacinth growth area at Zhushan Bay in Lake Taihu covered only 0.09% of the total water surface area of the lake. Hence, it was expected that the impact of water hyacinth on water quality would be minimal in the

Table 9.3.2-2. Shannon-Wiener (*H'*), Margalef (*d*), Gini-Simpson (*D*) and Peilou (J) indices in Cladocera, Copepoda and Rotifera in the sampled areas from August to October 2010.

Area		Cladocera				Copepoda				Rotifera			
		H'	*d*	*D*	*J*	*H'*	*d*	*D*	*J*	*H'*	*d*	*D*	*J*
Mean	WHA	1.08a	1.39a	0.56a	0.68a	0.65a	0.89a	0.34a	0.51a	0.10a	0.06a	0.06a	0.11a
	NWHA	0.93a	1.27a	0.48a	0.30a	0.59a	0.93a	0.33a	0.54a	0.79b	0.54b	0.42b	0.61b
	FWHA	0.85a	1.03a	0.46a	0.61a	0.43a	0.51a	0.27a	0.48a	0.37ab	0.20ab	0.24ab	0.41ab
Standard deviation	WHA	0.45	0.44	0.22	0.20	0.43	0.59	0.22	0.28	0.26	0.17	0.16	0.28
	NWHA	0.36	0.69	0.18	0.12	0.29	0.84	0.17	0.26	0.68	0.50	0.33	0.43
	FWHA	0.42	0.44	0.23	0.28	0.30	0.40	0.21	0.39	0.43	0.23	0.28	0.45

Notes: WHA refers to Water Hyacinth Area; ANWH refers to Area Near Water Hyacinth; AFWH refers to Area Far from Water Hyacinth; different lower case letters for a specific index represent significant differences ($p < 0.05$).

Table 9.3.2-3. Correlation analysis of physico-chemical parameters and zooplankton.

	WT	DO	pH	Chl-*a*	TN	NH_4^+	NO_3^-	TP	PO_4^{3-}
Cladocera	0.08	0.04	0.22	0.26	–0.01	0.03	–0.12	0.07	–0.09
Copepoda	–0.05	–0.11	0.10	0.19	0.38*	0.13	0.18	0.53**	0.06
Rotifera	0.12	0.16	0.10	0.03	–0.21	–0.13	–0.05	–0.13	0.07
Bosmina longirostris	–0.08	0.12	0.12	–0.44*	0.07	0.30	–0.10	0.01	0.23
Microcyclops varicans	–0.17	–0.01	0.10	0.22	0.49*	0.26	0.24	0.58**	0.09
Keratella quadrata	0.19	0.02	–0.07	–0.13	–0.19	–0.17	–0.20	–0.09	0.24
H' (Cladocera)	0.40*	–0.34	–0.17	–0.33	–0.16	–0.51**	–0.06	0.09	–0.43*
H' (Copepoda)	0.46**	–0.24	–0.26	–0.24	–0.30	–0.39*	–0.28	–0.19	–0.09
H' (Rotifera)	–0.04	0.36*	0.37*	0.21	–0.27	–0.17	0.29	–0.26	–0.23

Notes: WT = Water Temperature (°C); DO = Dissolved Oxygen (mg L^{-1}); Chl-*a* = Chlorophyll-*a* (mg L^{-1}); TN = Total Nitrogen (mg L^{-1}); TP = Total Phosphorus (mg L^{-1}); * and ** refer to $p < 0.05$ and $p < 0.01$.

Lake Taihu area, and the ecological engineering project at the lake only focused on biological diversity inside and outside the water hyacinth mat. The main contribution to water purification was assessed by determining the amount of nitrogen and phosphorus removed by harvesting water hyacinth biomass.

9.4.1 Macrozoobenthos at Zhushan Bay in Lake Taihu

Sample collection and analysis

Zhushan Bay (120°04'00"–120°04'40" E, 31°27'01"–31°27'22" N) is located northwest in Lake Taihu, Jiangsu Province, China (Fig. 9.4.1-1).

The ecological engineering project selected a demonstration area of 2 km^2 to grow water hyacinth in enclosures in the center of Zhushan Bay. A total of 33 sample sites (Fig. 9.4.1-1) in and around the enclosures were selected to investigate the impacts of water hyacinth on macrozoobenthos. Sites 1–12 referred to the areas far from water hyacinth; sites 13–24 referred to the area near water hyacinth; and sites 25–33 referred to the water hyacinth area. Water hyacinth plantlets were introduced to the area in July 2009. Samples were collected early each month from August to October 2009 using a 0.025-m^2

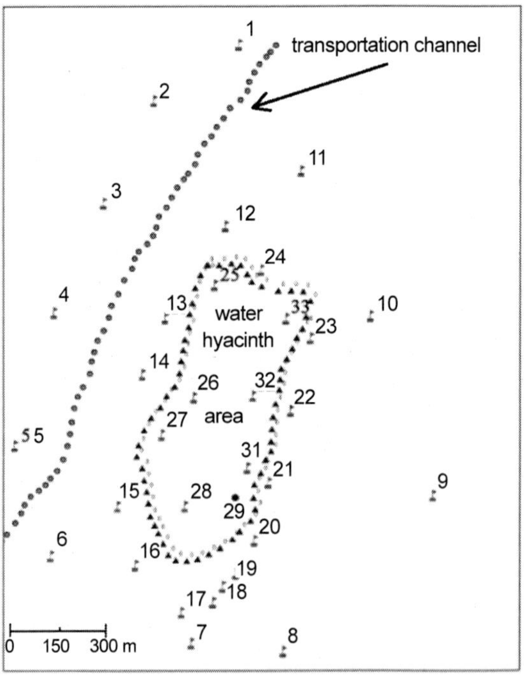

Fig. 9.4.1-1. Sampling sites at Zhushan Bay in Lake Taihu, site 1–12 referring to far from water hyacinth areas, site 13–24 referring to near water hyacinth areas, site 25–33 referring to water hyacinth areas (redrawn with permission).

Peterson sampler. All bottom samples were sieved through a 245-μm sieve. Specimens were manually sorted out from sediment on a white porcelain plate and preserved in 10% v/v formaldehyde solution. Macrozoobenthos and zooplankton were identified to the lowest feasible taxonomic level and counted under microscopes in the laboratory, weighed and calculated per square meter.

To estimate the biodiversity, Shannon-Wiener and Gini-Simpson indices were used as defined in formulas [9.3.1-1] and [9.3.1-3]. Index of Relative Importance (IRI) was defined to assess dominant species:

$$\mathrm{IRI} = (W + N) \times F \qquad [9.4.1\text{-}1]$$

where W is the biomass weight ratio of ith species to the total weight of the all species; N is the ratio of the ith species density (individual L^{-1}) to the total number of the species; F is the frequency of the ith species presence (Simenstad et al. 1991).

Macrozoobenthos composition

During the 3-month investigation, total of 120 samples were collected; and eight common species belonging to five families were identified (Table 9.4.1-1). These species appeared in all sampled areas.

Table 9.4.1-1. Species composition of macrozoobenthos in all sampled areas and the index of relative importance (IRI) (Liu et al. 2014).

Taxon	IRI	%	Taxon	IRI	%
Oligochaeta			Gastropoda		12.5%
Tubificidae		25%	*Bellamya aeruginosa*	113	
Limnodrilus hoffmeisteri	48		Bivalvia		25%
Rhyacodrilus sinicus	91		*Anodonta elliptica*	<1	
Diptera			*Unio douglaniae*[a]		
Chironomidae		25%	Clitellata		12.5%
Pelopia sp.	15		Hirudinea		
Chironomus plumosus	22		*Hirudo nipponica*	<1	

[a] Deleted from statistical analysis because it appeared less than twice.

Spatial and temporal changes in macrozoobenthos density

Gastropoda and Diptera larvae were found in all three sampled areas, whereas Oligochaeta rate of occurrence varied slightly. The average density and biomass weight of Gastropoda were highest in the water hyacinth area and were higher in the area near than far from water hyacinth. The density and biomass weight of Diptera larvae and Oligochaeta were higher in the area near than far from water hyacinth, but were lowest in the water hyacinth area (Table 9.4.1-2).

The dynamic changes in the average densities in each month showed that Oligochaeta density declined from August to October 2009 in the area far from water hyacinth, but in the area near water hyacinth it increased first and then declined (Fig. 9.4.1-2). The same occurred for the water hyacinth area.

The average density of Diptera increased from August to September, then declined in October in the area far from water hyacinth, but in the water hyacinth area it continuously declined from 1040 individuals m^{-2} in August to 449 individuals m^{-2}. The average density of Gastropoda showed a decline in the areas far from and near water hyacinth over the sampling period. Overall, macrozoobenthos declined in the area far from water hyacinth and increased in the area near water hyacinth during the monitoring period.

Table 9.4.1-2. Average density (individuals m^{-2}) and biomass (g m^{-2}) of macrozoobenthos in all sampled areas in Lake Taihu.

Taxon	Water hyacinth area			Area near water hyacinth			Area far from water hyacinth		
	%	Density	Biomass	%	Density	Biomass	%	Density	Biomass
Gastropoda	100	440	673	100	371	486	100	277	373
Oligochaeta	81	2409	2.4	100	4917	5.0	81	4630	4.8
Diptera	100	2058	4.6	100	5653	9.3	100	4043	7.3

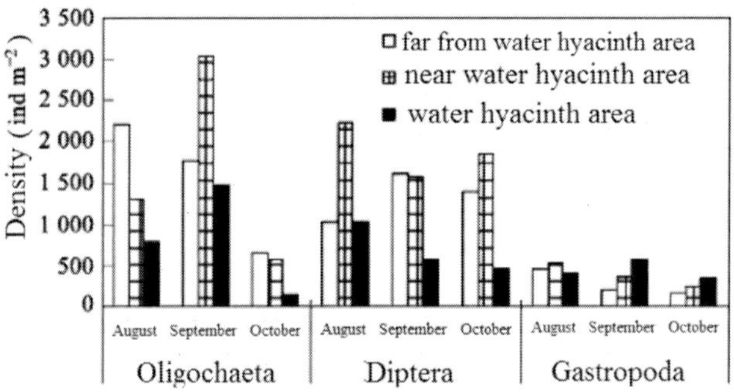

Fig. 9.4.1-2. Density changes of macrozoobenthos from August to October 2009 in the sampling areas in Lake Taihu (redrawn with permission).

Temporal and spatial changes in macrozoobenthos biomass and community structure

Given that the individual Gastropoda specimens generally have larger biomass than those in Oligochaeta and Diptera, even a small number of individual Gastropoda collected would have larger biomass than a relatively large number of Oligochaeta and Diptera specimens. In the area far from water hyacinth, the biomass of Oligochaeta and Gastropoda declined from August to October, while biomass of Diptera increased from August to September and then decreased in October (Table 9.4.1-3). In the area near water hyacinth, the biomass of Gastropoda declined from August to October, while Oligochaeta biomass increased from August to September and then decreased in October, and Diptera biomass declined from August to September and then increased in October. In the water hyacinth area, the biomass of Diptera declined from August to October, while biomass of Oligochaeta increased from August to September and then decreased in October, and Gastropoda biomass increased from August to September and then decreased slightly in October. In phytoremediation of eutrophic waters, species of Oligochaeta and Diptera may be regarded as indicators of the eutrophic status, whereas Gastropoda species are indicators of healthy aquatic environments.

Table 9.4.1-3. Average biomass (g m^{-2}) of macrozoobenthos from August to October 2009 at three sampling areas.

Sampling period	August			September			October		
Sampling group	O	D	G	O	D	G	O	D	G
Area far from water hyacinth	2.58	1.17	502	1.42	3.11	303	0.82	2.99	315
Area near water hyacinth	2.19	3.76	608	2.39	1.86	439	0.44	3.64	413
Water hyacinth area	0.75	1.93	507	1.49	1.77	854	0.18	0.90	657
Average for all areas	1.84	2.29	539	1.77	2.25	532	0.48	2.51	462

Notes: O = Oligochaeta; D = Diptera; G = Gastropoda.

Changes in biodiversity at the sampled areas

Shannon-Wiener and Gini-Simpson Biodiversity indices changed the same way as the density and biomass of the species in the three sampling areas, i.e., increased from August to September and then declined slightly in October. The two indices showed large variation and were higher in the water hyacinth area than the areas far from and near water hyacinth, especially in August and September (Fig. 9.4.1-3).

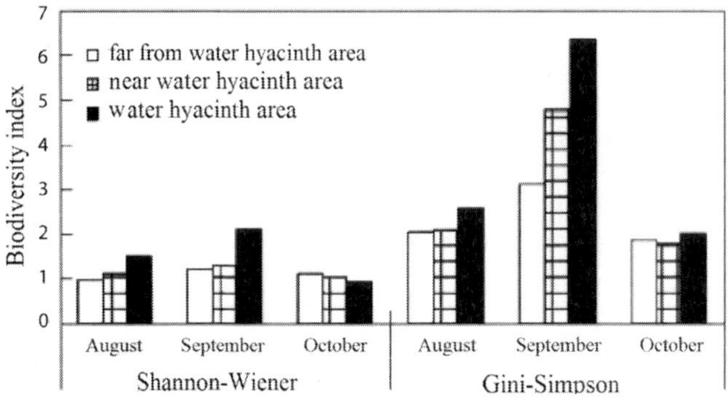

Fig. 9.4.1-3. Changes of biodiversity indexes of macrozoobenthos from August to October 2009 in the sampling areas (redrawn with permission).

9.4.2 Zooplankton at Zhushan Bay in Lake Taihu

Sample collection and analysis

The survey of zooplankton was conducted in the same areas (Fig. 9.4.1-1) and but different sampling schedule from 22 August to 11 November 2009. Zooplankton samples were collected from 0 to 0.5 m depth using a 1-L Patalas sampler. Samples were filtered through 35-μm mesh net, the zooplankton were preserved in a 5% v/v formaldehyde solution, identified and sorted to the genus level, and counted directly using a Nikon microscope at 100 x magnification.

Composition of zooplankton

Altogether 22 genera of zooplankton (protozoa not identified) were found, among which seven genera of Cladocera; five genera of Copepoda; and 10 genera of Rotifera (Table 9.4.2-1).

The species were evenly distributed in the areas of water hyacinth, near and far away with 19, 20 and 19 genera, respectively. Among the total genera, 17 were commonly distributed in all three sampled areas; only five genera showed a different distribution pattern in the three areas. The data indicated that the ecological engineering project using water hyacinth for phytoremediation of the large lake had little effects on the zooplankton community structure.

Changes in zooplankton density

The Cladocera average density in the water hyacinth area and nearby was significantly higher than in the area far from water hyacinth ($p < 0.05$). The average density of copepods showed an increasing (but non-significant) trend from the area far from water hyacinth to the area near water hyacinth and the water hyacinth area. The Rotifera density was significantly lower in the water hyacinth area than in the areas near and far from water hyacinth ($p < 0.05$). Total zooplankton density was higher in the water hyacinth and nearby areas compared with the area far from water hyacinth ($p < 0.05$, Table 9.4.2-2).

Temporal and spatial distribution of zooplankton

In the three sampled areas, the densities of Cladocera, Copepoda and Rotifera showed a similar pattern during August to November 2009, with no significant differences (Figs. 9.4.2-2a, 9.4.2-2b and 9.4.2-2c). The data suggested that the adverse effects of water hyacinth on zooplankton were very limited at Zhushan Bay in Lake Taihu if the population of the macrophyte can be controlled in properly-sized enclosures and harvested on time (Chen et al. 2012).

Table 9.4.2-1. List of taxa and dominant species in the sampled areas at Zhushan Bay in Lake Taihu from 22 August to 11 November 2009.

Taxon			Distribution			Dominant species (%)		
Zooplankton	**Family**	**Genus**	**FWHA**	**NWHA**	**WHA**	**FWHA**	**NWHA**	**WHA**
Cladocera	Bosminidae	*Bosmina Baird*	*	*	*	75	85	74
	Daphniidae	*Ceriodaphnia Dana*	+	+	+			
		Daphnia (D.s. str.)	+	+	+			
		Daphnia (D. carinata)	+	N	N			
	Moinidae	*Moina Baird*	+	+	+			
	Chydoridae	*Alona Baird*	+	+	+			
	Sididae	*Diaphanosoma Fischer*	N	+	+			
Copepoda	Oithonidae	*Limnoithona Burckhardt*	+	+	+			
	Centropagidae	*Sinocalanus Burckhardt*	N	+	+			
	Cyclopidae	*Mesocyclops Sars*	*	*	*	59	51	61
		Cyclops Müller	+	+	+			
		Microcyclops Claus	*	*	*	35	28	35
Rotifera	Lecanidae	*Monostyla*	*	*	*	22	6	20
		Lecane	+	+	+			
	Gastropodidae	*Ascomorpha*	+	+	+			
	Lindiidae	*Lindia*	+	N	N			
	Testudinellidae	*Filinia*	+	+	+			
	Philodinidae	*Rotaria*	+	+	+			
	Trichocercidae	*Trichocerca*	N	+	N			
	Brachionidae	*Brachionus*	*	*	*	46	57	49
		Keratella	*	*	*	23	28	26
		Lepadella	+	+	+			
Total	15 (6 + 3 + 7)	22 (7 + 5 + 10)	19	20	19			

Note*: dominant genus; +: present; N: not present; AFWH: area far from water hyacinth; ANWH: area near water hyacinth; WHA: water hyacinth area.

Table 9.4.2-2. Average density (individuals L^{-1}) of zooplankton in the three sampled areas from 22 August to 11 November 2009 at Zhushan Bay in Lake Taihu.

Zooplankton	Water hyacinth area	Area near water hyacinth	Area far from water hyacinth
Cladocera	$57.6\pm4.^{3}$a	51.2 ± 2.8^{a}	29.5 ± 6.3^{b}
Copepoda	4.5 ± 1.2^{a}	3.7 ± 0.8^{a}	3.2 ± 0.4^{a}
Rotifera	15.7 ± 1.9^{b}	20.5 ± 1.9^{a}	22.8 ± 2.6^{a}
Total	77.8 ± 5.8^{a}	75.4 ± 4.0^{a}	55.5 ± 7.8^{b}

Note: Different low case letters indicate significant differences among the areas.

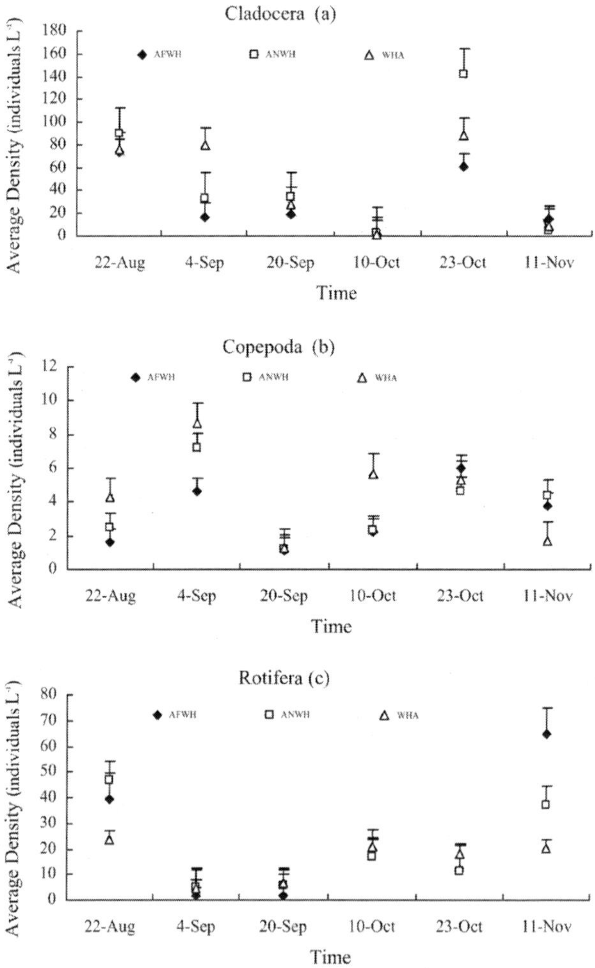

Fig. 9.4.2-1. Average density (mean + SE) of zooplankton from 22 August to 11 November 2009 in the area far from water hyacinth (AFWH), area near water hyacinth (ANWH) and water hyacinth area (WHA) (redrawn with permission).

Biodiversity indexes

Shannon-Wiener (H') and Gini-Simpson (D) indices for each zooplankton group (Cladocera, Copepoda and Rotifera) were not statistically different in the sampling period (Table 9.4.2-3). Hence, the data suggested that both the biodiversity and the stability of the zooplankton community were not significantly influenced by the confined growth of water hyacinth in Lake Taihu.

Table 9.4.2-3. Diversity indices of zooplankton sampled from 22 August to 11 November 2009 in the area far from water hyacinth (AFWH), the area near water hyacinth (ANWH) and the water hyacinth area (WHA).

Zooplankton	Shannon-Wiener index (H')			Gini-Simpson index (D)		
	FWHA	NWHA	WHA	FWHA	NWHA	WHA
Cladocera	0.34	0.24	0.39	0.61	0.45	0.58
Copepoda	0.40	0.46	0.29	0.60	0.79	0.48
Rotifera	0.57	0.57	0.61	1.05	1.04	1.11

9.5 Discussion and conclusions of the impact of water hyacinth on water quality and biodiversity

This case study investigated the water quality at Baisan Bay and Cao Hai in Lake Dianchi, and macrozoobenthos and zooplankton at Zhushan Bay in Lake Taihu and Baishan Bay in Lake Dianchi. Lake Taihu and Lake Dianchi are two large shallow lakes with hypereutrophic status; hence, the results from this case study may not be applicable to deep lakes and less eutrophic lakes and reservoirs, but the principles would be sound in general. During the investigation, the assessment of macrozoobenthos and zooplankton at Cao Hai and evaluation of water quality at Zhushan Bay were not done for various reasons. Further research is warranted also because of the complexity of the management of noxious weeds, water quality improvement and maintenance, and ecosystem management.

Summary of water quality at Baishan Bay

In the water hyacinth area, dissolved oxygen concentration decreased compared with the control area, but was still above the critical level for most species of fish and other aquatic animals. One important concern using water hyacinth for phytoremediation in eutrophic waters is that the water hyacinth mat may decrease dissolved oxygen concentration to cause damage to aquatic ecosystems. Dissolved oxygen in water is controlled primarily by photosynthesis of algae and submerged macrophytes plus oxygen exchanges at the air-water interface (Hunt and Christiansen 2000). Water hyacinth mat can impact all the primary contributors by blocking incident sunlight via covering the air-water interface, and reducing wind speed and waves. Also, decomposition of intercepted detritus and algae may result in consumption of a relatively large amount of oxygen, further depressing dissolved oxygen concentration (Meerhoff et al. 2003). Another situation occurring in eutrophic waters is the stratification of dissolved oxygen concentration in the water column because sunlight penetration into water is relatively low, restricting photosynthesis to the surface layer and causing oxygen depression in the lower layers of the water column (Wang et al. 2010). The results of this experiment suggest that the location and the size of water hyacinth mat versus the open space between enclosures should be carefully managed in order to optimize dissolved oxygen concentration in water with heavy algal load to avoid depression of dissolved oxygen. For example, in a 0.4-hectare test pond with water hyacinth coverage of less than 25% of the water surface, dissolved oxygen concentration was above the critical level for fish (McVea and Boyd 1975).

A decrease of pH in the water hyacinth area at the Baishan Bay was influenced by the biological nature of the macrophyte and the geology of the rock base (prevalence of carbonates in the calcium–magnesium–sodium material, with a pH of 8–10) (Institute of Environmental Science 1992). A decrease in water pH by the growth of water hyacinth may have resulted from intensive respiration of water hyacinth roots releasing carbon dioxide into the water, while low incident sunlight penetration in the water column resulted in relatively low photosynthesis and poor absorption of carbon dioxide from water. Water hyacinth is a good water pH stabilizer (Giraldo and Garzón 2002) because in acidic aquatic environments, the water pH may be increased towards neutral by the growth of water hyacinth.

In the water hyacinth area in this case study, the concentrations of total phosphorus, orthophosphate and total nitrogen were higher than those in the areas near and far from water hyacinth. If taking the hydraulic and wind conditions into consideration, and the data from chlorophyll-*a* and COD_{Mn} analysis, it may be suggested that water hyacinth roots trapped a lot of suspended particles and algae that were decomposed, causing a release of nutrients into the water (Kim and Kim 2000).

Although the dissolved oxygen concentration was above the critical level during the investigation period, there was depression of dissolved oxygen under the water hyacinth mat. This situation may not only impact aquatic animals, but also biological processes in the sediment, potentially increasing the nutrient release (Zhu et al. 2009). In the application of phytoremediation, the water hyacinth mat size, gap between mats, and harvesting strategies need to be further investigated for the best remediation result.

The case study found an interesting phenomenon, i.e., NH_4^+ concentration in the water hyacinth area and nearby was lower than in the area far from water hyacinth, while NO_3^- concentration showed opposite distribution. This finding may be caused by water hyacinth biology: (1) preferential absorption of ammonium over nitrate, with ammonium absorption occurring at two-fold greater rate than nitrate (Rommens et al. 2003); and (2) microbes accumulated on the water hyacinth roots may enhance transformation of ammonium to nitrate (Gao et al. 2012).

Water hyacinth can effectively decrease nitrogen and phosphorus concentrations in the surface water of relatively shallow lakes (<2.5 meters). In good conditions, ammonium can be totally eliminated (concentration down to 0 or below the detection limit) during the active growth of water hyacinth.

Phosphorus can be removed in excess of what was initially present in the surface water (it should be borne in mind that phosphorus may be released from the sediment when its concentration is very low in the overlying water). This phenomenon occurs in shallow lakes, such as Cao Hai in Lake Dianchi, and may not be relevant in deep lakes.

Successful phytoremediation relies on confining water hyacinth to designed locations as well as complete and timely harvesting of the huge fresh biomass of water hyacinth. The experiment at Cao Hai suggested there are benefits in effectively expanding water hyacinth population during the growing season (May to November for Cao Hai) to increase the biomass of water hyacinth population followed by harvesting as early as possible to control the population at a suitable size and thus reduce the management and equipment costs.

Macrozoobenthos at Baishan Bay

During the experimental period, 18 species of macrozoobenthos were collected from Baishan Bay: 14 in the water hyacinth area, 10 in the nearby area and six in the area far from water hyacinth. Margalef, Gini-Simpson and Shannon-Wiener indices were significantly higher in the water hyacinth area than the areas near and far away. The number of species and the biodiversity indices appeared not to support the hypothesis that water hyacinth negatively impacts biodiversity of macrozoobenthos. The reason may be that water hyacinth has complex effects on macrozoobenthos, including improvement of water quality to provide better ambient environment for macrozoobenthos, and providing shelter and food resources for macrozoobenthos. For example, Gastropoda (*Radix swinhoe*) and Crustacea (Decapoda, *Caridina* sp. and *Gammaridae* spp.) were present only in the water hyacinth area. This result was consistent with the literature report from Lake Okeechobee, Florida, USA (O'Hara 1967). Other researchers reported that macroinvertebrates such as Gastropoda and Arachnida were associated mostly with the water hyacinth areas in Lake Chivero, Zimbabwe (Brendonck et al. 2003), or that higher density and biodiversity of macroinvertebrates were associated with water hyacinth mats compared with the areas of open water and submerged macrophytes in Lake Chapala, Mexico (Villamagna 2009).

Concentration of dissolved oxygen was lower in the water hyacinth area than the areas nearby and far away. The extent of depression of dissolved oxygen concentration by water hyacinth is important. During the experiment in Lake Dianchi, the special design of the ecological engineering project and appropriate management strategies resulted in dissolved oxygen concentration maintained above 3.8 mg L^{-1}. However, whether or not such design and strategies would play the same roles at other locations needs to be further researched, with the case study providing only a general theoretical principle to underpin practice.

In hypereutrophic lakes, macrozoobenthos were largely pollutant-tolerant species that can survive at low dissolved oxygen concentration (Ellis 2011). The analysis of composition of functional feeding groups at the Baishan Bay also revealed that collector-gatherers were mainly associated with water hyacinth. The phenomenon may imply that collector-gatherers were adapted to the food sources associated with water hyacinth: plant detritus, roots and trapped algae. However, in the three monitoring areas, more functional feeding groups (including shredders, scrapers and parasites) appeared in the water hyacinth area than in the areas nearby and far away. This may imply that strategic phytoremediation design and controlled growth of water hyacinth may have positive impacts on abundance and biodiversity of aquatic fauna communities.

Zooplankton at Baishan Bay in Lake Dianchi

A total of 28 species in the area near water hyacinth was higher than the total number of species in the water hyacinth area and far from water hyacinth area. This result was consistent with the conclusion about the water quality being good in the area near water hyacinth (see earlier). The good water quality in the area near water hyacinth may also have contributed to the densities of Cladocera and Copepods gradually increasing from August to October (Fig. 9.3.2-1).

The densities of Cladocera and Copepoda, and the Shannon-Wiener, Gini-Simpson, Margalef and Peilou indices were not significantly different in the three sampling areas. This result may imply that short-term ecological engineering project has a minimal impact on Cladocera and Copepoda, which is consistent with the literature reports (Meerhoff et al. 2003, Chen et al. 2012). Meerhoff et al. (2003) found that micro-crustaceans abundance and biodiversity index were not significantly different between the submerged plant area and the water hyacinth area in hypertrophic Lake Rodó (34°55' S 56°10' W), Uruguay.

The case study found that water hyacinth significantly affected rotifer community structure, especially the dominant species in different monitoring areas. In the water hyacinth area, rotifer density and the Shannon-Wiener, Margalef, Gini-Simpson and Peilou indices were significantly lower than those in the near and far away areas. Literature also reported that rotifer density was lower in the water hyacinth area than the open area in the backwaters of the Delhi Segment of the Yamuna River (Arora and Mehra 2003). Large crustacean zooplankton can prey on rotifers to reduce their population (Yang et al. 2008). In this case study, the density of large crustacean zooplankton was lower in water under the water hyacinth mat than in the open water nearby and far away, which implies that the low rotifer density in the water hyacinth area may not have been caused by the crustacean zooplankton. In natural conditions, a lot of environmental factors such as incident sunlight, turbidity, water temperature, algal population and composition, dissolved oxygen and available food sources may influence the rotifer density and community structure (Villamagna and Murphy 2010). The actual reasons for the lower rotifer population in the water hyacinth area reported here needs to be further investigated.

In this case study, the active water hyacinth growing period (August to October) was selected to investigate zooplankton population dynamics as influenced by the presence of macrophyte since the period before July was the establishment stage of water hyacinth. The results suggested that using water hyacinth for phytoremediation may not cause damage to zooplankton community if the growth area, size of the water hyacinth mat and biomass can be controlled effectively. However, the size of the mat, the period of active growth, and the open space between water hyacinth mats need to be determined taking into account local climate, hydraulic patterns and the extent of eutrophic status because they all impact the management strategy. The basic principle is to design small enclosures and leave large enough gaps (such as twice the mat size) for initial set up.

Macrozoobenthos at Zhushan Bay in Lake Taihu

During the sampling period in different areas, macrozoobenthos were mainly Tubificidae in Oligochaeta (*Limnodrilus hoffmeisteri* and *Rhyacodrilus sinicus*), Chironomidae in Diptera (*Pelopia* sp. and *Chironomus plumosus*) and Gastropoda (*Bellamya aeruginosa*). Given that *Limnodrilus hoffmeisteri* and Chironomidae larvae can survive well in heavily polluted aquatic environments, they are often the dominant species indicating the contamination status (Riley et al. 2007, Kazanci and Girgin 1998). In the study presented here, Tubificidae (*Limnodrilus hoffmeisteri*) and the Chironomidae larvae were more abundant in the areas far from and near water hyacinth compared with the water hyacinth area in August and September, but significantly declined in October 2009, and also declined continuously from August to October in the water hyacinth area. The results should be interpreted together with the local environmental conditions and the settings for the confined growth of water hyacinth. Although the experiment area was set at the center of Zhushan Bay, it was still subjected to wave (wind) action and hydraulic current to mix the water layers, minimizing the negative effects on dissolved oxygen concentration.

The Gastropoda (*Bellamya aeruginosa*) biomass and density increased from August to September, and then decreased slightly in October 2009 in the water hyacinth area compared to the areas far away and near water hyacinth in Zhushan Bay, Lake Taihu. Gastropoda are ecologically important in aquatic environments. Adults of *Bellamya aeruginosa* live mainly at the bottom and feed on benthic algae, bacteria and organic debris and have adaptability ecological niches (Cai et al. 2009). It is assumed that dense fibrous root system of water hyacinth can intercept algae and detritus, creating favorable conditions and food resources for Gastropoda. The dense mat of the macrophyte may also modulate hydraulic patterns to make the environment more suitable to Gastropoda (Zhu 2007, Yuan et al. 2008) in addition to the suitable water temperature (30ºC) at the time of sampling.

In October, the biomass and density of Gastropoda and the dominant species of Oligochaeta and Chironomidae in Diptera all declined in October in all three sampled areas. This phenomenon may be caused by the changes in aquatic environment at Zhushan Bay in Lake Taihu. The water temperature and dissolved oxygen at the region decreased in October due to seasonal changes and increased decomposition of water hyacinth detritus and intercepted algae. The measured dissolved oxygen was as low as 3.2 mg L^{-1} in October 2009. Although the exact reasons for the decreased macrozoobenthos in the region may need further investigation, it is fairly certain that the confined growth of water hyacinth did not cause damage to biodiversity of macrozoobenthos during the experimental period.

Biodiversity and zooplankton at Zhushan Bay in Lake Taihu

The zooplankton biodiversity indices showed the same pattern as Gastropoda macrozoobenthos. The Shannon-Wiener and Gini-Simpson indices for Cladocera, Copepoda and Rotifera were not significantly different during the monitoring period among different sampled areas. These results from the investigation demonstrated that using water hyacinth as a biological agent for phytoremediation in eutrophic waters did not cause a significant impact on biodiversity and community structure of zooplankton.

Conclusions

This case study shows the results of the ecological engineering project using water hyacinth in two typical hypereutrophic lakes focusing on water quality and biodiversity on macrozoobenthos and zooplankton. The results showed that water hyacinth can effectively remove nitrogen and phosphorus and other pollutants from the water. However, the impact of water hyacinth on the water quality depends on the balance of input and output of plant nutrients, the growth period, the population size of water hyacinth compared to the total water surface area or hydraulic retention time in a particular water body. For example, at the Baishan Bay and Zhushan Bay, a small quantity of water hyacinth mat did not result in any water quality change in the lakes. On the other hand, at Cao Hai in Dianchi Lake, the water quality improved significantly due to the large coverage (~50% of the total water surface) of water hyacinth. The impact of water hyacinth on macrozoobenthos was mostly positive; and on zooplankton was not significant. This is because water hyacinth can improve water quality and provide food resources and shelter for aquatic fauna. Another important reason is that in this case study, the growth of water hyacinth was confined and completely harvested and processed, so that the negative impacts on dissolved oxygen were well controlled.

A good management system in such ecological engineering project means increased investment and technology inputs. For this reason, the most appropriate methods would vary for different locations, including the size of water hyacinth mat, gaps between mats, growth period, initial population size, time and methods of harvesting and processing, equipment, etc. Different water bodies would have different sources of pollutants, hydraulic properties and local climate (incident sunlight, wind, primary production pattern, etc.). All those factors influence the outcomes of an ecological engineering project. The case study presented here may provide theoretical underpinning, particularly regarding the effects of small water hyacinth coverage on macrozoobenthos and zooplankton in relatively small bays in lakes. At large coverage of 50% at Cao Hai in Lake Dianchi, the impact of water hyacinth on benthic and planktonic fauna was unfortunately not investigated, which should not be implied to mean unimportance; rather, the investigation was interrupted for various reasons. The preliminary investigation found that the various types of pollution at Cao Hai in Dianchi Lake were quite serious; the benthic fauna either did not exist or was represented only by the pollution indicator *Limnodrilus hoffmeisteri* and Chironomidae larvae.

By evaluating the biology of water hyacinth, water quality improvement, biodiversity of macrozoobenthos and zooplankton as well as reclamation of water resources, the ecological engineering using water hyacinth for phytoremediation was potentially practicable. The present study can be viewed as pioneering pollution control and remediation in practice.

References cited

Arora, J. and N. K. Mehra. 2003. Species diversity of planktonic and epiphytic rotifers in the backwaters of the Delhi Segment of the Yamuna River, with remarks on new records from India. *Zoological Studies* 42(2): 239–247.

Babourina, O. and Z. Rengel. 2011. Nitrogen removal from eutrophicated water by aquatic plants. In: Eutrophication: causes, consequences and control, ed. A. A. Ansari, S. S. Gill, G. R. Lanza, and W. Rast, 355–372. Dordrecht, Netherlands: Springer Netherlands.

Brendonck, L., J. Maes, W. Rommens, N. Dekeza, T. Nhiwatiwa, M. Barson et al. 2003. The impact of water hyacinth (*Eichhornia crassipes*) in a eutrophic subtropical impoundment (Lake Chivero, Zimbabwe). II. species diversity. *Archiv für Hydrobiologie* 158(3): 389–405.

Cai, Y., Z. Gong and B. Qin. 2009. Standing crop and spatial distribution pattern of mollusca in Lake Taihu, 2006–2007. *Journal of Lake Sciences* (*China*) 21(5): 713–719 (In Chinese with English Abstract).

Chen, H. G., F. Peng, Z. Y. Zhang, G. F. Liu, W. Da Xue, S. H. Yan et al. 2012. Effects of engineered use of water hyacinths (*Eichhornia crassipes*) on the zooplankton community in Lake Taihu, China. *Ecological Engineering* 38(1): 125–129.

Dai, S. and G. Che. 1987. Removal of some heavy metals from wastewater by waterhyacinth. *Environmental Chemistry* 6(2): 43–50 (In Chinese with English Abstract).

Ellis, A. T. 2011. Invasive species profile water hyacinth, *Eichhornia crassipes*. *University of Washington*. Seattle, WA, USA. http://depts.washington.edu/oldenlab/wordpress/wp-content/uploads/2013/03/ Eichhoria-crassipes_Ellis.pdf.

Elser, J. J., M. E. S. Bracken, E. E. Cleland, D. S. Gruner, W. S. Harpole, H. Hillebrand et al. 2007. Global analysis of nitrogen and phosphorus limitation of primary producers in freshwater, marine and terrestrial ecosystems. *Ecology Letters* 10(12): 1135–1142.

Fan, C. and M. Aizaki. 1997. Effects of aerobic and anaerobic conditions on exchange of nitrogen and phosphorus across sediment-water interface in Lake Kasumigaura. *Journal of Lake Sciences (China)* 9(4): 337–342 (In Chinese with English Abstract).

Gao, Y., N. Yi, Z. Zhang, H. Liu, L. Zou, H. Zhu et al. 2012. Effect of water hyacinth on N_2O emission through nitrification and denitrification reactions in eutrophic wate. *Acta Scientiae Circumstantiae* 32(2): 349–359 (In Chinese with English Abstract).

Giraldo, E. and A. Garzón. 2002. The potential for water hyacinth to improve the quality of Bogota River water in the Muña Reservoir: comparison with the performance of waste stabilization ponds. *Water Science & Technology* 45(1): 103–110.

Han, S., S. Yan, Z. Wang, W. Song, H. Liu, J. Zhang et al. 2009. Harmless disposal and resources utilizations of Taihu Lake blue algae. *Journal of Natural Resources* 24(3): 431–438 (In Chinese with English Abstract).

Huang, Y., H. Wen, J. Cai, M. Cai and J. Sun. 2010. Key aquatic environmental factors affecting ecosystem health of streams in the Dianchi Lake watershed, China. *Procedia Environmental Sciences* 2: 868–880.

Hunt, R. J. and I. H. Christiansen. 2000. *Understanding dissolved oxygen in streams. In Information Kit*. 1st ed. Townsville Qld, Australia: CRC Sugar Technical Publication (CRC Sustainable Sugar Production).

Institute of Environmental Science. 1992. *Survey on eutrophication of Lake Dianchi*. 1st ed. Kunming, China: Kunming Science and Technology Press.

Kang, Z., J. Tang and H. Liu. 2012. Eco-compensation based on cost-benefit analysis of nitrogen and phosphorus removal in Taihu Lake region. In: Advances in China farming system research 2012, ed. Farming System Committee, Chinese Agricultural Association, 182–187 p. (In Chinese). Beijing, China: China Agricultural Science and Technology Press.

Kazanci, N. and S. Girgin. 1998. Distribution of oligochaeta species as bioindicators of organic pollution in Ankara stream and their use in biomonitoring. *Turkish Journal of Zoology* 22: 83–87.

Kim, Y. and W. Kim. 2000. Roles of water hyacinths and their roots for reducing algal concentration in the effluent from waste stabilization ponds. *Water Research* 34(13): 3285–3294.

Liu, G., S. Han, J. He, S. Yan and Q. Zhou. 2014. Effects of ecological purification engineering of planting water hyacinth on macro-benthos community structure. *Ecology and Environmental Sciences* 23(8): 1311–1319 (In Chinese with English Abstract).

McVea, C. and C. E. Boyd. 1975. Effects of waterhyacinth cover on water chemistry, phytoplankton, and fish in ponds. *Journal of Environmental Quality* 4(3): 375–378.

Meerhoff, M., N. Mazzeo, B. Moss and L. Rodríguez-Gallego. 2003. The structuring role of free-floating versus submerged plants in a subtropical shallow lake. *Aquatic Ecology* 37(4): 377–391.

MEP-PRC. 2002. Environmental quality standards for surface water (GB3838-2002). Baijing, China: Ministry of Environmental Protection of The Peoples's Republic of China.

O'Hara, J. 1967. Invertebrates found in water hyacinth mats. *Quarterly Journal of the Florida Academy of Science* 30(1): 73–80.

Riley, C., S. Inamdar and C. Pennuto. 2007. Use of benthic macroinvertebrate indices to assess aquatic health in a mixed-landuse watershed. *Journal of Freshwater Ecology* 22(4): 539–551.

Risgaard-Petersen, N. and K. Jensen. 1997. Nitrification and denitrification in the rhizosphere of the aquatic macrophyte *Lobelia dortmanna* L. *Limnology and Oceanography* 42(3): 529–537.

Rommens, W., J. Maes, N. Dekeza, P. Inghelbrecht, T. Nhiwatiwa, E. Holsters et al. 2003. The impact of water hyacinth (*Eichhornia crassipes*) in a eutrophic subtropical impoundment (Lake Chivero, Zimbabwe). I. water quality. *Archiv für Hydrobiologie* 158(3): 373–388.

Simenstad, C. A., C. D. Tanner, R. M. Thom and L. L. Conquest. 1991. *Estuary habitat assessment protocol*. Seattle, Washington, U.S.A: U.S. Environmental Protection Agency.

Smith, V. 2003. Eutrophication of freshwater and coastal marine ecosystems: a global problem. *Environmental Science and Pollution Research* 10(2): 126–139.

Surridge, B. W. J., A. L. Heathwaite and A. J. Baird. 2007. The release of phosphorus to porewater and surface water from river riparian sediments. *Journal of Environmental Quality* 36(5): 1534–44.

Villamagna, A. M. 2009. Ecological effects of water hyacinth (*Eichhornia crassipes*) on Lake Chapala, Mexico. Ph.D. Thesis, Virginia Polytechnic Institute and State University, Blacksburg, Virginia, USA.

Villamagna, A. M. and B. R. Murphy. 2010. Ecological and socio-economic impacts of invasive water hyacinth (*Eichhornia crassipes*): a review. *Freshwater Biology* 55 (2): 282–298.

Wang, F. S., C. Q. Liu, M. H. Wu, Y. X. Yu, F. W. Wu, S. L. Lu et al. 2009. Stable isotopes in sedimentary organic matter from Lake Dianchi and their indication of eutrophication history. *Water, Air, & Soil Pollution* (199): 159–170.

Wang, Z., B. Xiao, X. Wu, X. Tu, Y. Wang, X. Sun et al. 2010. Linear alkylbenzene sulfonate (LAS) in water of Lake Dianchi - spatial and seasonal variation, and kinetics of biodegradation. *Environmental Monitoring and Assessment* 171(1-4): 501–12.

Wang, Z., Z. Zhang, Y. Han, Y. Zhang, Y. Wang and S. Yan. 2012. Effects of large-area planting water hyacinth (*Eichhornia crassipes*) on water quality in the bay of Lake Dianchi. *Chinese Journal of Environmental Engineering* 6(11): 3827–3832 (In Chinese with English Abstract).

Wang, Z., Z. Zhang, J. Zhang, Y. Zhang, H. Liu and S. Yan. 2012. Large-scale utilization of water hyacinth for nutrient removal in Lake Dianchi in China: the effects on the water quality, macrozoobenthos and zooplankton. *Chemosphere* 89(10): 1255–1261.

Wang, Z., Z. Zhang, Y. Zhang, J. Zhang, S. Yan and J. Guo. 2013. Nitrogen removal from Lake Caohai, a typical ultra-eutrophic lake in China with large scale confined growth of *Eichhornia crassipes*. *Chemosphere* 92(2): 177–183.

Xu, H., H. W. Paerl, B. Qin, G. Zhu and G. Gao. 2010. Nitrogen and phosphorus inputs control phytoplankton growth in eutrophic Lake Taihu, China. *Limonology and Oceanography* 55(1): 420–432.

Yang, G., B. Qin, G. Gao, G. Zhu, X. Tang and X. Wang. 2008. Comparative study on seasonal variations of community structure of rotifer in different lake areas in Lake Taihu. *Environmental Science* 29(10): 2963–2969.

Yuan, M., B. Huang, X. Qiu, H. Xu and Q. Chen. 2008. Effects of water hyacinth mat on hydraulic dynamic. *Guangdong Water Resources and Hydropower* 2(1): 7–10 (In Chinese).

Zhang, Z., Y. Gao, J. Guo and S. Yan. 2014. Practice and reflections of remediation of eutrophicated waters: a case study of haptophyte remediation of the ecology of Dianchi. *Journal of Ecology and Rural Environment* 30(1): 15–21 (In Chinese with English Abstract).

Zhang, Z. Y., J. C. Zheng, H. Q. Liu, Z. Z. Chang, L. G. Chen and S. H. Yan. 2010. Role of *Eichhornia crassipes* uptake in the removal of nitrogen and phosphorus from eutrophic waters. *Chinese Journal of Eco-Agriculture* 18(1): 152–157 (In: Chinese with English Abstract).

Zhao, H. W., D. S. Mavinic, W. K. Oldham and F. A. Koch. 1999. Controlling factors for simultaneous nitrification and denitrification in a two-stage intermittent aeration process treating domestic sewage. *Water Research* 33(4): 961–970.

Zheng, J., Z. Chan, L. Chen, P. Zhu and J. Shen. 2008. Feasibility studies on N and P removal using water hyacinth in Taihu Lake region. *Jiangsu Agricultural Science* 3: 247–250 (In Chinese).

Zhu, H. 2007. Research on hydrodynamic behavior of flow through *Eichhornia crassipes* in ecological watercourse, MSc. Thesis, Department of Environmental Hydraulic Science. Hohai University.

Zhu, J., H. Li and P. Wang. 2009. The impact of environmental factors on COD, TN, TP release from sediment. *Technology of Water Treatment* 35(8): 44–49 (In Chinese with English Abstract).

Part Four

Utilization of Water Hyacinth Biomass as Natural Resource

CHAPTER 10

Utilization of Biomass for Energy and Fertilizer

X. M. Ye

10.1 Introduction

Water hyacinth is one of the plants with the highest production capacity due to its biological characteristics of strong photosynthesis adapted to wide range of Photosynthetically-Active Radiations (PAR), rapid growth and prolific reproduction (Abbasi and Ramasamy 1999). Literature revealed that, although water hyacinth is a C3 plant, its maximum photosynthetic rate and light saturation point were much higher than those of typical C3 plant, such as rice (Zheng et al. 2011). Water hyacinth is adapted to a wide range of photosynthetically-active radiation from 20 to 2458 μmol m^{-2} s^{-1}, enabling the species to make maximal use of solar energy to produce and rapidly accumulate organic matter (biomass) and have a growth rate that is comparable to or superior than that of C4 plant, such as corn (Li et al. 2010).

Under suitable conditions, the number of water hyacinth plants could double in 5 days, and the fresh biomass could reach 270–720 tonnes ha^{-1} yr^{-1} (Li et al. 2011, Patil et al. 2014). The results of experiments conducted in Lake Taihu (31°27' N 120°4' E) and Nanjing (32°02' N 118°52' E) showed that water hyacinth grew rapidly in water and with an initial stocking rate of 0.06 kg m^{-2} dry weight, the biomass could increase to 0.14 kg m^{-2} after 7 days, 0.29 kg m^{-2} after 14 days and 1.22 kg m^{-2} after 42 days. When the density reached 0.49–0.86 kg m^{-2}, water hyacinth achieved the fastest growth rate of 0.053 kg m^{-2} d^{-1} (Zheng et al. 2011). The report on monitoring growth of water hyacinth at Caohai in Lake Dianchi (24°45' N 102°36' E) revealed that the largest growth rates of water hyacinth were 0.759 kg m^{-2} d^{-1} fresh weight

50 Zhong Ling Street, Nanjing, China.
Email: yexiaomei610@126.com

in Dongfeng dam and 0.602 kg m^{-2} d^{-1} in Laogan fishpond at 24°C (Zhang et al. 2014). The growth rates of water hyacinth in Dongfeng dam and Laogan fishpond are substantially higher than that in Lake Taihu due to initial density of standing crops being much higher at 2.25 kg m^{-2} than that in Lake Taihu. If the water hyacinth biomass in natural water body is not effectively controlled and timely harvested, it would expand on the water surface very quickly and can clog rivers and block irrigation channels; finally, upon dying the biomass would rot and pollute water and degrade water ecosystems. Given this invasive species covers large geographical area and consequently causes huge economic, social and ecological damage; it is a challenge to develop technologies for utilizing water hyacinth biomass. This challenge is not only to minimize the damage caused by water hyacinth, but also to mitigate the eutrophication of water bodies as well as to recover nutrients, reclaim water resources, achieve carbon sequestration and produce bio-energy by using integrated management strategies.

The chemical composition of water hyacinth revealed it is rich in carbon, nitrogen, phosphorus, potassium and other inorganic and organic components so that the water hyacinth biomass could be used to produce animal feed, organic fertilizer and bio-energy. Economic analysis showed that water hyacinth as a feedstock biomass could be economically produced at approximately 40 US$ per tonne dry mass including harvest, transportation and dewatering pretreatment using improved equipments and innovative technology (Hronich et al. 2008). This analysis suggested that utilization of water hyacinth biomass for bio-energy production may have a huge market potential.

Utilization of the massive water hyacinth biomass has been the aspiration over the past half century; many studies were performed to improve both equipment and management strategies, but few resulted in profitable commercial operation (Gettys et al. 2009). By carefully studying biology and ecology of water hyacinth, it can be revealed that the main reasons for failure of commercial utilization of water hyacinth are based on the following: (1) Water hyacinth fresh biomass contains 94–95% water, reducing its commercial value and creating difficulties in the process of harvest, transportation and storage. During the commercial operation, storage of raw materials is necessary at a production line, but fresh water hyacinth cannot meet the requirements for storage due to its high water content making it easy to decay; (2) As a single substrate for biogas fermentation, the physical properties and material composition of water hyacinth do not make for the best raw materials for energy production; (3) There is low efficiency of biogas production from water hyacinth; and (4) As an invasive species, water hyacinth should be strictly controlled and managed due to the ecological safety reasons. Considerable progress in designing special equipments for harvest and dehydration of fresh water hyacinth biomass improved the efficiency and made the harvest and dehydration feasible on a large scale, such as 350 tonnes in one day by one machine (Hronich et al. 2008), at a low cost and a target moisture less than 70% in the residual water hyacinth biomass. This makes dehydrated water hyacinth biomass valuable as raw material for industrial production.

10.2 Technologies for utilizing water hyacinth biomass for energy production

Water hyacinth fulfills all the criteria to become an ideal feedstock for bio-energy production and has unique advantages: (1) perennial and growing on water surface rather than on land; (2) availability at naturally growing in tropical and subtropical climate zones and at manual introduced by phytoremediation project after harvest under well managed locations without additional chemical fertilizer or pesticides as in cropping fields; (3) rapid reproduction after repeated population reduction (being harvested) and (4) high yields (Wilkie and Evans 2010). The energy produced per hectare by water hyacinth is equivalent to the energy released by 19,755 kg of standard coal, or 2.2 times the energy equivalent of rice, 1.7 times that of corn, 2.5 times that of oilseed rape and 2.1 times that of sweet potato (Zheng et al. 2011). The ways of utilizing water hyacinth biomass as energy crop include carbonization, incineration, briquetting, ethanol production by hydrolysis fermentation, and hydrogen or methane production by anaerobic fermentation.

1) *Carbonization*

   Carbonization involves three stages, gasification followed by pyrolysis and carbonization. The final products are bio-active charcoal and by-product gas that could be used as an energy source during the production process. The carbonization of water hyacinth biomass previously had two major problems: high moisture content of fresh water hyacinth and high ash content of dried water hyacinth. Both problems led to high production cost and low calorific value of the charcoal product and directly reduced the attractiveness of this technology for commercial application. Additionally, the high investment requirement and technological complexity also hamper utilization, especially in developing countries (Thomas and Eden 1990). Efficiently dewatering the fresh biomass solved the first problem, but the second problem of high ash content remains. Although the calorific value of unit volume could be increased by compression, high ash content makes it less advantageous on commodity market for fuel. Given the product is bio-active charcoal, it may find other applications such as soil amendments or composting for organic fertilizers or for growth media, etc.

2) *Incineration*

   This utilization method refers to directly burning dried water hyacinth biomass and is quite commonly used in developing countries/regions for supplemental energy in daily life. The drying is usually under the sun, and, hence, the moisture content of fresh water hyacinth after drying is inconsistent. In general, even though the moisture content of the biomass may be lowered to less than 15%, its calorific value would not exceed 1.3 GJ m^{-3} (Thomas and Eden 1990), which is lower than the calorific value of wood at 9.8 GJ m^{-3}. For this reason, the incineration of dried water hyacinth biomass is not attractive. In addition, this utilization method could do little to hundreds of millions tonnes of water hyacinth biomass worldwide.

3) *Briquetting*

Briquetting refers to the method of compressing dried biomass after shredding and sifting to produce solid fuel (Thomas and Eden 1990). Drying may also be under the sun because dehydration would not be sufficient to meet the requirements for compression and solid fuel production. The calorific value of solid fuel produced by compressing dried biomass was 8.3 GJ m^{-3} and almost equivalent to charcoal at 9.6 GJ m^{-3}. Solidification molding technology has strict requirements for moisture content of raw materials, generally 10–15% (Wu and Ma 2003). This technology demands huge space, labor and time as well as specific local weather (dry and sunny). Besides, the cost of transporting large amount of biomass may not be recovered in the value of the product: solid fuel which is basically burnt and the ash disposed of. The cost of this technology may prevent the product to be competitive on the market compared to coal or gasoline. However, in a particular case, if water hyacinth was used for remediation of heavy metal contamination and organic pollution, this method would be a good solution because ash has a very small volume and is easy to dispose by burying.

4) *Liquid fuel production*

The cellulose of water hyacinth could be used as the carbohydrate to produce liquid fuels including ethanol through hydrolysis and fermentation. This technology was pioneered in the late 1980's, with ethanol produced from water hyacinth biomass by hydrolysis (Kahlon and Kumar 1987). Given that ethanol is a valuable product, there are an increasing number of studies on ethanol production using water hyacinth biomass. For example, hemicellulose acid hydrolysate was utilized as substrate to produce ethanol using xylose-fermenting yeast (*Pichia stipitis* NRRL Y-7124); and the ethanol yield was significantly improved (by 84%) through boiling and overliming up to pH 10.0 with solid $Ca(OH)_2$ in combination with sodium sulfite. However, the presence of acetic acid in the hydrolysate decreased the ethanol yield considerably (Nigam 2002). In another ethanol production experiment, the technology of simultaneous saccharification and fermentation using a recombinant bacterial stain *Escherichia coli* KO11 was applied to water hyacinth biomass as substrate in ethanol production; ethanol productivity of 0.14–0.17 g g^{-1} dry matter was achieved (Mishima et al. 2008).

Kumar et al. (2009) tested ethanol production by pretreating water hyacinth biomass with dilute acid, produced ethanol by hydrolysis and obtained a yield of 0.43 g ethanol g^{-1} solid hemicellulose. About 73% of xylose was converted to ethanol, and ethanol productivity was about 0.18 g L^{-1} h^{-1}. Furthermore, Aswathy et al. (2010) optimized cellulase and β-glucosidase loading and adding surfactants to improve saccharification; they obtained 71% of saccharification from water hyacinth biomass by acid or alkali pretreatments at 95°C, then with redesigning enzyme blend of cellulases after cooling and washing to pH neutral. The consequent fermentation of the enzymatic hydrolysate using the common baker's yeast (*Saccharomyces cerevisiae*) yielded ethanol concentration of 4.4 g L^{-1}. The above experiment suggested that pre-treatment of water hyacinth biomass for ethanol production is necessary and requires a relatively high temperature and strong acid/alkali pretreatments given that water hyacinth has low sugar and high lignocellulose contents; hence, energy cost is relatively high, making it difficult to achieve a positive energy balance (Thomas and Eden 1990). Therefore, there is still a long way to go before commercial production can be implemented.

5) *Hydrogen production*

Hydrogen represents desirable and renewable clean energy because the product of its combustion is only water. Hydrogen can be produced biologically by algae, bacteria and archaea; the product is called biohydrogen and is quite common in nature. However, biohydrogen is a chemically active component often involved in other complex reactions to form stable chemical components such as methane (Kovács et al. 2000). The main process in biohydrogen production is driven by hydrogenase that loses its function in the presence of oxygen (Morra et al. 2015). Hydrogenase is present in many algal species such as green algae *Chlamydomonas reinhardtii, Scenedesmus obliquus* and others (Hemschemeier 2005). Hydrogenase also exists in many bacteria and archaea strains such as *Rhodobacter sphaeroides, Rhodopseudomonas palustris,* and *Rhodobacter capsulatus* (Rakhely and Kovacs 1996, Laguna et al. 2011, Abo-Hashesh et al. 2013). Although water electrolysis as well as thermochemical, photochemical, photocatalytic and phototelectrochemical processing can produce hydrogen (Antonopoulou et al. 2008), the simplest process to produce hydrogen is Acidogenic Anaerobic Digestion (AAD) or dark fermentation (Das and Veziroğlu 2001). This process uses organic waste including water hyacinth biomass.

The content of crude fiber in water hyacinth biomass reached 46% and that of crude protein 18%; the biomass also contained various amino acids (Xie et al. 1999). These biochemical components are critical to biohydrogen fermentation typically involving two steps: hydrolysis and acidogenesis. The first step breaks down complex organic polymers to simple soluble organic compounds; the second step converts the soluble organic compounds into Volatile Fatty Acids (VFAs), hydrogen, carbon dioxide and other intermediates (Angeriz-Campoy et al. 2015). The first step often involves pretreatment by heating, alkalis and acids in order to increase efficiency of hydrolysis.

In laboratory conditions biohydrogen production using water hyacinth pseudo-lamina, float, subfloat and stolons as substrate was better than using mixed substrate including water hyacinth root and rhizome. Applying alkali (NaOH) pretreatment was better than diluted sulfuric acid (H_2SO_4) pretreatment before hydrolysis of the substrate. Fermentation reaction temperature of 35°C was more beneficial than 55°C in biohydrogen production (Zhou et al. 2007). After alkali (NaOH) pretreatment and hydrolysis of water hyacinth substrate, the pH value during the fermentation was controlled at 6; the biohydrogen yield was 50 mL g^{-1} dry substrate with a maximum production rate 0.48 mL h^{-1} g^{-1} dry substrate.

Apart from substrate pretreatment and temperature and pH control during the fermentation, other factors such as specific bacterial strain, physical properties and chemical composition of substrate (floatability and heavy metal content) are also important. With improved conditions and bacterial inoculation, high biohydrogen yield of 116 mL g^{-1} total solids was obtained in laboratory conditions, with mean biohydrogen concentration of 65% in the biogas product (Cheng et al. 2006), even though the research failed to identify the biological taxonomy of the specific bacterial strain(s). The above experiment suggested it was preferable to remove water hyacinth roots to prevent inhibiting effects on hydrogen-producing bacteria. Recently, biohydrogen production using water hyacinth biomass and genetically engineered microorganisms coupled with advanced innovative technologies showed a lot of promise (Morra et al. 2015). However, biohydrogen commercial production on a large scale is still some time away due to economic and technical reasons; all reports available so far cover the laboratory conditions that need to be carefully integrated and scaled up to industrial size, including production equipment, storage, transportation and utilization of hydrogen for energy generation.

10.3 Methane production by anaerobic fermentation

Anaerobic fermentation using organic waste as a substrate and methanogenic bacteria as biological agents to produce methane represent mature technology. Water hyacinth either in a mixture with other types of organic waste or as the only component is a good substrate for methane fermentation. The product from fermentation is biogas that is a mixture of carbon dioxide, methane and other trace gases. Although the fermentation product is burnable as energy source, it must be refined to obtain the final product (pure methane). Many researchers suggested that using water hyacinth to produce methane was an important management strategy for utilization of water hyacinth biomass (Xu et al. 2008, Rezania et al. 2015). Water hyacinth contained much moisture, rich crude protein, cellulose, hemicelluloses and other organic substances with the C:N 10–30:1. As water hyacinth had relatively low lignin, the cellulose and hemicelluloses were easily degraded to produce good substrate for anaerobic fermentation compared with agricultural organic waste such as straw. A detailed chemical composition of water hyacinth biomass from different publications was summarized in Table 10.3-1. It showed large differences in the composition of water hyacinth biomass sourced from different locations, presumably caused by different growth stages and growth periods at sampling and the effects of ambient environment on the composition of water hyacinth biomass.

Table 10.3-1. Chemical composition (mg g^{-1}) of water hyacinth dried biomass from different sources.

Composition	References					
	(Poddar et al. 1991)	(Abdelhamid and Gabr 1991)	(Chanakya et al. 1993)	(Patel et al. 1993)	(Qian et al. 2011)	(Cheng et al. 2013)
Organic matter	836	743	835	—	—	—
Cellulose	256	195	340	178	177–278	270
Hemicellulose	184	334	180	434	200–344	203
Lignin	99	93	264	78	114–130	100
Crude Fate	16	35	—	—	19–29	9
Crude Protein	163	200	—	119	84–202	210
Ash	164	257	—	202	134–292	208
Nitrogen	28	—	—	—	13–32	—
Phosphorus	5	5	—	—	3–6	—
Calcium	23	6	—	—	—	—
Magnesium	—	2	—	—	—	—
Potassium	24	—	—	—	24–43	—

10.3.1 Methane production potential of water hyacinth biomass

Water hyacinth biomass in either fresh or semi-dried state has been used as substrate in anaerobic fermentation for biogas production for almost half a century; early reported biogas yield was 400 mL g^{-1} volatile solids at 60% methane concentration (240 mL methane g^{-1} volatile solids) with a bioconversion rate equaling 47% of the substrate (Hanisak et al. 1980).

However, other experiments yielded only 190 mL methane g^{-1} volatile solids (Chynoweth et al. 1993) being 30% lower than the methane yield reported by Hanisak (1980). Chanakya et al. (1993) used fresh and air-dried samples of water hyacinth biomass as substrate in the solid-liquid-phase and batch modes to ferment at temperatures 21–27°C for 300 days, yielding potential biogas for fresh (291 mL g^{-1} total solids or 348 mL g^{-1} volatile solids) and air-dried samples (245 mL g^{-1} total solids or 292 mL g^{-1} volatile solids). Another research team compared water hyacinth biomass and straw and suggested that the potential biogas production was higher from water hyacinth biomass than straw, reaching 400 mL g^{-1} total solids (Jiao et al. 1986); however, the authors failed to provide the methane concentration data in their results. Variable methane yields reported by different groups stimulated the interest in theoretical work on the potential methane yield from water hyacinth biomass.

Matsumura (2002) calculated that the chemical structure of water hyacinth biomass was averaged to $C_6H_{12}O_{6.8}$ and, based on this chemical structure of water hyacinth, further calculated conversion efficiencies for methane at 14.8% (w/w) and carbon dioxide at 40.5% (w/w), concluding that the production volume of biogas (methane + carbon dioxide) from water hyacinth biomass through anaerobic fermentation was 413 mL g^{-1} total solids, out of which 207 mL g^{-1} should be pure methane. In practice, the potential is difficult to estimate, especially in large-scale industry fermentation.

Compared with other special energy crops, such as corn, sugar beet and pastures, water hyacinth biomass showed a higher production potential. The annual growth of water hyacinth was reported to be 30–60 tonnes of dry biomass per hectare. Calculated on an average annual output of 45 tonnes dried biomass per hectare and an average potential biogas yield of 340 mL g^{-1}, each hectare of water hyacinth could generate 15,300 m^3 biogas per year, which was 2.5 times of corn and much higher than the production potential of other energy crops (Table 10.3.1-1).

Table 10.3.1-1. Unit yield of biomass and productive rate of biogas in different energy crops.

Energy crops	**Dry biomass yield (t ha^{-1})**	**Biogas yield (L kg^{-1})**	**Annual biogas yield (m^3 ha^{-1} yr^{-1})**	**Ref.**
Whole-plant silage corn (*Zea mays* L.)	15	390	5879	(Cheng et al. 2011)
Barley (*Hordeum vulgare* L.) straw silage	4	189	720	
Sugar beet (*Beta vulgaris* L.) silage	14	430	6173	
Clover (*Trifolium* spp.)	3	335	906	
Sunflower (*Helianthus annuus* L.)	22	225	5055	
Water hyacinth (*Eichhornia crassipes*)	45	340	15,300	(Ye et al. 2011)

Water hyacinth in different growth stages and ambient environment conditions would have different chemical composition, especially lignin content, resulting in large differences in biogas production potential. Water hyacinth had the gas production potential of 336 mL g^{-1} total solids and 517 mL g^{-1} volatile solids in the tillering stage, but only 231 mL g^{-1} total solids and 266 mL g^{-1} volatile solids in the slow-growth period during winter (Ye et al. 2009). This experiment partially explained the variations in the reported data and also confirmed the theoretical calculations done by Matsumura (2002).

The growth environment of water hyacinth also affects its biogas production potential. Qian et al. (2011) reported that the biomass collected from Caohai (Lake Dianchi), where the water body had the highest concentrations of nitrogen and phosphorus, had the highest biogas production of 391 mL g^{-1} total solids (499 mL g^{-1} volatile solids) whereas the biomass collected from Baishan Bay (Lake Dianchi), with the lowest concentrations of nitrogen and phosphorus, had the lowest biogas production of 289 mL g^{-1} total solids (334 mL g^{-1} volatile solids). Verma et al. (2007) tested the influence of brass and chrome on biogas production using water hyacinth biomass as substrate collected from phytoremediation of brass-contaminated and chrome-contaminated waters and reported biogas production at 1.82 and 0.89 μg g^{-1} dry biomass, respectively. High concentration of copper and chromium in water hyacinth biomass significantly reduced biogas production, but microelements at low concentration may be beneficial to the biogas production by methanogens. Even though water hyacinth growing in water with low-level heavy metal contamination may still be a good source of biomass for methane production, the slurry resulting from production needs to be disposed appropriately.

To increase the biogas yield or conversion rate from water hyacinth biomass or to decrease fermentation time, a pretreatment is very important. Similar to biohydrogen production, the pretreatment often involves heating and alkali and acid dissolutions or combinations of methods in sequence. The pretreatment could improve the efficiency of hydrolysis of cellulose and hemicelluloses, especially regarding break-down of lignin-cellulose and lignin-hemicellulose structures. Patel et al. (1993) studied the effects of alkali pretreatment (NaOH-adjusted pH 11 at 121°C for 1 hour) of water hyacinth biomass powder (0.3 mm) on biogas production, and reported that the biogas productivity was increased by 60%, with the methane content of 62–64%. Gao et al. (2013) investigated pretreatment using 50 g 1-N-butyl-3-methylimidazolium chloride plus 10 g dimethyl sulfoxide to treat 3.0 g water hyacinth biomass at 120°C for 2 hours, increasing biogas production by 98% compared with untreated water hyacinth biomass and increasing the methane content from 53 to 68% of biogas. Xia (2014) studied biogas production by continuous anaerobic digestion of a mixture of water hyacinth and rice straw (1:4 on a dry-mass basis) with or without silage pretreatment and obtained yield increases of 67% in biogas and 139% in methane compared with the digestion of the mixture without silage treatment.

The above examples illustrated that water hyacinth biomass, or mixtures of water hyacinth biomass with other crop residues or other organic waste are highly feasible for production of bio-energy. Various pretreatments such as heating, biological (silage), chemical (alkali, acid and ionic liquids) may significantly increase methane yield and decrease fermentation time. However, these methods have their specific area of application. For example, the heating method requires a high energy input. If there is no specific heating source, it is difficult to scale it up to commercial production. The biological (silage) method needs large space and a relatively long time (1 to 4 days). The chemical pretreatment requires consumption of chemicals and needs additional step(s) to recover the processing chemicals in order to avoid secondary pollution.

A relatively new and promising method by applying ionic liquid to treat water hyacinth biomass (Mora-Pale et al. 2011) also has its limitations such as the cost of ionic liquids, although as high as 95% recovery of ionic liquid after fermentation process can be achieved together with the solvent reusability (up to four times). Another limitation of the ionic liquid method is that the potential toxicology of the chemicals involved is not clear (Masten 2004), although many reports claimed the organic solvents are environmentally friendly ("green solvents") due to heat and chemical stability and low volatility (TCI 2007).

10.3.2 Commercial methane fermentation using water hyacinth biomass

As mature bio-energy technology, anaerobic fermentation has a potential for large-scale commercial production worldwide, but is still in an early development stage. Due to the specific physical properties of water hyacinth biomass, it would be difficult to directly input and output bulk materials because of a potential reactor blockage (Abbasi et al. 1992, Malik 2007). In order to overcome the disadvantages of water hyacinth biomass, such as floating on water surface during fermentation and blocking inlet and outlet pipes, many studies innovated special technologies to improve anaerobic fermentation in large-scale industrial production.

Two-phase fermentation of water hyacinth

Annachhatre and Khanna (1987) used alkali to pre-treat water hyacinth biomass combined with whole-cell immobilization technology and invented a two-step system (acidification phase and methane phase) to obtain a higher biogas production (0.44 L g^{-1} total solids) than in batch technology, but the scale was still of the laboratory type. Zhou et al. (1996) solved the problems of water hyacinth biomass floating on the surface of a reactor and blocking the pipes by two-phase anaerobic technology with the first anaerobic solid acid phase and the second flow biofilm-production methane phase, yielding an average biogas production of 100 mL g^{-1} (fresh water hyacinth) with methane content up to 73–83%; however, the scale was still laboratory size and the data could not be compared with other studies because fresh water hyacinth has a wide range of total solids and water content.

Sharma et al. (1999) designed field-scale (10 m^3) batch fermentation reactors that consisted of an alkali pretreatment reactor, an acidification reactor and an anaerobic digestion reactor, achieving stable biogas production rate of 0.31 m^3 kg^{-1} d^{-1} (with methane concentration 65%) from a sun-dried mixture of three parts water hyacinth biomass (70% water content) and one part sugarcane press mud (70% water content). Abbasi and Ramasamy (1999) commercialized methane production from water hyacinth biomass in an acidification reactor that could be stirred continuously, used some livestock excrements as inoculum in the fermentation process, and brought volatile organic acid generated from fermentation into an up-flow anaerobic filter reactor and obtained biogas production rate of 0.38 m^3 kg^{-1} d^{-1} (with a methane concentration 60%), which was quite comparable with the yield obtained in the laboratories. In another case, Chen et al. (2007) designed a two-stage reaction tank (including acidification and gas production parts) at a field scale (20 m^3 total volume); it operated for 80 days with an average biogas production rate of 0.31 m^3 kg^{-1} total solids d^{-1}. All examples described above of anaerobic fermentation technology using water hyacinth biomass as substrate had a complex design and procedure, long-term fermentation period, low gas production rate, large space requirement and high investment, making it difficult to achieve industrial utilization of water hyacinth biomass for commercial bio-energy production.

Fermentation using mixture of water hyacinth biomass and other organic waste

In methane production, high yields are often obtained by optimizing fermentation conditions such as pH, temperature, C:N ratio, inoculums and digestion reactor design, among which C:N ratio and digestion reactor design are the most important. Water hyacinth biomass has a wide range of C:N ratios (10–30:1). The best methane production is often obtained at C:N ratio around 15 (Shanmugam and Horan 2009). The mixture of water hyacinth biomass with other organic waste is often used to adjust C:N ratio.

Regarding digestion reactor design, much research has been devoted to feed-batch (Sharma et al. 1999), two-stage or multiple phase anaerobic digestion reactors (Chanakya et al. 1992, Kivaisi and Mtila 1998). However, investors prefer a digestion system with simple design, low investment and simple operation for industrial application. At present, 90% of anaerobic biogas production plants in Europe operate using continuous single-phase fermentation digesters (Bouallagui et al. 2005). Given that single-phase anaerobic digestion of water hyacinth biomass had slow speed and low conversion rate, fermentation of water hyacinth biomass mixed with other organic wastes with high content of volatile solids could significantly increase the speed and conversion rate of anaerobic digestion. El-Shinnawi et al. (1989) compared the methane production rate using rice straw, corn and cotton stalks and water hyacinth biomass as substrate, each enriched with partially digested cattle dung to C:N ratio of 30 and obtained the highest methane yield of 0.24 m^3 kg^{-1} added volatile solids[1] from water hyacinth biomass mixture in a laboratory-scale (2.5 liter) reactor.

Lu et al. (2010) tested mixtures of water hyacinth biomass with pig manure at field scale (300 m^3 reactor) and obtained higher biogas production from the mixture than the pig manure alone; the 46% biogas production was obtained from the mixture of 15% water hyacinth biomass with 85% pig manure at C:N ratio 17, but the report only presented data on daily biogas production and failed to express biomass production rate in units of either total solids or volatile solids for comparison with other experiments. Chen et al. (2008) concluded that mixture of pig manure and water hyacinth biomass could promote biogas production by 34% compared with only water hyacinth biomass or by 420% compared with only pig manure, with the mixture having the highest biogas production at 0.35 m^3 g^{-1} volatile solids and biological conversion rate[2] of 0.44.

The technology using different source of materials combined with water hyacinth biomass has three purposes: (1) adjust C:N ratio to improve methane yield; (2) change physical properties of raw material to have better operational processes during the fermentation, especially for continuous feeding equipment; and (3) use different source of materials for methane production as long as the biogas yield is not reduced. For example, municipal solid waste combined with 5% (w/w) water hyacinth biomass and pre-treated with 1.0% (w/w) $Ca(OH)_2$ could reduce hydraulic retention time by 7 days during the fermentation process for each production cycle (70 days) and also increase biogas yield by 33%, methane content by 1.1% and volatile solid degradation rate by 12% (Wang 2013).

Separate fermentation technology using juice and residues of water hyacinth

In the research on large-scale remediation of water pollution in Lake Taihu using water hyacinth as a biological agent, Jiangsu Academy of Agricultural Science, China, implemented dehydration of harvested water hyacinth biomass

---

[1] Because not all volatile solids in substrate are converted to methane, the conversion rate is often expressed as consumed volatile solids/added volatile solids.

[2] Biological conversion rate = (actual methane produced)/(theoretical methane yield).

and utilization of both extruded juice and biomass residue separately to assist rapid and efficient disposal of huge amounts of water hyacinth biomass. In the process, one tonne fresh water hyacinth biomass could produce 0.8 tonne juice with COD content 12–19 g L^{-1} and total N 1.3–1.4 g L^{-1} plus 0.2 tonne residue with dry matter content about 25% (i.e., 50 kg dry organic matter). Thus, dehydration could reduce biomass volume by 80%. The biogas production potential showed that biogas yield can be 327 mL g^{-1} COD from juice and from residue of water hyacinth biomass 398 mL g^{-1} total solids or 445 mL g^{-1} volatile solids (Ye et al. 2010). Solid-state single-phase anaerobic fermentation for the residue of water hyacinth biomass could have a biogas production 0.6 m^3 m^{-3} d^{-1} (Ye et al. 2011), which was higher than that of using mixture of fresh water hyacinth biomass and pig manure as substrate in two-phase anaerobic fermentation (Chen et al. 2007).

Ye et al. (2011) compared biogas production using chopped fresh water hyacinth biomass and extruded juice as feedstock in two continuous stirred-tank reactors. Using chopped fresh water hyacinth as a substrate required 27 days of hydraulic retention and had biogas production rate of 267 mL g^{-1} volatile solids and biogas production efficiency of 0.61 m^3 m^{-3} d^{-1} with average methane content 58%. In contrast, extruded juice only required 2.4 days of hydraulic retention and had biogas production rate of 231 mL g^{-1} COD and biogas production efficiency of 1.4 m^3 m^{-3} d^{-1} with an average methane content of 66%. This example showed that by separating juice from fresh water hyacinth biomass, the hydraulic retention time during fermentation was reduced 10-fold and the biogas production efficiency was significantly increased two-fold (Ye et al. 2012).

Hydraulic retention time and production efficiency are the two most important properties for industry-scale fermentation. For instance, treating 1 tonne of fresh water hyacinth biomass per day requires an anaerobic reactor of 27 m^3 volume, but only 2-m^3 anaerobic reactor is required for fermentation of extruded water hyacinth juice as feedstock so that the same volume of fermentation reactor could treat much more water hyacinth biomass in practice and greatly reduce investment in purchasing reactors (Table 10.3.2-1). Meanwhile, the water hyacinth dehydration process removed 88% suspended detritus in the water hyacinth juice so that the slurry after juice fermentation could be directly transported in pipeline to crop fields for application as liquid fertilizer. Water hyacinth residue after dehydration is also suitable for solid-state single-phase fermentation to produce bio-energy, or to be used as silage or compost.

The technology on utilizing water hyacinth biomass for renewable and sustainable energy production is only one of the multiple targets in the integrated management strategy to solve problems, including invasive weed control, eutrophic waters remediation, water resources reclamations, retrieval of nutrients leached from farmland and carbon fixation and sequestration

Table 10.3.2-1. Comparison of two anaerobic fermentation technologies for using water hyacinth biomass and extruded water hyacinth juice (Ye et al. 2012).

Technical parameters	**Fermentation using fresh water hyacinth biomass**	**Fermentation using extruded water hyacinth juice**
Hydraulic retention time (day)	27	2.4
Daily treatment volume per m^3 reactor	37 kg	415 kg water hyacinth juice (obtained from dehydration of 520 kg fresh water hyacinth)
Daily biogas production (m^3 d^{-1})	0.6	1.4
Average methane content (%)	58	66
Pretreatment energy consumption	Particle size <1 cm, energy consumption of about 8 kW t^{-1}	Particle size ≤5 cm, grinding energy consumption of 6.0 kW t^{-1}, energy consumption of dehydration of about 2.08 kW t^{-1}
Stirring energy consumption	High energy consumption	Low energy consumption
Residue treatment	Difficult, with a need to separate solids and liquid	Easy to be treated, biogas slurry could be transported in pipelines, the residue after juice extrusion could be used to make compost

(Fig. 10.3.2-1). In a commercial-scale project for exploring the multi-target integrated management strategy in Lake Taihu region (China), harvesting boats, extruding equipment (Fig. 10.3.2-2), large-scale fermentation reactors, and a 100-W power generator (Fig. 10.3.2-3) were designed for daily processing

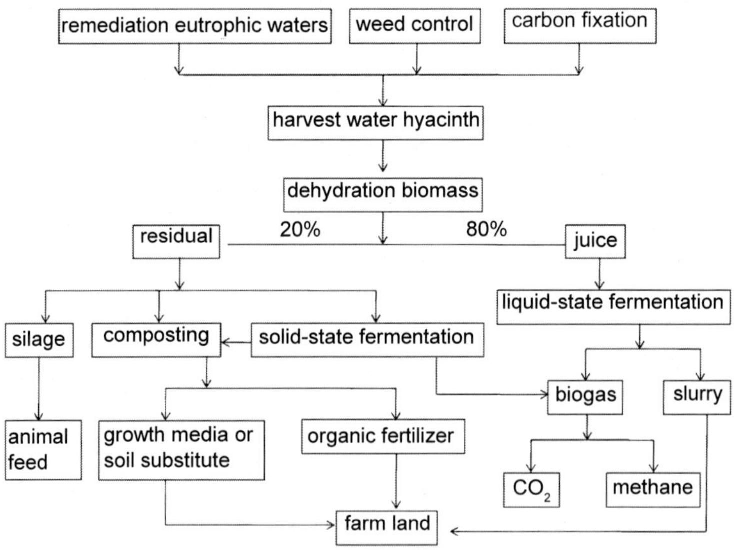

Fig. 10.3.2-1. Technical diagram of comprehensive utilization of water hyacinth in Wujin, Changzhou city, China.

Fig. 10.3.2-2. Dehydration equipments to separate water hyacinth juice and residues (Photo by Xiaomei Ye 2011).

Fig. 10.3.2-3. The 2000-m^3 biogas fermentation reactors for using water hyacinth juice and a 100-W power generator running on purified methane (Photo by Xiaomei Ye 2011).

of 800 tonnes of fresh water hyacinth biomass and production of 2000 m^3 biogas using fresh water hyacinth juice as substrate.

10.4 Technology of water hyacinth composting and application

10.4.1 Water hyacinth directly used as green manure

Using fresh water hyacinth biomass as green manure may be a simple way of utilization by directly covering the farmland or directly spreading on soil and then plowing down to mix in the nutrients (nitrogen, phosphorus and potassium) and sometimes for its high moisture content (Lu et al. 2003). Green manure keeps soil moisture and improves soil structure; it has been practised over several centuries in tea and fruit gardens, and many field crops including rice, sorghum, onion, carrot, corn, pea, potato and soybean (Jiao et al. 1986). Sheng et al. (2009) tested green manure application using fresh water hyacinth biomass (equivalent to 4500 kg dry matter per hectare) and showed that the application could increase the contents of soil-available nitrogen and phosphorus compared with the application of the same amount of fertilizer nitrogen and phosphorus. Liu et al. (2011) dried water hyacinth biomass under the sun and then applied to farmland at 4500 kg dry matter per hectare and showed increased content of nitrogen and potassium in plant tissue in different growth stages of rice, and promoted early grain maturation. It was also reported that water hyacinth as a potassium source mixed with chemical fertilizer and applied to sandy soil could increase wheat and barley production and improve product quality (increasing mineral and crude protein contents) comparing with single application of chemical fertilizer (Mawlys and Zanouly 1999). The recommended application rate of water hyacinth as green fertilizer was 20–30 tonnes of fresh weight per hectare. However, this method was not suitable for mechanized operation and large-scale disposal and utilization of water hyacinth because of high transportation costs and relatively low efficiency.

10.4.2 Water hyacinth compost

Composting of water hyacinth biomass is a feasible method to utilize water hyacinth biomass on a large scale or in commercial operation. Water hyacinth biomass after dehydration can be easily transported, utilized and composted to ensure that most of nitrogen, phosphorus and potassium were preserved in organic fertilizer. Montoya et al. (2013) composted water hyacinth biomass at temperatures above 57°C to inactivate water hyacinth seeds, which made the organic fertilizer safe for application in a range of ecosystems. The moisture content of raw material that is suitable for compost fermentation was about 75% for water hyacinth biomass; the moisture content and texture of water hyacinth residue from dehydration could meet the requirements regarding physical properties for compost fermentation. Suitable moisture content and texture created conditions for increasing direct contact between microorganisms and water hyacinth and improved biodegradation efficiency, thus to obtain better compost maturity.

High quality compost could be produced through indoor fermentation using a mixture of dehydrated water hyacinth biomass and pig manure in piles with a height of about 1 meter, being turned once every 2 days (Luo et al. 2014). Shi et al. (2012) investigated the effect of water content at 65, 70, 75 and 80% (w/w) in the process of composting using mixture of dehydrated water hyacinth biomass and rice straw and concluded that the water content of 75% was the best to achieve increased efficiency (time in days) and quality of compost, and decrease environmental impact such as lowering ammonia volatilization and carbon dioxide release. Each 100 tonnes of water hyacinth residues after dehydration (5 tonnes of dry matter) combined with 9.9 tonnes of straw (8.4 tonnes of dry matter) could produce 18 tonnes of organic fertilizer.

Goyal et al. (2005) monitored the changes in biological and chemical parameters during the composting process for four types of organic waste: mixture of bagasse and cow dung, dehydrated sludge, water hyacinth biomass and poultry wastes; they found that water hyacinth and mixture of bagasse and cow dung had a higher nitrogen loss because of ammonia volatilization within the first 30 days, and suggested that the C:N ratio could be used as a maturation index for composted organic fertilizers. In order to reduce N loss at high temperatures during composting of water hyacinth biomass, mixing 2% (w/w) ammonia-synergist $Mg(OH)_2$:H_3PO_4:H_2O at 3:12:85 in organic waste before composting can increase N retention by 33% by minimizing ammonia volatilization and denitrification (Haihou Wang et al. 2011).

Vermicomposting of 25% water hyacinth biomass + 75% cow dung (dry based) preserved nitrogen, phosphorus and potassium better than composting cow dung or water hyacinth alone (Gupta et al. 2007, Yu et al. 2010). Gajalakshmi et al. (2001) explored the vermicomposting using earthworms (*Eudrilus eugeniae*) and suggested that 950 g mixture (85% water hyacinth and 15% cow dung w/w dry base) in each 10-day run with 250 healthy earthworms per reactor (3 liters) of an vermireactor could significantly (5.6-fold) increase production efficiency.

The above examples indicated that mixture of water hyacinth biomass with organic waste from animal production produced quality organic fertilizers by either composting or vermicomposting, decreased CO_2 release and increased nitrogen, phosphorus and potassium retention during the composting process; in addition, organic fertilizer could also decrease CO_2 emission from soil during crop production and increase organic carbon content in soil (Shi et al. 2012). Other experiments also showed that water hyacinth composted into organic fertilizer increased grain yield of rice up to 15 tonnes per hectare (Sharma and Mittra 1990) and increase yield of corn and sesame seeds and promote crop uptake of micronutrients (Abdel-Sabour et al. 2001, Oroka 2012).

Water hyacinth compost was used as manure in nursery ponds for rearing larvae of Indian major carp, *Labeo rohita,* and increased survival in the ponds 186% compared with that in the ponds without any treatment (Sahu et al. 2002). Chukwuka and Omotayo (2008) found that applying water hyacinth compost and *Tithonia diversifolia* green manure in the ratio of 3:1 into soil not only effectively improved nutrient status of soil, but also replaced chemical fertilizer. Oroka (2012) reported that application of water hyacinth-based vermicompost and inorganic fertilizer together significantly increased productivity of groundnut (*Arachis hypogaea* L.) and cassava (*Manihot esculenta* Crantz).

Large-scale production of organic fertilizer using water hyacinth biomass was successfully explored in the pilot project to demonstrate annual production of 10 thousand tonnes of organic fertilizer (Fig. 10.4.2-1) using water hyacinth biomass, especially using dehydrated water hyacinth residues mixed with pig

Fig. 10.4.2-1. Organic fertilizer production facilities to utilize dehydrated water hyacinth biomass with production capacity of 10 thousand tonnes per year (Photo by Xiao-mei Ye 2011).

manure. After composting for 60 days at ambient temperature of 15–20°C, the mixture of water hyacinth biomass and pig manure turned into quality fertilizer that conformed with the Chinese standards for organic fertilizer quality (Luo et al. 2014). The cost of dehydration and composting for each tonne of fresh water hyacinth was US$2.8.

10.4.3 Biogas slurry and biogas residue fertilizer from anaerobic fermentation

The anaerobic fermentation of either water hyacinth juice or residue produces waste water and slurry, which are high-quality organic fertilizers that contain almost all nutrients needed to support plant growth (Gunnarsson and Petersen 2007) and have the promoting effects on plant growth. The biogas slurry from fermentation of water hyacinth juice contains 1.0–1.5 kg N t^{-1}, 0.42–0.68 kg P_2O_5 t^{-1} and 0–4.8 kg K_2O t^{-1}. Compared with chemical fertilizer application, substituting 25% of chemical fertilizer by biogas slurry resulted in a 20-g increase in weight of single peach (*Prunus persica*) fruit, an increase in production per tree of 18.4 kg, 6.3% higher fruit soluble solids, and 14% higher sugar content, with the sugar-acid ratio reaching 56 (Wang et al. 2013). Applying biogas slurry to substitute 50, 75 and 100% of chemical fertilizer nitrogen in production of lettuce (*Lactuca sativa*) increased yield by 9.1, 16 and 10%, respectively, and decreased nitrate content by 14, 14 and 11%, respectively, compared with the 100% chemical fertilizer nitrogen treatment. Applying biogas slurry to substitute 75% of chemical N fertilizer yielded the highest content of amino acid, soluble sugar and vitamin C (Jidong Wang et al. 2011). Xue et al. (2011a) tested soaking seeds of Chinese vegetable *Brassica rapa* in biogas slurry from fermentation of water hyacinth juice and reported an increase in root system activity and an increase in crude protein content and a reduction in nitrite content. Producing *Brassica rapa* using biogas slurry from water hyacinth juice fermentation instead of chemical fertilizer not only increased yield, but also enhanced the metabolic cycle of reduced ascorbate-glutathione (AsA-GSH) and anti-oxidation capability (Xue et al. 2011b).

The biogas slurry from fermentation water hyacinth biomass also showed a good antimicrobial effect. Xue et al. (2010) investigated the inhibitory effects of biogas slurry plus eugenol on three types of pathogen fungi (*Fusarium graminearum, Phytophthora capsici* and *Fusarium oxysporum*) *in vitro* and reported positive interaction that enhanced the antifungal activity; this mixture could generate a synergistic effect and had a significantly larger antimicrobial effect than that of applying eugenol or biogas slurry individually. Therefore, there may be a prospect of developing the biogas slurry from fermentation of water hyacinth biomass or juice as a new biological antifungal agent.

The biogas residue of water hyacinth is also a superior organic fertilizer. Shi et al. (2011) compared the effects of four different kinds of biogas residue from water hyacinth biomass fermentation on the production and quality of eggplant (*Solanum melongena*) and found that the yield increased 15–31% (w/w) with applications of 15, 22, 30, or 37.5 tonnes biogas residue per hectare compared with application of 37.5 tonnes of black earth per hectare. Meanwhile, the content of available phosphorus and potassium in soil was higher in treatments with biogas residue than the treatment with black earth.

10.4.4 Production of culture substrate from dehydrated biomass of water hyacinth

Culture medium or soil substitute in agriculture is rich in organic matter and nutrients for supporting seed germination, seedling nurseries or soil less cropping and greenhouse vegetable production. The residues of water hyacinth biomass after dehydration could be easily turned into good materials for culture substrate due to its rich cellulose, hemicelluloses and ash content. Zhou et al. (2014) reported that the seedling-raising substrate prepared from water hyacinth residues improved cucumber (*Cucumis sativus*) root system quantity, stem thickness, plant height, and single plant fresh weight compared with control.

Culture substrate prepared from water hyacinth residues is also widely used for culturing edible mushrooms. Klibansky et al. (1993) reported that using water hyacinth mixed 50/50 (w/w) with sugarcane residues as culture substrate to produce mushrooms (*Pleurotus ostreatus*) increased yield two-fold. Nageswaran et al. (2003) found that sawdust supplemented with 25% of water hyacinth residues increased oyster mushroom (*Pleurotus sajor-caju*) production by 20%. Chen et al. (2010) prepared culture medium from anaerobic fermentation of water hyacinth residue mixed with 50% w/w sawdust for production of mushrooms (*Pleurotus geesteranus*), increasing yield by 23% compared with other culture substrates.

References cited

Abbasi, S. A., P. C. Nipaney and E. V. Ramasamy. 1992. Use of aquatic weed Salvinia (*Salvinia-Molesta*, Mitchell) as full/partial feed in commercial biogas digesters. *Indian Journal of Technology* 30: 451–457.

Abbasi, S. A. and E. V. Ramasamy. 1999. *Biotechnological methods of pullution control*. Hyderabad, Indian: University Press Ltd.

Abdelhamid, A. M. and A. A. Gabr. 1991. Evaluation of water hyacinth as a feed for ruminants. *Archiv für Tierernaehrung* 41(7-8): 745–756.

Abdel-Sabour, M. F., H. I. Abdel-Shafy and T. M. Mosalem. 2001. Heavy metals and plant-growth-yield as affected by water hyacinth compost applied to sandy soil. *Environment Protection Engineering* 27(2): 43–53.

Abo-Hashesh, M., N. Desaunay and P. C. Hallenbeck. 2013. High yield single stage conversion of glucose to hydrogen by photofermentation with continuous cultures of *Rhodobacter capsulatus* JP91. *Bioresource Technology* 128: 513–517.

Angeriz-Campoy, R., C. J. Álvarez-Gallego and L. I. Romero-García. 2015. Thermophilic anaerobic co-digestion of organic fraction of municipal solid waste (OFMSW) with food waste (FW): enhancement of bio-hydrogen production. *Bioresource Technology* 194: 291–296.

Annachhatre, A. P. and P. Khanna. 1987. Methane recovery from water hyacinth through whole-cell immobilization technology. *Biotechnology and Bioengineering* 29(7): 805–818.

Antonopoulou, G., H. N. Gavala, I. V. Skiadas, K. Angelopoulos and G. Lyberatos. 2008. Biofuels generation from sweet sorghum: fermentative hydrogen production and anaerobic digestion of the remaining biomass. *Bioresource Technology* 99(1): 110–119.

Aswathy, U. S., R. K. Sukumaran, G. L. Devi, K. P. Rajasree, R. R. Singhania and A. Pandey. 2010. Bio-ethanol from water hyacinth biomass: an evaluation of enzymatic saccharification strategy. *Bioresource Technology* 101(3): 925–30.

Bouallagui, H., Y. Touhami, B. R. Cheikh and M. Hamdi. 2005. Bioreactor performance in anaerobic digestion of fruit and vegetable wastes. *Process Biochemistry* 40(3-4): 989–995.

Chanakya, H. N., S. Borgaonkar, G. Meena and K. S. Jagadish. 1993. Solid-phase biogas production with garbage or water hyacinth. *Bioresource Technology* 46(3): 227–231.

Chanakya, H. N., S. Borgaonkar, M. G. C. Rajan and M. Wahi. 1992. Two-phase anaerobic digestion of water hyacinth or urban garbage. *Bioresource Technology* 42(2): 123–131.

Chen, B., Y. Zhao, W. Cao, J. Lan and J. Wang. 2007. Research and application of anaerobic fermentation of water hyacinth. *Environmental Pollution and Control* 29(6): 455–458 (In Chinese with English Abstract).

Chen, G., Z. Zheng, X. Zou and S. Yang. 2008. Effects of swine feces on anaerobic digestion of water hyacinth. *China Environmental Science* 28(10): 898–903 (In Chinese with English Abstract).

Chen, X., Z. Jiang, X. Chen, J. Lei, B. Weng and Q. Huang. 2010. Use of biogas fluid-soaked water hyacinth for cultivating *Pleurotus Geesteranus*. *Bioresource Technology* 101(7): 2397–2400.

Cheng, J., H. Pan, F. Qi, J. Zhou, J. Liu and K. Cen. 2006. Influence factors on bio-H_2 production from digested sludge and hyacinth by co-fermentation. *Journal of Wuhan University of Technology* 28(2): 209–214 (In Chinese with English Abstract).

Cheng, J., A. Xia, H. Su, W. Song, J. Zhou and K. Cen. 2013. Promotion of H_2 production by microwave-assisted treatment of water hyacinth with dilute H_2SO_4 through combined dark fermentation and photofermentation. *Energy Conversion and Management* 73: 329–334.

Cheng, X., W. Zhu and Z. Cui. 2011. Research achievements on dedicated energy crops for biogas in the EU countries. *Renewable Energy Resources* 29(5): 133–136 (In Chinese with English Abstract).

Chukwuka, K. S. and O. E. Omotayo. 2008. Effects of tithonia green manure and water hyacinth compost application on nutrient depleted soil in South-Western Nigeria. *International Journal of Soil Science* 3(2): 69–74.

Chynoweth, D. P., C. E. Turick, J. M. Owens, D. E. Jerger and M. W. Peck. 1993. Biochemical methane potential of biomass and waste feedstocks. *Biomass and Bioenergy* 5 (1): 95–111.

Das, D. and T. N. Veziroğlu. 2001. Hydrogen production by biological processes: a survey of literature. *International Journal of Hydrogen Energy* 26(1): 13–28.

El-Shinnawi, M. M., M. N. El-Din and S. A. El-Shimi. 1989. Biogas production from crop residues and aquatic weeds. *Resources, Conservation and Recycling* 3(1): 33–45.

Gajalakshmi, S., E. V. Ramasamy and S. A. Abbasi. 2001. Assessment of sustainable vermiconversion of water hyacinth at different reactor efficiencies employing *Eudrilus eugeniae* kinberg. *Bioresource Technology* 80(2): 131–5.

Gao, J., L. Chen, K. Yuan, H. Huang and Z. Yan. 2013. Ionic liquid pretreatment to enhance the anaerobic digestion of lignocellulosic biomass. *Bioresource Technology* 150: 352–358.

Gettys, L. A., W. T. Haller and M. Bellaud. 2009. *Biology and control of aquatic plants: a best management practices handbook.* 2nd ed. Marietta GA, USA: Aquatic Ecosystem Restoration Foundation.

Goyal, S., S. K. Dhull and K. K. Kapoor. 2005. Chemical and biological changes during composting of different organic wastes and assessment of compost maturity. *Bioresource Technology* 96(14): 1584–1591.

Gunnarsson, C. C. and C. M. Petersen. 2007. Water hyacinths as a resource in agriculture and energy production: a literature review. *Waste Management* 27(1): 117–29.

Gupta, R., P. K. Mutiyar, N. K. Rawat, M. S. Saini and V. K. Garg. 2007. Development of a water hyacinth based vermireactor using an epigeic earthworm *Eisenia foetida*. *Bioresource Technology* 98(13): 2605–2610.

Hanisak, M. D., L. D. Williams and J. H. Ryther. 1980. Recycling the nutrients in residues from methane digesters of aquatic macrophytes for new biomass production. *Resource Recovery and Conservation* 4(4): 313–323.

Hemschemeier, A. C. 2005. The anaerobic life of the photosynthetic alga *Chlamydomonas reinhardtii*. Ph.D. Thesis, Department of Biology. Ruhr-University Bochum, Germany.

Hronich, J. E., L. Martin, J. Plawsky and H. R. Bungay. 2008. Potential of *Eichhornia crassipes* for biomass refining. *Journal of Industrial Microbiology and Biotechnology* 35(5): 393–402.

Jiao, B., R. S. Gu and X. S. Zhang. 1986. *Green manure in China*. Beijing, China: Agriculture press.

Kahlon, S. S. and P. Kumar. 1987. Simulation of fermentation conditions for ethanol production from water hyacinth. *Indian Journal of Ecology* 14: 213–217.

Kivaisi, A. K. and M. Mtila. 1998. Production of biogas from water hyacinth (*Eichhornia crassipes*) (Mart) (Solms) in a two-stage bioreactor. *World Journal of Microbiology & Biotechnology* 14: 125–132.

Klibansky, M. M., M. Mansur, I. Gutierrez and L. Gonzalez. 1993. Production of *Pleurotus ostreatus* mushrooms on sugar cane agrowastes. *Acta Biotechnol* 13(1): 71–78.

Kovács, K. L., C. Bagyinka, L. Bodrossy, R. Csáki, B. Fodor, K. Győrfi et al. 2000. Recent advances in biohydrogen research. *European Journal of Physiology* 439(1): 81–83.

Kumar, A., L. K. Singh and S. Ghosh. 2009. Bioconversion of lignocellulosic fraction of water-hyacinth (*Eichhornia crassipes*) hemicellulose acid hydrolysate to ethanol by *Pichia stipitis*. *Bioresource Technology* 100 (13): 3293–3297.

Laguna, R., F. R. Tabita and B. E. Alber. 2011. Acetate-dependent photoheterotrophic growth and the differential requirement for the Calvin-Benson-Bassham reductive pentose phosphate cycle in *Rhodobacter sphaeroides* and *Rhodopseudomonas palustris*. *Archives of Microbiology* 193(2): 151–154.

Li, X., W. Cong, C. Ren, J. Sheng, P. Zhu, J. Zheng et al. 2011. Photosynthetic productivity and the potential of carbon sink in cultivated water hyacinth (*Eichhornia crassipes*) in Taihu Lake. *Jiangsu Journal of Agricultural Sciences* 27(3): 500–504 (In Chinese with English Abstract).

Li, X., C. Ren, M. Wang, J. Sheng and J. Zheng. 2010. Response of photosynthesis of leaves to light and temperature in *Eichhornia crassipes* in Jiangsu Province. *Jiangsu Journal of Agricultural Sciences* 26(5): 943–947 (In Chinese with English Abstract).

Liu, H., L. Chen, P. Zhu, J. Sheng, Y. Zhang and J. Zheng. 2011. Effects of hyacinth mulching on dry matter production and distribution of rice (*Oryza sativa* L.). *Chinese Journal of Applied Environmental Biology* 17(4): 521–526 (In Chinese).

Lu, J., L. Zhu, G. Hu and J. Wu. 2010. Integrating animal manure-based bioenergy production with invasive species control: a case study at Tongren pig farm in China. *Biomass and Bioenergy* 34(6): 821–827.

Lu, L., N. Su and Y. Sun. 2003. Low input, high output, multi-purpose water hyacinth. *Jilin Animal Husbandry And Veterinary Medicine* 12: 26–27 (In Chinese).

Luo, J., L. Liu, T. Wang, H. Liu, Y. Gao, Z. Zhang et al. 2014. Study on fermentation conditions of water hyacinth and pig manure co-composting. *Jiangsu Agricultural Sciences* 42(6): 336–339 (In Chinese).

Malik, A. 2007. Environmental challenge vis a vis opportunity: the case of water hyacinth. *Environment International* 33(1): 122–38.

Masten, S. A. 2004. Ionic liquid: review of toxicological literature. North Carolina, USA: Integrated Laboratory Systems, Inc.

Matsumura, Y. 2002. Evaluation of supercritical water gasification and biomethanation for wet biomass utilization in Japan. *Energy Conversion and Management* 43(9-12): 1301–1310.

Mawlys, M. D. and I. Zanouly. 1999. The application of water hyacinth as a supplemental source of K for wheat and barley grown on a sandy soil. *Assiut Journal of Agricultural Sciences (Egypt)* 30(2): 73–82.

Mishima, D., M. Kuniki, K. Sei, S. Soda, M. Ike and M. Fujita. 2008. Ethanol production from candidate energy crops: water hyacinth (*Eichhornia crassipes*) and water lettuce (*Pistia stratiotes* L.). *Bioresource Technology* 99(7): 2495–2500.

Montoya, J. E., T. M. Waliczek and M. L. Abbott. 2013. Large scale composting as a means of managing water hyacinth (*Eichhornia crassipes*). *Invasive Plant Science and Management* 6(2): 243–249.

Mora-Pale, M., L. Meli, T. V. Doherty, R. J. Linhardt and J. S. Dordick. 2011. Room temperature ionic liquids as emerging solvents for the pretreatment of lignocellulosic biomass. *Biotechnology and Bioengineering* 108(6): 1229–1245.

Morra, S., F. Valetti, V. Sarasso, S. Castrignanò, S. J. Sadeghi and G. Gilardi. 2015. Hydrogen production at high Faradaic efficiency by a bio-electrode based on TiO_2 adsorption of a new [FeFe]-hydrogenase from *Clostridium perfringens*. *Bioelectrochemistry* 106: 258–262.

Nageswaran, M., A. Gopalakrishnan, M. Ganesan, A. Vedhamurthy and E. Selvaganapathy. 2003. Evaluation of waterhyacinth and paddy straw waste for culture of oyster mushrooms. *Journal of Aquatic Plant Management* 41(2): 122–123.

Nigam, J. N. 2002. Bioconversion of water-hyacinth (*Echhornia crassipes*) hemicellulose acid hydrolysate to motor fuel ethanol by xylose-fermenting yeast. *Journal of Biotechnology* 97: 107–116.

Oroka, F. O. 2012. Water hyacinth-based vermicompost on yield, yield components, and yield advantage of cassava+groundnut intercropping system. *Journal of Tropical Agriculture* 50(1-2): 49–52.

Patel, V., M. Desai and D. Madamwar. 1993. Thermochemical pretreatment of water hyacinth for improved biomethanation. *Applied Biochemistry and Biotechnology* 42(1): 67–74.

Patil, J. H., M. AntonyRaj, B. B. Shankar, M. K. Shetty and B. P. P. Kumar. 2014. Anaerobic co-digestion of water hyacinth and sheep waste. *Energy Procedia* 52: 572–578.

Poddar, K., L. Mandal and G. C. Banerjee. 1991. Studies on water hyacinth (*Eichhornia crassipes*) —chemical composition of the plant and water from different habitats. *Indian Veterinary Journal* 68: 833–837.

Qian, Y., X. Ye, Z. Chang, J. Du and J. Pan. 2011. Methane production characteristics of water hyacinth from different water areas. *China Environmental Science* 31(9): 1509–1515 (In Chinese with English Abstract).

Rakhely, G. and K. L. Kovacs. 1996. Plating hyperthermophilic archea on solid surface. *Analytical Biochemistry* 243(1): 181–183.

Rezania, S., M. Ponraj, M. F. M. Din, A. R. Songip, F. M. Sairan and S. Chelliapan. 2015. The diverse applications of water hyacinth with main focus on sustainable energy and production for new era: an overview. *Renewable and Sustainable Energy Reviews* 41: 943–954.

Sahu, A. K., S. K. Sahoo and S. S. Giri. 2002. Efficacy of water hyacinth compost in nursery ponds for larval rearing of Indian major carp, *Labeo rohita*. *Bioresource Technology* 85(3): 309–11.

Shanmugam, P. and N. J. Horan. 2009. Optimising the biogas production from leather fleshing waste by co-digestion with MSW. *Bioresource Technology* 100 (18): 4117–4120.

Sharma, A. R. and B. N. Mittra. 1990. Response of rice to rate and time of application of organic materials. *The Journal of Agricultural Science* 114(03): 249.

Sharma, A., B. G. Unni and H. D. Singh. 1999. A novel fed-batch digestion system for biomethanation of plant biomasses. *Journal of Bioscience and Bioengineering* 87(5): 678–682.

Sheng, J., J. Zheng, L. Chen, P. Zhu and X. Xue. 2009. Absorption of water nutrients by hyacinth and its application in wheat production. *Journal of Agro-Environment Science* 28(10): 2119–2123 (In Chinese with English Abstract).

Shi, L. L., M. X. Shen, Z. Z. Chang, H. H. Wang, C. Y. Lu, F. S. Chen et al. 2012. Effect of water content on composition of water hyacinth *Eichhornia crassipes* (Mart.) Solms residue and greenhouse gas emission. *Chinese Journal of Eco-Agriculture* 20(3): 337–342 (In Chinese with English Abstract).

Shi, L., W. Xu, S. Wang, L. Yang, G. Shi, R. Liu et al. 2011. Effect of biogas residue of *Eichhornia crassipes* on yield and quality of eggplant. *Journal of Changjiang Vegetables* 14: 41–43 (In Chinese with English Abstract).

TCI. 2007. Material safety data sheet: D3240. Portland OR, USA: TCI America.

Thomas, T. H. and R. D. Eden. 1990. Water hyacinth—a major neglected resource. In: Energy and Environment: Into the 1990's—Proceedings of the 1st World Renewable Energy Congress, ed. A. A. M. Sayigh, 2096–2096. Reading, UK: Pergamon.

Verma, V. K., Y. P. Singh and J. P. N. Rai. 2007. Biogas production from plant biomass used for phytoremediation of industrial wastes. *Bioresource Technology* 98(8): 1664–1669.

Wang, H., M. Shen, Z. Chang, C. Lu, F. Chen, L. Shi et al. 2011. Nitrogen loss and technique for nitrogen conservation in high temperature composting of hyacinth. *Journal of Agro-Environment Science* 30(6): 1214–1220 (In Chinese with English Abstract).

Wang, J., Y. Cao, Zh. Chang, Y. Zhang and H. Ma. 2013. Effects of combined application of biogas slurry with chemical fertilizers on fruit qualities of *Prunus persica* L. and soil nitrogen accumulation risk. *Plant Nutrition and Fertilizer Science* 19(2): 379–386 (In Chinese with English Abstract).

Wang, J., H. Ma, X. Gao, X. Xu, Y. Ning, H. Zhang et al. 2011. Effects of different ratios of biogas slurry of water hyacinth (*Eichhornia crassipes*) substitute chemical nitrogen fertilizers on growth and quality of lettuce (L. *sativa*). *Soils* 43(5): 787–792 (In Chinese with English Abstract).

Wang, L. 2013. Experimental study on dry anaerobic co-digestion of municipal solid waste and water hyacinth. MSc. Thesis, Department of Chemical Engineering. South China University of Technology, Guangzhou, China.

Wilkie, A. C. and J. M. Evans. 2010. Aquatic plants: an opportunity feedstock in the age of bioenergy. *Biofuels* 1(2): 311–321.

Wu, C. and L. Ma. 2003. *Modern technologies of biomass energy utilization*. Beijing, China: Chemical Industry Press.

Xia, Y. 2014. Effect of Pretreatments of Water Hyacinth (*Eichhornia crassipes*) and Rice Straw on Biogas Production via Anaerobic Digestion. Ph.D thesis, Zhe Jiang University.

Xie, P., X. Zhou, J. Yang, Y. Zhu, Y. Zhang, J. Zhang et al. 1999. Study on broiler chickens of feeding water hycacinth in Lake Dianchi. *Feed Industry* 20(4): 26–28 (In Chinese).

Xu, Z., Y. Gao and S. Wang. 2008. Review on comprehensive and utilization of water hyacinth. *Resources and Environment in the Yangtze Basin* 17(2): 201–205 (In Chinese with English Abstract).

Xue, Y., H. Feng, Z. Shi, S. Yan and J. Zheng. 2011a. Effect of biogas slurry of *Eichhornia crassipes* on the seedling quality of Chinese cabbage. *Pratacultural Science* 28(4): 687–692.

Xue, Y. 2011b. Dynamic changes in the growth and the AsA-GSH circulation metabolism of Chinese cabbage treated with a biogas slurry of water hyacinth. *Acta Prataculturae Sinica* 20(3): 91–98 (In Chinese with English Abstract).

Xue, Y., S. Zhou, Z. Hi, S. Yan and J. Zheng. 2010. The combination of water hyacinth biogas slury and eugenol controls fungi. In: Congress of the 11th Soil Microorganism, 6th Soil Biology and Biochemistry and 4th Microbial Fertilizers, ed. Chinese Society of Soil Science, 189 (In Chinese). Chang Sha, China: Chinese Society of Soil Science.

Ye, X., Z. Chang, Y. Qian, P. Zhu and J. Du. 2012. Comparison of biogas production effiency of anaerobic digestion using water hyacinth and its juice from solid-liquid separation as feedstock. *Transactions of the CSAE* 28(4): 208–214 (In Chinese with English Abstract).

Ye, X., J. Du, Z. Chang, Y. Qian, Y. Xu and J. Zhang. 2011. Anaerobic digestion of solid residue of water hyacinth. *Jiangsu Journal of Agricultural Sciences* 27(6): 1261–1266 (In Chinese with English Abstract).

Ye, X., L. Zhou, S. Yan, Z. Chang and J. Du. 2010. Anaerobic digestion of water hyacinth juice in CSTR reactor. *Fujian Journal of Agricultural Science* 25(1): 100–103 (In Chinese with English Abstract).

Ye, X., L. Zhou, S. Yan, Z. Chang and B. Gao. 2009. Studies on the anaerobic digestion of water hyacinth. *Jiangsu Journal of Agricultural Sciences* 25(4): 787–790 (In Chinese with English Abstract).

Yu, J., Z. Chang and R. Li. 2010. Microbial and physicochemical properties of the mixture of water hyacinth residue and excrement through vermicomposting. *Jiangsu Journal of Agricultural Sciences* 26(5): 970–975 (In Chinese with English Abstract).

Zhang, Z., H. Qin, H. Liu, X. Li, X. Wen, Y. Zhang et al. 2014. Effect of large-scale confined growth of water hyacinth improving water quality of relatively enclosed eutrophicated waters in Caohai of Lake Dianchi. *Journal of Ecology and Rural Environment* 30(3): 306–310 (In Chinese with English Abstract).

Zheng, J., J. Sheng, Z. Zhang, X. Li, Y. Bai and P. Zhu. 2011. Ecological function of hycinth and its utilization. *Jiangsu Journal of Agricultural Sciences* 27(2): 426–429 (In Chinese with English Abstract).

Zhou, J., Q. Feng, J. Cheng, J. Liu and K. Cen. 2007. Studies on the hydrogen production from hyacinth fermentation. *China Environmental Science* 27(1): 141–144 (In Chinese with English Abstract).

Zhou, X., H. Wang, L. Shi, M. Shen, C. Lu, T. Wu et al. 2014. Study on the substitution effect of *Eichhornia crassipes* on cucumber seedling substrate. *Chinese Agricultural Science Bulletin* 30(25): 201–206 (In Chinese with English Abstract).

Zhou, Y., X. Kong, L. Hao and Q. Fu. 1996. Two-phase anaerobic digestion of water hyacinth. *China Biogass* 14(3): 8–12 (In Chinese with English Abstract).

CHAPTER 11

Utilization of Water Hyacinth Biomass for Animal Feed

Y. F. Bai[1] and *J. Y. Guo*[2,*]

11.1 Introduction

In the integrated water hyacinth management system, utilization of water hyacinth biomass as animal feed on large scale in commercial production is desirable because animal feed always fetch good financial return on the market. After the problems in technical and economic aspects of harvesting and dehydration are solved, utilization of water hyacinth biomass for animal feed can be practised on large scale. Based on the triple bottom line of global development, phytoremediation with water hyacinth must generate social, economic and environmental benefits simultaneously. Offsetting costs with benefits is important for sustainable development. Potential challenges exist in linkages between the large-scale management of water hyacinth on one side and the sustainable development of agriculture based on recycling nutrients or bio-energy production or silage and feed production, among which silage and feed production may be selected as the number one priority.

Over the last century, many ideas on utilization of water hyacinth have been advanced all over the world. The biomass of water hyacinth was utilized as animal feed or leaf protein for cattle, sheep, geese, pigs, etc. (in either fresh or silage form). However, the linkages between management system and water hyacinth biomass utilization have been difficult to establish, partly because the technical aspects of biomass (nutrient) value vary with the harvest time and location, partly because heavy metal contents vary spatially and temporally.

[1] 50 Zhong Ling Street, Nanjing, China.
Email: blinkeye@126.com
[2] 5 Armagh Way, Ottawa, Canada.
* Corresponding author: guoj1210@hotmail.com

The changes in quality and safety of the water hyacinth fresh biomass represent problems in achieving stability in large-scale commercial animal production.

The dehydration technology to achieve water content of less than 64% in water hyacinth bagasse was appropriate only for using biomass as a silage or feed ingredient. Special feed formulae from water hyacinth were developed for herbivores such as lambs (Bai et al. 2010), fattening geese and cattle (Tham et al. 2013). A goat could intake 2153 grams of formulated water hyacinth silage per day and could gain up to 122 g d^{-1} in body weight. A goose could intake 744 g of silage d^{-1} and could gain up to 26 g d^{-1} in body weight (Yun-feng Bai et al. 2011). Formulated water hyacinth can be good feed for cattle with the highest digestibility of 72% for organic matter, 75% for crude protein, 62% for neutral detergent fiber and 49% for acid detergent fiber (Tham and Udén 2013). After 8-hours sun drying of the water hyacinth fresh biomass, the resulting dry matter had a crude protein content of 180 g kg^{-1}, and leaves and shoots had greater digestibility than the whole plant (Aboud et al. 2005). Sunday (2002) recommended the utilization of sun-dried water hyacinth biomass at 40% of goat feed. It is clear that dehydrated fresh biomass of water hyacinth has a significant potential in feeding herbivores.

There is always a requirement for cheap and nutrient-rich feed for general animal production beyond the herbivores and omnivores. To assess its nutritional potency for poultry, ducks were fed 50 g fresh water hyacinth per day for a month; they exhibited better daily feed intake and higher egg-laying capacity in treatment group compared to the control group (Lu et al. 2008). Mangisah et al. (2010) evaluated the composition and nutritional attributes of *Aspergillus niger*-fermented leaves of water hyacinth for Tegal ducks. Fermentation enhanced the crude protein digestibility, true metabolizable energy and nitrogen retention. Konyeme et al. (2006) tested the nutritive value of various percentages of water hyacinth to catfish *Clarias gariepinus* fingerlings. To minimize feed costs and achieve good yield in aquaculture, 40% supplementation of water hyacinth in fish meal was recommended. Saha and Ray (2011) reported that fish feed containing 30% water hyacinth leaf meal fermented by *Bacillus subtilis* and lactic acid bacteria can be used successfully. The rohu (*Labeo rohita*) fingerlings fed fermented leaf meal showed 25% more weight gain, 33% higher protein efficiency ratio and three times higher apparent net protein utilization than fingerlings fed reference diet (Patel 2012). Unformulated fish feeds containing water hyacinth were also reported (Li and Zhang 1992, Sahu et al. 2002). Other examples on swine and broiler rearing were also reported. Broiler diets with 3–7% dry water hyacinth, did not have negative effects on the average daily gain, feed efficiency and carcass performance (Xie et al. 1999). For fattening swine diets, supplement of 5–15% water hyacinth biomass in feed did not affect daily gain, feed intake during growth and the final product in dressing percentage, fat thickness, and loin eye area compared to the control (Cui et al. 2004).

Water hyacinth can also be used as roughage to reduce the shortage of animal feed. Abdelhamid and Gabr (1991) tested a crude protein content of 200 g kg^{-1} DM and got a good result in using water hyacinth biomass as roughage for ruminants. Tag EI-Din (1992) reported that up to 30% of the roughage (bean straw) can be substituted with dried water hyacinth without a loss in growth rate of sheep. However, using water hyacinth as only roughage can greatly reduce the average daily weight gain because crude protein content of 90 g kg^{-1} DM is the minimum in the fodder for ruminants.

The above literature reported the results using largely untreated or sun-dried water hyacinth biomass. The water hyacinth tissues contain intercellular spaces filled with air, which soak up water while the animals are digesting. That leads to excessive water consumption making the animals feel replete, but having little material of nutritional value in their rumens (Bolenz et al. 1990). Bolenz et al. (1990) also observed under a microscope sharp needles formed of calcium oxalate in water hyacinth tissue and assumed that these needles could damage the digestive tract of animals, if not dissolved by the digestive acid. To avoid these problems, Bolenz et al. (1990) suggested the following preparation of the water hyacinths used for feeding animals: (i) chop the tissues to eliminate air and negate the capacity to absorb water, and (ii) separate the soluble tissue components (juice) by pressing and centrifuging. The bagasse could be washed with acid to eliminate acid-soluble calcium oxalate followed by processing to ruminant fodder. These suggestions could be put into practice by using a harvesting shredder and dehydration. However, the acid wash procedure may be replaced by fermentation. For successful fermentation the pH needs to be lowered (<4), which can be achieved by adding sugar. The fresh water hyacinth biomass contains about 0.52% w/w fermentable sugar, so an addition of 0.4% w/w sugar was enough to reach the desired pH value (Gunnarsson and Petersen 2007).

Although literature reported good results from adding water hyacinth biomass to feed for pigs, chicken and fish, the nutritional physiology pointed that monogastric animals such as pigs and chicken (without rumen digestion as in cattle and sheep, and without a cecum as in grazing livestock such as horse and donkey) do not have the capacity to digest water hyacinth biomass with high fiber content. Therefore, feeding pigs and chicken with water hyacinth is still a topic open to discussion.

11.2 Feed quality assurance and risk assessment

The literature suggests that water hyacinth biomass could be a good potential for animal feed based on the nutrient value of the dehydrated water hyacinth biomass. However, the source of the biomass may vary from time to time because most public water bodies are subject to variable pollution. The most important pollutants to affect the feed quality and safety are herbicides, antibiotics, pesticides and heavy metals.

11.2.1 Chemical composition of water hyacinth biomass

The reported values on chemical composition vary widely due to inconsistent harvesting (e.g., at different growing stages) and variable harvesting methods and growth conditions (Poddar et al. 1991). Generally, at a mature stage, biomass has more fiber and carbohydrates, as do water hyacinth roots in comparison with leaves. The macrophyte growing in waters with low or unbalanced nutrient concentrations may have less crude protein content than that growing in waters with high and balanced nutrient concentrations. When the utilization of the fresh biomass of water hyacinth becomes one of the important targets for the integrated management system, the harvest method and schedule are important to ensure good quality feed from the biomass (Table 11.2.1-1).

Dried biomass of water hyacinth has crude protein content from 104 to 155 g kg^{-1}; regarding quality, it is good roughage, equivalent to grass, but lower than alfalfa and can be improved by mixing with other ingredients. The biomass is slightly higher in ash and ADF than standard and may have negative effects on feed digestibility. However, the quality can be much improved by adjusting the harvesting schedule and selecting the right growth stage of the macrophyte. For example, the biomass (dry matter) harvested in July and August 2010 in Lake Taihu had gross energy 13.47 MJ kg^{-1}, ash 191 g kg^{-1}, phosphorus 4.1 g kg^{-1} and crude protein 198 g kg^{-1} dry matter, whereas at the same place in October, it had gross energy 12.92 MJ kg^{-1}, ash

Table 11.2.1-1. Chemical composition of dried water hyacinth biomass (g kg^{-1} DM).

Items	Whole plant	Roots	Stems and leaves	Roots	Stems	Leaves
OM	841					
Ash	159	192	165	182	158	149
CP	122	107	98	104	155	113
CF				292	204	283
ADF	432	381	231			
NDF	729	673	554			
GE	15.5					
Fat				10	11	21
Ca	32	13	27	8	18	20
P	3.8	4.4	5.3	2.1	3.3	3.0
Mg	3.1			3.1	8.8	3.8
K	33			14	43	67
References	(Baldwin et al. 1975)	(Li 2006)		(Yu 1989)		

Note: DM = Dry Matter; OM = Organic Matter; CP = Crude Protein; CF = Crude Fiber; ADF = Acid Detergent Fiber; NDF = Neutral Detergent Fiber; GE = Gross Energy (MJ g^{-1}); Ca = Calcium; P = Phosphorus; Mg = Magnesium; K = Potassium.

233 g kg^{-1}, phosphorus 3.6 g kg^{-1} and crude protein 159 g kg^{-1} dry matter showing an increase in ash but a decrease in GE, P and CP content. The nutritional quality changed after dehydration. The over-dried biomass harvested from a pond in Nanjing, China in July and August, and the same growth stage had gross energy 14.34 MJ kg^{-1}, ash 148 g kg^{-1}, phosphorus 4.5 g kg^{-1}, crude protein 209 g kg^{-1}, ADF 276 g kg^{-1} and NDF 481 g kg^{-1} dry matter, but biomass dehydrated by a roller press had gross energy 15.15 MJ kg^{-1}, ash 108 g kg^{-1}, phosphorus 1.9 g kg^{-1}, crude protein 147 g kg^{-1}, ADF 429 g kg^{-1} and NDF 628 g kg^{-1} dry matter. The experiment revealed that, during dehydration, soluble components were significantly decreased and the ADF and NDF were increased. The ADF and NDF are important for herbivores.

11.2.2 Effects of heavy metals on the safety of feed from water hyacinth biomass

Water hyacinth has a strong capacity to adsorb, absorb and accumulate heavy metals such as cadmium (Cd), chromium (Cr), lead (Pb) and mercury (Hg), and metalloids such as arsenic (As) and fluorine (F) (Zhou et al. 2005, Shi and Zhao 2007). Roots have three times as high as the capacity to hold and transport the toxic elements than the other parts and tissues. For this reason, water hyacinth growing in industry waste water and in waters surrounding mining areas may be unsafe to utilize as feed.

Li (2006) investigated different parts of water hyacinth growing in domestic wastewater and reported that the contents of Cd, Cr, Pb, As and F were all under the critical limits of the feed standard (GB13078-2001 2001) in all plant parts, but the root content of Hg exceeded the standard limit. This investigation implied that: (1) the biomass harvested from general wastewater (other than heavily polluted industry wastewater) generally may be safe for feed; (2) roots may be removed for safety reasons regardless of whether toxic elements are under the standard limits or not; (3) standards differ from country to country and from region to region so specific investigations are needed; and (4) the toxic elements may be accumulated in particular organs of animals, suggesting that a particular toxic element with content in feed below the standard limit may still transfer and accumulate in specific animal parts to exceed the standard limits for human consumption.

The contents of the toxic elements in the water hyacinth biomass are the main concern in utilization of the biological resources, not only for feed, but also for organic fertilizer and growth substrate in agriculture. Some elements such as iron and calcium may also affect the anaerobic fermentation process of biogas production and the formula of substrate for mushroom culture. The utilization of the biomass for feed needs to follow very detailed working procedures such as investigating and monitoring water chemistry in the water hyacinth growing regions, harvesting plan and schedule, dehydration, and biomass quality and analytical chemistry especially regarding toxic elements.

11.3 Water hyacinth biomass ensilage and modulation

After dehydration, the bagasse of water hyacinth at 60–70% water content is suitable for silage and can be processed on an industrial scale. Silage of vegetative biomass for feed has been successfully used in animal production for over 100 years. To apply this technology to water hyacinth biomass, fermentable sugar content of the bagasse needs to be modulated to above 0.9% w/w before silage production using standard technology: anaerobic conditions in the presence of *Lactobacillus* spp. or other lactic fermentation bacteria. Given that silage technology uses an anaerobic lactic acid fermentation process, the production of water hyacinth silage needs to establish an anaerobic environment by compacting and sealing. The optimum moisture for anaerobic lactic fermentation is 60 to 70%. Higher water content may promote growth of harmful bacteria, but lower moisture may make the fermentation process prolonged and incomplete. The fermentable sugar at 0.9% w/w can bring down the pH <4 quickly to ensure a good quality silage with respect to low contents of ammonium and butyric acid and high content of lactic acid (Zhuang et al. 2007).

Water hyacinth silage is not only a feed processing method, but also the long-term storage method. Conventional methods are silage silos, silage towers, trench silage and silage bags, all using anaerobic conditions, compacting and sealing. For this reason, the advances in technology, equipment and accessories and new methods of modulation may promote the utilization of water hyacinth biomass for feed.

Zhuang et al. (2008) tested silage method with water hyacinth biomass at two water contents of 71 and 40% with three types of additives: fermented green juice, cellulase and the mixture of fermented green juice and cellulase, and reported the significant quality improvements of silage by fermented green juice and the mixture of fermented green juice and cellulase at both water contents. The soluble carbohydrates in silage were increased by cellulase addition. In another experiment on silage methods, additives were extended to fermented green juice, sugar, formic acid, ammonium formate, fermented green juice+sugar, and no additive (control). The results revealed that all the additives significantly decreased pH and concentrations of ammonia, and the quality was improved especially by the treatment with fermented green juice+sugar (Zhuang et al. 2007). The above-mentioned two experiments reported that the dry matter recovery from silage averaged as high as 98%, which was better than for corn stalks, alfalfa and Napier grass (*Pennisetum purpureum*) (Zhuang et al. 2006).

Li et al. (2007) investigated the effects of *Lactobacillus* ($0, 1 \times 10^6, 1 \times 10^8$ CFU kg^{-1} dry matter) and mixture of enzymes (0, 3000, 6000 IU kg^{-1} dry matter) on the quality of silage using biomass of water hyacinth leaves and stolons, and reported that the control treatment had higher pH and higher contents of ammonium and butyric acid and lower content of lactic acid than the other treatments. The quality was positively related to the amount of added *Lactobacillus* and enzymes.

Tham et al. (2013) tested different additives (molasses, rice bran, inoculants of fermented vegetable juice and their combinations) with water hyacinth biomass for silage production. The results showed that pH values decreased to 4.0 in all treatments in 3 days, except in control and the treatment with inoculum of fermented vegetable juice. Butyric and lactic acids and ammonia concentration were maintained at good quality level in all treatments, except in control and the treatment with inoculum of fermented vegetable juice due to low water soluble carbohydrates.

El-Sayed (2003) investigated the nutritive value of nine isonitrogenous (35% w/w crude protein) and isocaloric (GE 18.8 KJ g^{-1}) diets containing water hyacinth biomass for Nile tilapia fingerlings. Wheat bran at 10 and 20% w/w in the nine test diets were substituted by fresh dry water hyacinth, molasse-fermented water hyacinth, cow rumen content-fermented water hyacinth and yeast-fermented water hyacinth. The results revealed three conclusions: (1) control treatment (no water hyacinth inclusion) had the best performance regarding weight gain, specific growth rate and feed utilization efficiency than all water hyacinth-containing diets; (2) different fermentation methods of water hyacinth (added at 10% w/w) did not significantly affect growth rate and feed utilization efficiency, but these parameters were better at 10 than 20% w/w water hyacinth inclusion in the diets; and (3) at 20% w/w inclusion, best feed utilization efficiency was obtained by molasse-fermentation method compared to other fermentation methods. The results implied that water hyacinth may not be suitable for fish diet, and if used, the fermentation method may only affect diets containing more than 20% w/w of water hyacinth biomass.

Saha and Ray (2011) studied the nutritive value of nine isonitrogenous (30% w/w crude protein) and isocaloric (18.23 kJ g^{-1}) test diets incorporating water hyacinth leaf meal for rohu fingerlings (*Labeo rohita*). Fish meal in the reference diet (RD) was replaced by water hyacinth leaf meal at 20, 30 and 40% w/w in nine test diets. The water hyacinth leaf meal for substitution was pretreated at 37°C for 15 days by one of the two fermentation methods: (1) *Bacillus subtilis* CY5 (isolated from *Cyprinus carpio*); or (2) *B. megaterium* CI3 (isolated from *Ctenopharyngodon idella*). The test diet prepared with *Bacillus subtilis* CY5 fermentation method was fed to fish fingerlings with supplement of a commercial preparation of lactic acid bacteria (*Lactobacillus acidophilus*). The results showed that crude fiber, cellulose and hemicellulose contents and the anti-nutritional factors (tannin and phytic acid) were significantly decreased, and free amino acids and fatty acids increased, in the fermented water hyacinth leaf meal. The response to different diets (growth, feed conversion ratio, protein efficiency ratio, and apparent net protein utilization) from best to worst was 30% leaf meal substitute fermented with *B. megaterium* CI3, followed by 20% leaf meal substitute fermented with *B. megaterium* CI3, then 20% substitute fermented with *Bacillus subtilis* CY5, and the fish meal-based reference diet. The experiment concluded that water hyacinth leaf meal fermented with fish gut bacteria showed good activity of extracellular enzymes and can be recommended as a dietary ingredient for rearing *Labeo rohita* fingerlings at as high as 40% substitution of fish meal without any adverse effect on growth rate but a significant reduction in feed cost.

The conflict in conclusions from the reports by El-Sayed (2003) and Saha and Ray (2011) may lie in the composition of water hyacinth biomass, bacteria used for fermentation, feed preparation processes, the methods of feeding the fish and the fish species in the experiment.

11.4 Case studies on large scale utilization of water hyacinth silage

The above discussion indicated that the utilization of water hyacinth silage including the methods of ensilage and modulation, types of animals to be fed, digestibility and nutritional value had a long history of research in laboratories around the global. However, large scale utilization of water hyacinth silage in commercial goat production is very rare, partly because water hyacinth silage is not commonly produced and because of the risks mentioned above, and partly due to relatively high ash content in the water hyacinth bagasse and lower digestibility after treatment compared with high-quality forage silage (Baldwin et al. 1975); in addition, there are high harvest costs and high water content in fresh water hyacinth biomass (Chen and Zhuang 2009). Furthermore, large scale case studies are also important to understand standard procedures, risks and application methods, and evaluate production processes and potential economic values of utilizing water hyacinth silage in the commercial system.

In an integrated management system designed to solve multidimensional problems of noxious weed control, remediation of eutrophication water bodies, and production of bio-energy, feed and organic fertilizers, the two ecological engineering projects in Lake Taihu and Lake Dianchi in China were executed to produce large quantity of water hyacinth bagasse and turn it into silage to feed goats; the utilization of water hyacinth bagasse silage for goat production was expected to increase the economic value of water hyacinth biomass on the commercial market. The project selected two commercial animal production systems for demonstration. One was a goat production farm at Luhe in Nanjing City, Jiangsu province, China, with a production scale of 1000 weaned goats (Fig. 11.4-1). The other was a commercial geese production farm in Chan Zhou City, Jiangsu Province, China, with 10,000 geese (Fig. 11.4-2).

11.4.1 Feeding water hyacinth silage to goats in commercial production

Use of water hyacinth as silage to feed ruminants was reported as early as the 1970s in China, mainly due to a lack of forage and increased demand for commercial goat meat production at that time. The production scale and the ensilage of water hyacinth were all at a farmer (individual) scale. With the social

Fig. 11.4-1. Water hyacinth silage as feed for goat production.

Fig. 11.4-2. Water hyacinth silage as feed for goose production.

and economic development, and population increase and industrialization, the animal production in China changed rapidly from individual scale to large commercial production scale, and utilization of water hyacinth silage became rare. However, eutrophication in aquatic system has caused the explosive growth of water hyacinth naturally in the southern part of China due to subtropical climate being naturally suitable for the weed (Ding et al. 2001).

Literature reported problems in feeding water hyacinth bagasse to goats because the silage contained high proportion of cellulose and hemicellulose; hence, improved fermentation methods and nutritional modulation were necessary to enhance water hyacinth bagasse as a good source of energy feed for ruminants (Mukherjee and Nandi 2004). For this reason, evaluation of silage fermentation methods and diet modulation, and an assessment of heavy metal risks were undertaken using the commercial goat production.

Quality of water hyacinth silage and dietary composition

The silage fermentation of water hyacinth bagasse (69% water content) was performed by adding 15.8% corn flour and 7.2% vinegar residue and sealing in a thick plastic bag (0.28 m^3) at average room temperature 25.8°C for 15 days. The pH values of the silage were 3.73, 3.62, 3.42 and 3.46 at 15, 20, 25 and 30 days, respectively, which showed a successful fermentation process. The other quality related properties of the silage were assessed on day 30. The results showed that the quality of water hyacinth silage was lower than that of alfalfa but better than that of corn stalks (Table 11.4.1-1).

Literature also suggests that water hyacinth silage is not good enough for animal feed, but can be mixed with other feed (modulation) to improve dietary nutrient utilization (Tag EI-Din 1992), with the percentage of daily feed being less than 50% (Abdelhamid and Gabr 1991) or less than 30% (Tag EI-Din 1992) or 40% (Sunday 2002). The case study adapted 50% water hyacinth silage (dry matter) mixed with 18.5% corn flour, 8% soybean residue, 10% vinegar residue, 12.5% wheat bran and the pre-mix[1]; the control treatment was commercial feed with 50% corn stalk silage (dry matter) mixed with the same other ingredients as specified above for the water hyacinth treatment.

Table 11.4.1-1. Comparison of quality of silage from water hyacinth, alfalfa and corn stalks (g kg^{-1}, except for GE).

Silage	GE (MJ kg^{-1})	CP	P	Ash	NDF	ADF	HC	Cellulose	Lignin	Ref.
WH	16.62	125	2.1	57	540	250	290	168	83	(Yunfeng Bai et al. 2011)
Alfalfa	—	195	—	87	460	360	—	—	79	(Martin et al. 2004)
Corn		88		51	540	270	—	—	23	(Martin et al. 2004)

Note: WH = Water Hyacinth; GE = Gross Energy; CP = Crude Protein; P = Phosphorus; NDF = Neutral Detergent Fiber; ADF = Acid Detergent Fiber; HC = Hemicellulose; — = data missing.

[1] Pre-mix consisted 0.26% w/w salt, 0.13% w/w dicalcium phosphate, 0.26% w/w limestone powder, 0.13% w/w vitamin and mineral premix.

Water hyacinth silage digestibility

Feed digestibility is a very important property of commercial feed for animal production and closely linked with profitability output. The case study randomly selected 7-month old male goats at ~25 kg individual weight to test feed digestibility by collecting all fecal output (Schroeder 2004). The calculations were performed by the following formulae:

Apparent digestibility X_i (g kg^{-1}) = $(X_{in}-X_f)/X_{in} \times 1000$ [11.4.1-1]

Apparent retainability of a nutrient Y_i (g kg^{-1}) = $(Y_{in}-Y_f-Y_u)/X_{in} \times 1000$ [11.4.1-2]

Apparent digested nutrient XX_i (kg) = $XX_{in}-XX_f$ [11.4.1-3]

Apparent retained nutrient YY_i (kg) = $YY_{in}-YY_f-YY_u$ [11.4.1-4]

Apparent digestible energy GE (MJ kg^{-1}) = $(E_{in}-E_f)/DM_{in} \times 1000$ [11.4.1-5]

Apparent energy digestibility AED (KJ MJ^{-1}) = $(E_{in}-E_f)/E_{in} \times 1000$ [11.4.1-6]

Biological nitrogen value BV (g kg^{-1}) = $N_{re}/(N_{in}-N_f) \times 1000$ [11.4.1-7]

Where X_i is apparent digestibility of nutrient *i*; X_{in} is nutrient *i* intake (kg); X_f is nutrient *i* in feces (kg); Y_i is apparent retainability of nutrient *i*; Y_{in} is nutrient *i* intake (kg); Y_f is nutrient *i* in feces (kg); Y_u is nutrient *i* in urine (kg); XX_i is apparent digested nutrient *i*; XX_{in} is nutrient *i* intake (kg); XX_f is nutrient *i* in feces (kg); YY_i is apparent retained nutrient *i*; YY_{in} is nutrient *i* intake (kg); YY_f is nutrient *i* in feces (kg); YY_u is nutrient *i* in urine (kg); GE is apparent digestible energy; E_{in} is energy intake (MJ); E_f is energy in feces (MJ); DM_{in} is dry matter intake (kg); BV is biological value as the proportion of nitrogen retained; N_{re} is the nitrogen retained (kg); N_{in} is nitrogen intake (kg); N_f is nitrogen in feces (kg).

The results showed that apparent digestibility of nitrogen, biological value, dry matter, organic matter, NDF, ADF, and apparent digestible energy were generally lower in water hyacinth silage than control, but other properties were compatible in the two types of feed (Table 11.4.1-2).

In this case study, the whole-plant water hyacinth biomass was used for silage. The crude protein content of the biomass was 125 g kg^{-1}, which was lower (probably because roots were included as well) than the crude protein content in leaves and stolons (151–224 g kg^{-1} dry matter) (Li 2006, Zhang et al. 2010, Rodríguez et al. 2012), the crude protein content in animal diet and therefore crucial for incorporating water hyacinth silage into feed. Water hyacinth biomass may be incorporated into fish diet (up to 40% w/w dry matter) to achieve 350 g kg^{-1} crude protein level in feed (Konyeme et al. 2006). Another reason for lower digestibility in water hyacinth feed than the control may be influenced by the impact of dehydration process (necessary for utilization of the biomass) on the bagasse protein quality. However, in the

Table 11.4.1-2. Digestibility of water hyacinth silage for feeding fattening goats.

Properties	Control	WHS	Properties	Control	WHS
Intake nitrogen (kg)	17.8	13.9	Dry matter digestibility ($g\ kg^{-1}$ DM)	657^A	517^B
Feces nitrogen (kg)	5.4	5.3	Organic matter digestibility ($g\ kg^{-1}$ OM)	700^A	545^B
Urine nitrogen (kg)	7.0	5.3	Crude protein digestibility ($g\ kg^{-1}$ CP)	696^A	622^B
Retained nitrogen (kg)	5.3^A	3.2^B	Acid detergent fiber digestibility ($g\ kg^{-1}$ ADF)	513^A	350^B
Nitrogen retainability ($g\ kg^{-1}$ nitrogen)	298^A	230^B	Neutral Detergent Fiber digestibility ($g\ kg^{-1}$ NDF)	693^A	440^B
Biological nitrogen value ($g\ kg^{-1}$ nitrogen)	431^A	370^B	Intake Energy ($MJ\ d^{-1}$)	14.7^A	8.6^B
Intake calcium (kg)	9.9^A	7.4^B	Feces Energy ($MJ\ d^{-1}$)	4.8	4.2
Feces calcium (kg)	8.1^A	6.1^B	Apparent digested Energy ($MJ\ d^{-1}$)	9.9^A	4.4^B
Urine calcium (kg)	0.28^A	0.18^B	Apparent digestible energy ($MJ\ kg^{-1}$ DM)	10.7^A	8.1^B
Retained calcium (kg)	1.6^A	1.1^B	Apparent energy digestibility ($KJ\ MJ^{-1}$)	673^A	512^B
Calcium retainability ($g\ kg^{-1}$ calcium)	156	153			
Intake phosphorus (kg)	3.7^A	2.8^B			
Feces phosphorus (kg)	2.7^A	1.9^B			
Urine phosphorus (kg)	0.09	0.09			
Retained phosphorus (kg)	0.97	0.85			
Phosphorus retainability ($g\ kg^{-1}$ phosphorus)	265	282			

Note: WHS = Water Hyacinth Silage; Different superscript letters indicate significant differences between control and WHS.

process of dehydration, crude protein would move from plant tissues to water hyacinth juice, thus lowering the crude protein content as well as reducing the quality of crude protein in bagasse (Du et al. 2012).

Effects of heavy metals and metalloids in water hyacinth on safety of goat production

Water hyacinth biomass is strongly influenced by the heavy metal and metalloid pollutants in water, especially in regions with industries such as chemical and electronic product processing, mining and smelting. Therefore, it is essential to monitor the heavy metal and metalloid contents in the water hyacinth biomass before utilization. In this case study, 11 elements in water hyacinth biomass were assessed using Hydride Generation Atomic

Absorption Spectrometry (HGAAS) before the biomass incorporation into feed (Table 11.4.1-3, unpublished data).

Table 11.4.1-3 showed that elemental contents differ significantly among locations, between bagasse and whole water hyacinth, and between different harvest times. Generally, the contents of Ca, Fe, Cu, Mn and Zn were in a range suitable for feed application, especially Fe and Mn. In Lake Taihu, the contents of Cr (22–25 mg kg^{-1}) and As (3.2–5.0 mg kg^{-1}) exceeded the standard (GB13078-2001 2001). The contents of Hg (0.11–0.22 mg kg^{-1}) and As (4.4–10.1 mg kg^{-1}) in water hyacinth/bagasse from Lake Dianchi exceeded the standard. It should be borne in mind that the whole-plant water hyacinth biomass was used in the case study. However, heavy metals and metalloids accumulate more (1.5–21 times) in/on roots than leaves (Cordes et al. 2000, Yapoga et al. 2013) due to resistance mechanisms minimizing translocation of these elements in leaves and stolons (Chen 2011). The results further confirmed the importance of developing equipment to separate roots and leaves at harvest in order to reduce the risk of heavy metal and metalloid contamination in water hyacinth silage.

The risk of heavy metal and metalloid contamination extends from feed to accumulation of these contaminants in the different organs of goats. The results from the case study revealed that Pb and As were translocated predominantly into the liver, heart and kidneys, whereas Cr was accumulated preferentially in the lungs, although significant amounts also accumulated in the liver, heart and kidneys (Table 11.4.1-4).

Table 11.4.1-4 showed that Hg, Pb and As were well under the maximum permissible contaminant level for both the European Union and China, but the Cr content was at a very high level in all organs of goats and exceeded the maximum limits of food safety standard even in goats fed the commercial feed from the local market. Cr is a transition metal in Group 6 in the periodic table and has different oxidation states, most commonly appearing as trivalent and hexavalent Cr. The trivalent Cr is non-toxic at low dosage (with LD_{50} of 187 mg kg^{-1} as chromium nitrate in female rats), but the hexavalent Cr is recognized as a genotoxic carcinogen with LD_{50} of 13 mg kg^{-1} as sodium chromate in female rats (Agency for Toxic Substances and Disease Registry (ASTDR) 2012).

The trivalent Cr is an essential nutrient for humans. National Academy of Sciences in the US has established safe and adequate daily intake of trivalent Cr at 0.05–0.2 mg for adults. On average, adults in the United States take in 0.06–0.08 mg of trivalent Cr per day in food (Yu 2008). The biologically active form (organic trivalent Cr complex) is essential in sugar metabolism as Glucose Tolerance Factor (GTF), which functions by facilitating the interaction between insulin and its cellular receptor sites. The trivalent Cr supplementation can rapidly reverse many symptoms of Cr deficiency (Mertz 1993, Cohen et al. 1993, Sun et al. 2015), although direct evidence of trivalent Cr deficiency in humans is not adequate (Agency for Toxic Substances and Disease Registry (ASTDR) 2012). Cr was unlikely to accumulate in animal or human body (Seaborn and

Table 11.4.1-3. Element content of water hyacinth biomass (whole plant) grown in different locations (mg kg^{-1}, air-dried basis, HGAAS method).

Element	WH from pond	Bagasse from pond	WH from Lake Taihu[a]	WH from Lake Taihu[b]	WH from Lake Dianchi	Bagasse from Lake Dianchi	Standard
Fe	703	1226	1948	2602	1905	1605	
Mn	260	374	1012	645	914	557	
Cu	9.26	13.8	24.1	32.5	11.0	7.63	
Zn	49.8	83.5	123	138	35.5	28.2	
Ca	30118	34957	21861	31238	16657	12729	
K	85696	19006	92790	94703	13276	3394	
Mg	7759	5204	6713	8118	7137	3457	
Cr	1.45	1.90	22.2	24.8	-	1.03	≤10.0
Pb	0.02	0.05	0.01	-	-	-	≤5.0
As	0.69	1.06	5.04	3.19	10.1	4.41	≤2.0
Hg	0.05	0.04	0.04	0.03	0.22	0.11	≤0.1

Note: WH = Water Hyacinth; [a] Harvested from Lake Taihu in August; [b] Harvested from Lake Taihu in October; Standard is Standard for Feed (GB13078-2001 2001).

Table 11.4.1-4. Contaminants in different goat organs influenced by silage from water hyacinth grown at different locations (mg kg^{-1} DM).

Element	Organs	Control	Group 1	Group 2	EU standard[a]	China standard[b]
Cr	Liver	1.33[a]	5.76[b]	14.9[c]	—	≤1.0 mg kg^{-1}
	Kidneys	2.80[a]	7.04[b]	12.5[c]		
	Heart	2.42[a]	7.76[b]	14.8[c]		
	Lungs	3.85[a]	9.64[b]	17.3[c]		
	Muscles	3.35[a]	8.34[b]	13.4[c]		
Hg	Liver	0	0	0	≤1.0 mg kg^{-1}	≤0.05 mg kg^{-1}
	Kidneys	0	0	0		
	Heart	0	0	0		
	Lungs	0	0	0		
	Muscles	0	0	0		
Pb	Liver	0.042	0.049	0.045	≤0.1 mg kg^{-1}	≤0.5 mg kg^{-1}
	Kidneys	0.020[a]	0.043[b]	0.047[b]		
	Heart	0.010[a]	0.020[b]	0.009[a]		
	Lungs	0	0	0		
	Muscles	0	0	0		
As	Liver	0	2.1×10^{-3}	2.2×10^{-3}	—	≤0.5 mg kg^{-1}
	Kidneys	0	0	0		
	Heart	0	0	0		
	Lungs	0	0	0		
	Muscles	0	0	0		

Note: Control goats were fed commercial silage; Group 1 was fed water hyacinth silage from a pond at Nanjing, Jiangsu Province, China; Group 2 was fed water hyacinth silage from Lake Taihu, Jiangsu Province, China; EU represents European Union; Different superscript letters indicate significant difference between treatments; — represent data missing or unspecified; [a] Maximum permissible contaminant level (The Commission of the European Union 2006); [b] Maximum permissible contaminant level (GB2762-2012 2012).

Stoecker 1990). The stable trivalent Cr is excreted from a human body in a week, although some trivalent Cr in lungs may stay for years (Agency for Toxic Substances and Disease Registry (ASTDR) 2012).

The hexavalent Cr is a very strong oxidant and can easily be reduced to trivalent Cr in water hyacinth roots. Some plant species (e.g., *Prosopis laevigata*) may hyperaccumulate hexavalent Cr from a growth medium (Buendía-González et al. 2010). However, that report failed to explain the processes by which hexavalent Cr enters the root tissues from the growth medium and the chemical form of Cr that accumulated in the plant. Water hyacinth grown in a solution supplemented with hexavalent Cr accumulated non-toxic trivalent Cr in roots and shoots through catalyzing reduction of hexavalent Cr in the fine lateral roots, with trivalent Cr subsequently being translocated to leaf tissues (Lytle et al. 1998). However, the above assessment in the goat experiment failed to report the Cr state in water hyacinth silage because analytical tools (e.g., X-ray microscopy) were not available during the experiment. Silage fermentation is an anaerobic process, and an impact of anaerobic fermentation on the changes in Cr redox state is unclear and warrants further research.

The hexavalent Cr is transported into cells in human bodies via the sulfate transport mechanism, which does not transport trivalent Cr (Salnikow and Zhitkovich 2008). The toxicity of hexavalent Cr was not caused by the actual Cr (VI) itself, but the trivalent Cr produced from intracellular hexavalent Cr. Because Cr (III) has poor capacity to cross cellular membrane, the intracellular trivalent Cr is trapped inside the cells, inducing Cr-DNA adducts and leading to DNA damage and genetic lesions.

Given that the hexavalent Cr is toxic and therefore of concern for public heath, the environmental and health agencies established Maximum Contaminant Level (MCL) of Cr in drinking water and food. For example, China has the MCL standard for drinking water at 0.05 mg L^{-1} for hexavalent Cr (Ministry of Public Health 2006). World Health Organization (WHO) recommended a standard for drinking water at 0.05 mg L^{-1} for total Cr (WHO 2011). The United States Environmental Protection Agency has the MCL for total Cr at 0.1 mg L^{-1} (EPA 2011).

Impacts of water hyacinth silage dietary on goat production

The impacts of water hyacinth silage replacement in commercial feed were assessed on feed intake, daily weight gain, feed conversion rate and feed digestibility over 60 days at 15-day intervals. The results showed that the feed intake increased with goat growth over the measuring period, but the daily weight gain decreased (Table 11.4.1-5). The water hyacinth silage diet produced average daily weight gain of 104–145 g d^{-1} (Bai et al. 2010). The annual consumption of water hyacinth silage dry matter by 1000 fattening goats was 468 tonnes, which was equivalent to 2000 tonnes of fresh biomass, digestible energy 1135 billion Joules, and crude protein 11 tonnes (Bai et al. 2010).

In animal production, quality of forage and feeding management determine feed efficiency and economic return, especially with ruminants such as goats because the characteristics of the digestive tract of these animals do not allow them to take in large volumes of feed and/or to pass the feed faster through the digestive tract to compensate for poor-quality feed. The

Table 11.4.1-5. Feed intake, daily weight gain and feed conversion rate over time (Bai et al. 2010).

Items	15 days	30 days	45 days	60 days
Feed intake (g d^{-1})	1429	1585	1670	1754
Daily weight gain (g d^{-1})	145	104	109	110
Feed conversion rate (g d^{-1})	4.0	5.6	5.6	6.2

NDF content in good-quality feed is recommended at no less than 380 g kg^{-1} (Schroeder 2013). The recommendation is compatible with the water hyacinth feed mixture (NDF 399 g kg^{-1}), but higher than the control (NDF 369 g kg^{-1}) used in the case study (Yunfeng Bai et al. 2011). The crude protein content in good-quality feed should be no more than 230 g kg^{-1} (Schroeder 2013), which is much higher than the water hyacinth feed mixture (145 g kg^{-1}) and the control (124 g kg^{-1}) used in this case study (Yunfeng Bai et al. 2011). The quality assessment showed that both silages used in the case study were of poor quality judging from the crude protein content, which implies a need to improve the process of producing water hyacinth biomass and modulate it for animal feed.

11.4.2 Feeding water hyacinth silage to fattening geese

Geese have different digestive physiology from other poultry and can digest and use green fodder. Adding roughage to diet can reduce production costs without decreasing the productivity and economic return. Geese have adapted to various living environments and food sources in the long history of evolution and domestication, especially regarding unique physical structure of the digestive tract and digestive characteristics. They have a long and flat beak that is significantly different from a chicken's conical beak. The rounded tip, rough edges and serrated tuck of upper and lower beak can cut off grass and separate the solid particles from water and reserve feed in the mouth (Jin et al. 1994). Geese are herbivorous waterfowl eating plants, especially aquatic vegetation, and are adapted to passing green fodder fast through the digestive tract to compensate for low protein and energy content in poor-quality feed; hence, geese can feed on large amounts of poor-quality green fodder to meet their growth and metabolic needs. Geese have a strong capacity to use crude fiber and digest neutral detergent fiber up to 270 g kg^{-1} feed, acid detergent fiber up to 420 g kg^{-1} feed and hemicellulose up to 520 g kg^{-1} feed. The digestibility test showed a positive correlation between crude fiber digestibility and diet fiber (Han et al. 2014).

Water hyacinth silage diet

The water hyacinth silage was made the same way as for goat feed described above. A study was conducted to assess the standard production procedures and application methods, as well as to evaluate production potential and economic value of utilizing water hyacinth silage in commercial production of geese.

In geese production (as in other animal production systems) the feed quality and management determine the animal growth and development and thus the economic return. In this case study, the control diets were designed according to different growth stages (Table 11.4.2-1).

Table 11.4.2-1. Control diets composition (g kg^{-1}, except for metabolic energy).

Items	For age 0–4 weeks	For age 4–8 weeks	> 8 weeks of age	Water hyacinth silage
Corn meal	610	600	600	146
Soybean residue	280	220	170	45
Wheat bran	79	150	200	736 (water hyacinth[a])
Vinegar residue	0	0	0	70
Salt	4	4	4	2.6
Dicalcium phosphate	5	5	5	1.3
Minerals	12	11	11	2.6
Pre-mix[2]	10	10	10	1.3
Dry matter	900	899	899	360
Metabolic energy (MJ kg^{-1})	11.45	11.20	11.04	3.7
Crude protein	196	177	161	52
Calcium	6.7	6.3	6.2	2.9
Phosphate	4.5	4.6	4.7	1.3
Crude fiber	31.4	32.3	32.6	230

[a] water content 69%.

Impact of water hyacinth silage diet on anatomical changes in digestive tract of goslings

In commercial production of geese, the feed is mostly purchased on the market as meal or pellet. When feeding on silage, the volumes of feed are significantly different. A goose is a small animal compared to a cow or goat. The amount of nutrition in silage compared with commercial feed that a goose takes in is significantly different and may subsequently affect the growth and development. The case study investigated (i) the anatomical changes in the gastrointestinal development, and (ii) carcass traits during geese production to compare water hyacinth silage-based diet with the commercial feed purchased from the local market. The investigation selected 120 Yangzhou geese 4-week-old and randomly divided them into three treatment groups replicated four times, with 10 goslings in each replicate. The experimental period was 42 days. The control group was fed commercial feeds according to the growth stages. The two treatment feeds were made by replacing different percentages of commercial feed with water hyacinth silage: in Group I, 18% replacement for 5–8 weeks of age and 37% replacement for 9–10 weeks of age; in Group II, 37% replacement for 5–8 weeks of age and 55% replacement for 9–10 weeks of age.

[2] Pre-mix: antibiotics + multi-vitamin for poultry (0.6% w/w), trace elements and methionine (0.4% w/w).

Results showed that the geese ingested more water hyacinth silage test feeds than commercial feed throughout the experimental period (Table 11.4.2-2).

These findings confirmed the hypothesis that geese may eat a large volume of low quality feed and pass it through the digestive tract fast to compensate for higher water content and lower energy and crude protein levels in water hyacinth silage. This mechanism led to anatomical changes in gastrointestinal development, with the water hyacinth silage replacement inducing an increase in the gastrointestinal weight of individual goslings. The gastrointestinal weight and the relative weight in the group II treatment were higher than those in the group I and control treatments (Tables 11.4.2-3 and 11.4.2-4). The components of gastrointestinal weight and the relative weight of the group II treatment were significantly higher than those of the group I and the control treatments, except for the duodenum that was less affected compared to the group I and control treatments.

The gastrointestinal length may reveal the goose adaptability to water hyacinth silage replacement, which would influence decisions in a commercial production of geese. The experiment revealed that the length of jejunum, ileum and appendix was longer in the group II treatment than those in the group I and the control treatments ($p < 0.05$, Table 11.4.2-5).

Table 11.4.2-2. Effect of water hyacinth silage diet on daily feed intake for individual geese.

Group	5–6 weeks old (g individual^{-1})	7–8 weeks old (g individual^{-1})	9–10 weeks old (g individual^{-1})	Average (g individual^{-1})
Control	146	143b	156b	149
Group I	198	197	259	218
Group II	263	272a	368a	301

Note: Different letters in a column represent significant differences ($p < 0.05$) among the treatments.

Table 11.4.2-3. The effect of water hyacinth silage on the gastrointestinal weight (g) of individual geese (10-week old).

Treatment	Proventriculus	Gizzard	Duodenum	Jejunum	Ileum	Appendix
Control	8.4a	91a	6.9a	14.8a	12.0a	4.7a
Group I	9.1a	106b	7.7ab	17.5a	12.4a	5.2a
Group II	13.6b	127c	8.9b	25.1b	26.0b	6.8b

Note: Different letters in a column represent the significant differences among the treatments.

Table 11.4.2-4. The effect of water hyacinth silage on the relative weight of digestive tract of 10-week old geese (g kg^{-1} body weight).

Treatment	Proventriculus	Gizzard	Duodenum	Jejunum	Ileum	Appendix
Control	3.95a	42.67a	3.22	6.96a	5.76a	2.02a
Group I	3.57a	41.79a	3.03	6.86a	4.86a	2.03a
Group II	5.17b	49.55b	3.45	9.74b	10.06b	2.64b

Note: Different letters in a column represent the significant differences among the treatments.

The gastrointestinal relative length may also reveal the gosling adaptability to water hyacinth silage replacement. The experiment revealed that the relative length of ileum was longer in the group II treatment than that in the group I and the control treatments ($p < 0.05$, Table 11.4.2-6).

The length and weight of digestive tract may change with the growth stage, animal species and feed type. Generally, high fiber content tends to increase the length and weight of digestive tract. The gizzard and intestine are important functional organs to digest and absorb nutrients. The results of the case study (Tables 11.4.2-3 to 11.4.2-6) were consistent with the findings in the literature (Shao and Han 1992), but the changes are generally influenced by fiber sources, species, fiber balance, etc. (Zhao 2003).

Table 11.4.2-5. The effects of water hyacinth silage on the length (cm) of digestive tract of 10-week old geese.

Treatment	Duodenum	Jejunum	Ileum	Appendix
Control	29.8 ± 2.30	72.4 ± 5.67a	67.0 ± 4.25a	40.2 ± 4.94a
Group I	31.2 ± 4.54	78.3 ± 4.72a	72.7 ± 4.26a	41.0 ± 4.88a
Group II	31.6 ± 4.26	90.2 ± 5.04b	94.3 ± 5.38b	47.9 ± 4.99b

Note: Different letters in a column represent the significant differences among the treatments.

Table 11.4.2-6. The effects of water hyacinth silage on the relative length of digestive tract of 10-week old geese (cm kg^{-1} body weight).

Treatment	Duodenum	Jejunum	Ileum	Appendix
Control	14.0 ± 1.88	34.1 ± 3.51	31.7 ± 3.98a	19.1 ± 3.55b
Group I	12.3 ± 1.85	31.1 ± 3.98	28.7 ± 3.04a	16.1 ± 1.46a
Group II	12.2 ± 2.33	35.2 ± 3.03	36.8 ± 5.44b	18.5 ± 2.75ab

Note: Different letters in a column represent the significant differences among the treatments.

Effects of water hyacinth silage on goose growth and production

The experiment revealed that, during weeks 5 to 8, geese in the groups I and II gained greater daily body weight than the control. During weeks 9 to 10, group I gained greater daily body weight than group II ($p < 0.05$) and almost the same as the control (Table 11.4.2-7).

The results also revealed that the rates of eviscerated yield, half-eviscerated yield and abdominal fat rate were significantly lower in the group II treatment than those in group I and the control treatments (Table 11.4.2-8) (Bai et al. 2013).

The case study may suggest that water hyacinth silage after proper fermentation can be a good source of feed for large-scale geese production. Geese eating feed containing 37 to 55% water hyacinth silage showed better production performance or no significant difference than those eating commercial feed with no water hyacinth silage. The results from the case study were consistent with the literature (Zhou 2006). The theoretical assumption

Table 11.4.2-7. Effect of water hyacinth silage diet on body weight and average daily gain of goslings.

Group	4 weeks body weight (g individual^{-1})	6 weeks body weight (g individual^{-1})	8 weeks body weight (g individual^{-1})	10 weeks Body weight (g individual^{-1})	Average Day gain (g individual^{-1} d^{-1})
Control	1131	1705b	1993b	2265	27
Group I	1146	1948	2248	2530	33
Group II	1197	2060a	2353a	2313	27

Note: Different letters in a column represent the significant differences among the treatments; no letters, no differences.

Table 11.4.2-8. Impacts of water hyacinth silage on the carcass properties of geese (% of total carcass weight, except column two).

Treatment	Carcass of live weight (%)	Half-eviscerated	Eviscerated	Breast muscle	Leg muscle	Abdominal fat
Control	89	78b	68b	5.5	12.5	1.56b
Group I	89	79b	69b	6.6	12.4	1.63b
Group II	89	73a	65a	6.3	12.3	0.74a

Note: Different letters in a column represent the significant differences among the treatments; no letters, no differences.

was that geese ingested more feed that contained water hyacinth silage and/or achieved better feed utilization to achieve improved growth. The experiment also revealed the differences between treatments with 37 and 55% of commercial feed being replaced by water hyacinth silage, with the 37% treatment resulting in better performance regarding some parameters, which may indicate a need for balance between feed nutrients, and fiber and water contents in geese feed (Feng et al. 2007). The results of the case study also indicated that the water hyacinth silage had different effects at different growth stages, with geese ingesting more feed in the group II treatment at 9–10 weeks old compared with the group I treatment but gaining less body weight at the week 10 assessment. We can suggest that feed for the group II treatment failed to provide enough nutrients for geese. However, the case study did not provide sufficiently detailed data and analysis. Dietary supplement of water hyacinth silage at 37 to 55% of commercial feed did not adversely impact the performance of Yangzhou geese in commercial production, but significantly decreased the production cost (Wang et al. 2012).

References cited

Abdelhamid, A. M. and A. A. Gabr. 1991. Evaluation of water hyacinth as a feed for ruminants. *Archiv für Tierernaehrung* 41(7-8): 745–756.

Aboud, A. A. O., R. S. Kidunda and J. Osarya. 2005. Potential of water hyacinth (*Eicchornia crassipes*) in ruminant nutrition in Tanzania. *Livestock Research for Rural Development* 17(8): Art. #96.

Agency for Toxic Substances and Disease Registry (ASTDR). 2012. Toxicological profile for chromium. *ATSDR 2000*. Washington, DC. USA: U.S. Department of Health and Human Services.

Bai, Y., L. Jiang, L. Gao, S. Yan and Y. Tu. 2013. Growth performance, carcass traits and gastrointestinal development of Yangzhou gosling affected by water hyacinth silage diet. *Jiangsu Journal of Agricultural Sciences* 29(5): 1107–1113 (In Chinese with English Abstract).

Bai, Y., W. Zhou, S. Yan, J. Liu, H. Zhang and L. Jiang. 2010. Effects of feeding water hyacinth silage on lambs fattening. *Jiangsu Journal of Agricultural Sciences* 26(5): 1108–1110 (In Chinese with English Abstract).

Bai, Y. 2011. Ensilaging water hyacinth: effects of water hyacinth compound silage on the performance of goat. *Chinese Journal of Animal Nutrition* 23(2): 330–335 (In Chinese with English Abstract).

Bai, Y., J. Zhu, S. Yan, W. Wang, J. Liu, Y. Tu et al. 2011. Effects of feeding water hyacinth silage on fattening geese in large scale. *China Poultry* 33(7): 49–52 (In Chinese).

Baldwin, J. A., J. F. Hentges, L. O. Bagnall and R. L. Shirley. 1975. Comparison of pangolagrass and water hyacinth silages as diets for sheep. *Journal of Animal Science* 40(5): 968–971.

Bolenz, S., H. Omran and K. Gierschner. 1990. Treatments of water hyacinth tissue to obtain useful products. *Biological Wastes* 33(4): 263–274.

Buendía-González, L., J. Orozco-Villafuerte, F. Cruz-Sosa, C. E. Barrera-Díaz and E. J. Vernon-Carter. 2010. *Prosopis laevigata* a potential chromium (VI) and cadmium (II) hyperaccumulator desert plant. *Bioresource Technology* 101(15): 5862–7.

Chen, X. 2011. The physiological and biological characteristics of the water hyacinth (*Eichhornia crassipes*) in adapting to different growth conditions, MSc. Thesis, School of Life Science. Fujian Agriculture and Forestry University (In Chinese with English Abstract).

Chen, X. and Y. Zhuang. 2009. Advance in utilization of water hyacinth as feed resource. *Fujian Journal of Animal Husbandry and Veternary Medicine* 31(4): 29–31 (In Chinese).

Cohen, M. D., B. Kargacin, C. B. Klein and M. Costa. 1993. Mechanisms of chromium carcinogenicity and toxicity. *Critical Reviews in Toxicology* 23(3): 255–281.

Cordes, K. B., A. Mehra, M. E. Farago and D. K. Banerjee. 2000. Uptake of Cd, Cu, Ni and Zn by the water hyacinth, *Eichhornia crassipes* (Mart.) Solms from pulverised fuel ash (PFA) leachates and slurries. *Environmental Geochemistry and Health* 22(4): 297–316.

Cui, L., H. Xiao, L. Chen, Y. Xu, Y. Huang and J. Chu. 2004. Feeding effects of water hyacinth from Lake Dinshan on swine fattening. *Feed Industry* 25(3): 39–40 (In Chinese).

Ding, J., R. Wang, W. Fu and G. Zhang. 2001. Water hyacinth in China: its distribution, problems and control status. In: Proceedings of the second meeting of the global working group for the biological and integrated control of water hyacinth, Beijing, China, 9–12 October 2000, ed. M. H. Julien, M. P. Hill, T. D. Center, and J. Ding, 29–32. Beijing, China: Australian Centre for International Agricultural Research.

Du, J., Z. Chang, X. Ye, Y. Xu and J. Zhang. 2012. Pilot-scale study on dehydration effect of water hyacinth with different pulverization degree. *Transactions of the Chinese Society of Agricultural Engineering* 28(5): 207–212 (In Chinese with English Abstract).

El-Sayed, A. -F. M. 2003. Effects of fermentation methods on the nutritive value of water hyacinth for Nile tilapia *Oreochromis niloticus* (L.) fingerlings. *Aquaculture* 218(1-4): 471–478.

EPA. 2011. 2011 edition of the drinking water standards and health advisories, EPA 820-R-11-002. Washington, DC. USA: U.S. Environmental Protection Agency.

Feng, W., Z. Xie, W. Zhang, Z. Li, H. Liu, J. Yang et al. 2007. Goose digestibility and utilization of stalks. *China Animal Husbandry and Veterinary Medicine* 34(4): 35–36 (In Chinese).

GB13078-2001. 2001. Hygienical standard for feeds. Beijing, China: General Administration of Quality Supervision, Inspection and Quarantine of the People's Republic of China (AQSIQ).

GB2762-2012. 2012. National food safety standard - maximum levels of contaminants in food. *National Food Safety Standard*. Beijing, China: Ministry of Health of the People's Republic of China.

Gunnarsson, C. C. and C. M. Petersen. 2007. Water hyacinths as a resource in agriculture and energy production: a literature review. *Waste Management* 27(1): 117–29.

Han, J., D. Jiang, Z. Wang, H. Yang and K. Huang. 2014. Dietary crude fiber utilization of geese. *Chinese Journal of Animal Nutrition* 26(5): 868–876 (In Chinese with English Abstract).

Jin, G., J. Wang, Y. Wang and M. Zhu. 1994. Anatomy study on digestorius system in Wan-Xi white geese. *Journal of Anhui Agrotechnical Teachers College* 8(1): 54–56 (In Chinese with English Abstract).

Konyeme, J. E., A. O. Sogbesan and A. A. A. Ugwumba. 2006. Nutritive value and utilization of water hyacinth (*Eichhornia crassipes*) meal as plant protein supplement in the diet of *Clarias gariepinus* (Burchell, 1822) (Pisces: Clariidae) fingerlings. *African Scientist* 7(3): 127–133.

Li, J. Di. 2006. Nutritional evaluation and optimized utilization of water hyacinth for ruminants. Ph.D. Thesis, Department of Animal Nutrition and Feed Science. Zhe Jiang University.

Li, J., J. Liu, Y. Wu and J. Ye. 2007. Effects of lactobacillus and cellulase on fermentation characteristics of water hyacinth stem and leaf. *Grass Processing* 43(11): 27–30 (In Chinese with English Abstract).

Li, W. and Y. Zhang. 1992. Utilization of water hyacinth as fish feed. *Fisheries Science* 11(5): 1–4 (In Chinese).

Lu, J., Z. Fu and Z. Yin. 2008. Performance of a water hyacinth (*Eichhornia crassipes*) system in the treatment of wastewater from a duck farm and the effects of using water hyacinth as duck feed. *Journal of Environmental sciences* 20(5): 513–9.

Lytle, C. M., F. W. Lytle, N. Yang, J. H. Qian, D. Hansen, A. Zayed et al. 1998. Reduction of Cr(VI) to Cr(III) by wetland plants: potential for *in situ* heavy metal detoxification. *Environmental Science and Technology* 32: 3087–3093.

Mangisah, I., H. I. Wahyuni, S. Sumarsih and S. Setyaningrum. 2010. Nutritive value of fermented water hyacinth (*Eichornia crassipes*) leaf with *Aspergillus niger* in Tegal Duck. *Journal of Animal Production* 12(2): 100–104.

Martin, N. P., D. R. Mertens and P. J. Weimer. 2004. Alfalfa: hay, haylage, baleage, and other novel products. In: Proceedings of the Idaho Alfalfa and Forage Conference, ed. Retrieved from http://www.extension.uidaho.edu/forage/Proceedings/2004 Proceedings pdf/Alfalfa hay_haylage--Martin.pdf. Idaho, USA: University of Idaho Extension.

Mertz, W. 1993. Chromium in human nutrition: a review. *The Journal of Nutrition* 123(4): 626–633.

Ministry of Public Health. 2006. Standards for drinking water quality - GB5749-2006. Beijing, China: Ministry of Health, P.R. China.

Mukherjee, R. and B. Nandi. 2004. Improvement of *in vitro* digestibility through biological treatment of water hyacinth biomass by two *Pleurotus* species. *International Biodeterioration & Biodegradation* 53(1): 7–12.

Patel, S. 2012. Threats, management and envisaged utilizations of aquatic weed *Eichhornia crassipes*: an overview. *Reviews in Environmental Science and Bio/Technology* 11(3): 249–259.

Poddar, K., L. Mandal and G. C. Banerjee. 1991. Studies on water hyacinth (*Eichhornia crassipes*) – chemical composition of the plant and water from different habitats. *Indian Veterinary Journal* 68: 833–837.

Rodríguez, M., J. Brisson, G. Rueda and M. S. Rodríguez. 2012. Water quality improvement of a reservoir invaded by an exotic macrophyte. *Invasive Plant Science and Management* 5(2): 290–299.

Saha, S. and A. K. Ray. 2011. Evaluation of nutritive value of water hyacinth (*Eichhornia crassipes*) leaf meal in compound diets for rohu, *Labeo rohita* (Hamilton, 1822) fingerlings after fermentation with two bacterial strains isolated from fish gut. *Turkish Journal of Fisheries and Aquatic Sciences* 11(2): 199–207.

Sahu, A. K., S. K. Sahoo and S. S. Giri. 2002. Efficacy of water hyacinth compost in nursery ponds for larval rearing of Indian major carp, *Labeo rohita*. *Bioresource Technology* 85(3): 309–11.

Salnikow, K. and A. Zhitkovich. 2008. Genetic and epigenetic mechanisms in metal carcinogenesis and cocarcinogenesis: nickel, arsenic and chromium. *Chemical Research in Toxicology* 21(1): 28–44.

Schroeder, J. W. 2004. AS-1250 Forage nutrition for ruminants. Fargo, ND, USA: North Dakota State University Agriculture and University Extension.

Schroeder, J. W. 2013. AS-1251 Interpreting composition and determining market value. Fargo, ND, USA: North Dakota State University Agriculture and University Extension.

Seaborn, C. D. and B. J. Stoecker. 1990. Effects of antacid or ascorbic acid on tissue accumulation and urinary excretion of ^{51}chromium. *Nutrition Research* 10(12): 1401–1407.

Shao, C. and Z. Han. 1992. Studies on fibre components digestion of cecum in goose. *Journal of Nanjing Agricultural University* 15(4): 86–89 (In Chinese with English Abstract).

Shi, Z. and R. Zhao. 2007. Accumulation of Cd^{2+}, Zn^{2+} by water hyacinth. *Fisheries and Water Conservation* 27(4): 66–68 (In Chinese).

Sun, H., J. Brocato and M. Costa. 2015. Oral chromium exposure and toxicity. *Current Environmental Health Reports* 2(3): 295–303.

Sunday, A. D. 2002. The utilization of water hyacinth (*Eichhornia crassipes*) by West African dwarf (Wad) growing goats. *African Journal of Biomedical Research* 4: 147–149.

Tag EI-Din, A. R. 1992. Utilization of water-hyacinth hay in feeding of growing sheep. *Indian Journal of Animal Sciences* 62(10): 989–992.

Tham, H. T., N. Van Man and T. Pauly. 2013. Fermentation quality of ensiled water hyacinth (*Eichhornia crassipes*) as affected by additives. *Asian-Australasian Journal of Animal Sciences* 26(2): 195–201.

Tham, H. T. and P. Udén. 2013. Effect of water hyacinth (*Eichhornia Crassipes*) silage on intake and nutrient digestibility in cattle fed rice straw and cottonseed cake. *Asian-Australasian Journal of Animal Sciences* 26(5): 646–653.

The commission of the european communities. 2006. Commission regulation (EC) No. 1881/2006. Brussels, Belgium: The commission of the european communities.

Wang, X., Z. Wang and H. Yang. 2012. Effects of different metabolizable energy, crude protein and lysine levels on the growth performance of 5–10 weeks old Tangzhou gosling. *Feed Industry* 33(7): 22–26 (In Chinese with English Abstract).

WHO. 2011. Guidelines for drinking-water quality - 4th Edition. Geneva, Switzerland: World Health Organization.

Xie, P., X. Zhou, J. Yang, X. Zhu, Y. Zhang, J. Zhang et al. 1999. Feeding effects of water hyacinth from Lake Dianchi on production of broiler. *Feed Industry* 20(4): 26–28 (In Chinese).

Yapoga, S., Y. B. Ossey and V. Kouame. 2013. Phytoremediation of zinc, cadmium, copper and chrome from industrial wastewater by *Eichhornia crassipes*. *International Journal of Conservation Science* 4(1): 81–86.

Yu, D. 2008. Chromium toxicity - ATSDR - CSEM. Atlanta, GA, USA: Agency for Toxic Substances and Disease Registry.

Yu, Y. 1989. Nutrition and ensilage of water hyacinth. *Feed China* (1): 38–41 (In Chinese).

Zhang, Z. Y., J. C. Zheng, H. Q. Liu, Z. Z. Chang, L. G. Chen and S. H. Yan. 2010. Role of *Eichhornia crassipes* uptake in the removal of nitrogen and phosphorus from eutrophic waters. *Chinese Journal of Eco-Agriculture* 18(1): 152–157 (In Chinese with English Abstract).

Zhao, L. 2003. Study on utilization of different fiber resources for geese. MSc. Thesis, Department of Animal Nutrition and Feed Science. Jiling Agricultural University (In Chinese with English Abstract).

Zhou, S. 2006. Effects of dietary fiber on the digestive tract development, growth and physiological change of goose. MSc. Thesis, Department of Animal Nutrition and Feed Science. Huazhong Agricultural University (In Chinese with English Abstract).

Zhou, W., L. Tan, D. Liu, H. Yan, M. Zhao and D. Zhu. 2005. Research advances of *Eichhornia crassipes* and it's utilization. *Journal of Huazhong Agricultural University* 24(4): 423–428 (In Chinese with English Abstract).

Zhuang, Y., Y. Gao, W. Zhang and S. Wang. 2006. Effects of fermented green juice on the quality of corn straw silage. *Acta Ecologiae Animalis Domastici* 27(6): 70–73, 86 (In Chinese with English Abstract).

Zhuang, Y., W. Zhang, X. Chen and L. Xu. 2008. Effect of fermented green juice, cellulase and mixture on quality of water hyacinth silage. *Chinese Agricultural Science Bulletin* 24(5): 35–38 (In Chinese with English Abstract).

Zhuang, Y., W. Zhang, L. Zhang, X. Chen and D. Yao. 2007. Effect of additives on the quality of water hyacinth silage. *Chinese Agricultural Science Bulletin* 23(9): 32–35 (In Chinese with English Abstract).

CHAPTER 12

Economic Assessment and Further Research

Z. H. Kang[1] and *J. Y. Guo*[2,*]

12.1 Cost-benefit assessment of phytoremediation using water hyacinth

In a multidimensional management system such as water hyacinth management and control, the decision-making process faces more challenge than making decisions in a purely commercial or environmental process because of a combination of ecological and socio-economic targets focused on public concerns. Hence, during the decision-making process, not only information on natural science and technology is needed, but also information on the socio-economic expectations and cost-benefit details. The cost-benefit analyses are essential for evaluating the economic implications of phytoremediation using water hyacinth. The analyses can be useful for multidimensional strategic policy development and for assessing ecosystem services as influenced by the policies. Evidence on economic value can also be used in implementing the policy and in decisions about eutrophic ecosystem rebuilding and/or restoration (UNEP-WCMC 2011). Information on natural science and technology can only provide an estimate of whether the management program will meet the final targets; in addition, the socio-economic expectations and the cost-benefit analysis can shed additional light on the planned budgetary expenditure and possible financial returns.

From the point of view of the public, the management and control of water hyacinth and environmental protection all represent common good, some of which is invisible. The projects on this subject must be transparent

[1] 50 Zhong Ling Street, Nanjing, China.
Email: kzh_mm@126.com
[2] 5 Armagh Way, Ottawa, Canada.
* Corresponding author: guoj1210@hotmail.com

and assessed reliably in both scientific and socio-economic contexts. However, the implementation of cost-benefit analysis in phytoremediation faces many challenges:

1) The evaluation of ecosystem services together with natural science and technology processes that may not have the established market values (Yang 2013);
2) Methods to evaluate many ecosystem services in the traditional cost-benefit analysis may not be commonly agreed on (de Groot, Fisher et al. 2010); and
3) There may also be no common agreement on the methods to evaluate multidimensional management projects on nitrogen and phosphorus removal, water quality increase and waterway transportation improvement for the traditional cost-benefit analysis.

The advances in both natural and socio-economic sciences have promoted the understanding of multidimensional targets in socio-economic and ecological context by the system approach, such as the coupling-coordination analysis at the top level to estimate the system evolutionary direction and the cost-benefit based structure matrix model analysis at the subsystem level to evaluate natural and social processes together (Binder 2007, Bai et al. 2011, Schiller et al. 2014, Ding et al. 2015). Nevertheless, an agreement by all relevant disciplines (e.g., scientists in natural and technological aspects with economists in social and economic aspects) may be very difficult to achieve.

Coupling refers to the relationship and influence between two or more dynamic processes regardless of natural or socio-economic context. Coordination refers to approaching coexistence and co-development by adjustment and adaptation through dynamic changes of all subsystems (elements) or bringing in new elements into the system. The analyses bridge the gap in analytical methods and bring socio-economic and natural sciences together to reflect the final status of the system, targets of which are indispensable in guiding the decision-making process. However, the top-level analyses do not provide indication on innovative direction or any additional elements to improve the system-level performance, whereas the cost-benefit based structure matrix model analysis can provide evaluation of the improvement of subsystems or any additional elements being brought into the system. The methods employed to assess the management performance of multidimensional targets in phytoremediation using water hyacinth are mainly on the subsystem level for additional elements in the improved systems.

12.1.1 Methods for assessing performance of multidimensional phytoremediation projects

Traditionally, socio-economic development has often been achieved at the expense of environment and ecosystem degradation. However, this is not acceptable anymore in modern society with advanced natural science and technology as well as modern socio-economic sciences. History teaches lessons on how to couple and coordinate various processes and how to improve system performance by additional elements to achieve desired multidimensional targets in phytoremediation using water hyacinth. The major difficulties in applying cost-benefit based structure matrix model analysis in phytoremediation using water hyacinth are a lack of information on the global market prices, such as the price of removing one unit of nitrogen or phosphorus from eutrophic water or the price of improving water quality from surface water standard IV to standard III, not to mention global differences in defining the standards of water quality.

To solve these problems, the concepts of ecosystem services and ecosystem functions must be clearly defined. Ecosystem functions are natural processes of energy flow and nutrient cycling in biomes with the surrounding environment. Ecosystem services are defined as the benefit that the ecosystem can provide directly or indirectly to human welfare (Costanza et al. 1997). Ecosystem services and ecosystem functions may not be necessarily on a one-to-one correspondence. A single ecosystem service may be the product of two or more ecosystem functions, whereas a single ecosystem function may contribute to two or more ecosystem services. For example, mature water hyacinth actively produces offshoots to achieve growth (involving energy flow and nutrients assimilation), which is the function of ecosystems, while the removal of nitrogen and phosphorus from the waters by water hyacinth as well as the photosynthesis of water hyacinth during growth that supplies oxygen to atmosphere (benefits to human welfare) are ecosystem services. Ecosystem services provide very efficient, least-cost benefit to human life-supporting systems that are necessary for the survival of both humans and animals, although many of these are not fully acknowledged.

Ecosystem services are subject to degradation or improvement upon human interferences. For instance, when pollutants of nitrogen and phosphorus are drained to rivers and lakes, the clean water resources and aquatic recreation activities may be compromised. Therefore, understanding and evaluating ecosystem services are essential to assess the performance of phytoremediation projects. For decades, scientists in socio-economic and natural sciences have worked together to define the "value" of ecosystem services in a way that is applicable and understandable globally for both market and non-market benefits to human welfare (Holl and Howarth 2000), including the moderation and mitigation of extreme events and climate change, provisioning of food, water and recreation and many other services. The basic idea is to determine what it would cost if the service is replicated in technologically produced, artificial biomes, or to quantify ecosystem services by contingent valuation, thus replacing market price techniques. In case of non-market benefits, contingent valuation or "willingness-to-pay" techniques are often used (Pascual et al. 2010) although not without criticism (Holl and Howarth 2000). The "willingness to pay" survey is conducted to collect answers on how changes in the quantity or quality of various types of ecosystem services may impact human welfare and how much may be acceptable to pay. The basic theory behind this method is that any particular change in ecosystem services may result in costs or benefits regarding maintenance of human welfare. During the valuation, terminology of Total Economic Value (TEV) is often used and refers to both direct and indirect benefits derived from ecosystem services. For instance, if phytoremediation of eutrophic water provided a US$80-increment to the fish productivity of a lake, then the beneficiaries of this service should be willing to pay up to US$80. In addition to cultural value, if the lake offered non-market aesthetic, existence and recreational values of US$120, those receiving this non-market benefit should be willing to pay up to US$120. The TEV of ecological services in this case would be $200 of the lake. However, only the improvement in fish productivity would contribute to the monetary value of ecological services as the value derived through the markets. Costanza et al. (1997) synthesized a large volume of literature, presented the estimates of values for ecosystem services per unit area by biome, and grouped ecosystem services into 17 major categories (Table 12.1.1-1).

Enhanced quality of lakes, rivers and reservoirs can improve waste treatment by up to US$666 ha^{-1} yr^{-1}, with these values not showing up in any market. Although in many cases there is no obvious relationship between the value of ecosystem services and the current spending on the market, the benefits to humans are huge and indispensable (Costanza et al. 1997). Table 12.1.1-1 indicated that the total value of lakes, rivers and reservoirs is US$8599 ha^{-1} yr^{-1}, among which the remediation projects may provide about US$2388 ha^{-1} yr^{-1} in services such as recreation, food production and water supply. In another case study in Uganda, Kakuru et al. (2013) reported valuation of US$814 ha^{-1} yr^{-1} in wetland ecosystem services only in categories of fish spawning grounds, fish yields, water regulation/recharge and recreation/aesthetic values.

Research over the last several decades in the valuation of ecosystem services has elucidated their indispensable roles in human societies. Phytoremediation of eutrophic waters and management strategies are now widely accepted as integrating the "environmental/ecosystem development" and "socio-economic development" (de Groot, Alkemade et al. 2010). Thus, the assessment of phytoremediation projects based on the valuation of the ecosystem services reveals more meaningful information and knowledge. Investments in phytoremediation, ecosystem rebuilding or restoration and sustainable ecosystem development are increasingly seen as a win-win situation that generates considerable benefits to humans (de Groot, Alkemade et al. 2010).

Table 12.1.1-1. Categories of ecosystem services and their valuation (Costanza et al. 1997).

Ecosystem services	Total value per hectare (US$ ha^{-1} yr^{-1})	Example biomes
Gas regulation	1341	All biomes
Climate regulation	684	All biomes
Disturbance moderation	1779	All biomes
Water regulation	1115 (5445)[a]	All biomes (Lakes, rivers and reservoirs)
Water supply	1692 (2117)[a]	All biomes (Lakes, rivers and reservoirs)
Erosion control and sediment retention	676	All biomes
Soil formation	63	All biomes
Nutrient cycling	17,076	All biomes
Waste treatment	2277 (666)[a]	All biomes (Lakes, rivers and reservoirs)
Pollination	117	All biomes
Biological control	417	All biomes
Refugia	124	All biomes
Food production	1388 (41)[a]	All biomes (Lakes, rivers and reservoirs)
Raw materials	721	All biomes
Genetic resources	79	All biomes
Recreation	815 (230)[a]	All biomes (lakes, rivers and reservoirs)
Cultural	3015	All biomes

Note: [a] values in parentheses represent the values for the biomes in parentheses in the rightmost column.

12.1.2 Estimation of phytoremediation compensation on removals of nitrogen and phosphorus

In the above discussion, the valuation of ecosystem services in biomes of rivers, lakes and reservoirs was represented as the total economic value in the several service categories (water resources, wastewater treatment, food production and recreation activities). The removal of nitrogen and phosphorus is only part of these services, albeit the first and essential step in improving the water quality; hence, using the total economic valuation in assessing phytoremediation management may not be appropriate.

Pagiola et al. (2004) reported a case study in water conservation in New York City, USA. New York City obtains its water supply from watershed in Catskill Mountains and requires about 1.1 billion cubic meters per day. After system analysis of the cost of alternative strategies to meet the city's requirement, the government chose a non-point pollution control program with initial investment of US$1.5 billion, saving US$8–10 billion on a water filtration option at the 2004 market value. The best option showed a rate of US$1.35 per cubic meter of water. In other literature, Söderqvist et al. (2005) reported a cost-benefit analysis on alternative strategies to reduce eutrophication in the Stockholm Archipelago. In the study, water quality improvement to increase water transparency by 1 m was desired to achieve both ecosystem health and human recreational benefits in the area. The management target required annual reduction of 2725 tonnes of nitrogen loading through a combination of management strategies including increased sewage water treatment and reduced fertilizer application in the surrounding area. The total costs of these measures were estimated to be US$7.45 million per year at the 2004 market value. The cost of the reduction was translated to US$2735 t^{-1} nitrogen. Furthermore, the benefits of such reduction were estimated to be about US$7.8 million per year for recreational income and US$65.4 million per year for total benefits. The cost-benefit ratio was estimated at about 8:1 or better. Although this case study in a highly-populated area may not be applicable to less populated areas, it provides good reference information for valuation of the phytoremediation projects using water hyacinth. In South Africa, the price of supplying water was estimated at about US$0.25 m^{-3} at the 2010 market value (de Lange and van Wilgen 2010). The examples discussed above showed the costs of eutrophic management in aquatic ecosystems ranging from US$0.25 to 1.35 per cubic meter of water supplied, showing wide differences among regions and the types of ecosystem services, indicating that the cost-benefit analysis should be location- and site-specific.

Liu et al. (2011) suggested formulating the ecological compensation policy in the process of eutrophic management, with financial sources from ecosystem service beneficiaries; in addition, phytoremediation targets and compensation standards were suggested to be based on the cost-benefit analysis of phytoremediation. Kang et al. (2012) applied cost-benefit analysis of phytoremediation using water hyacinth in Lake Taihu region of China and suggested compensation of US$4800 for the removal of one tonne nitrogen or nitrogen and phosphorus combined. In the analysis, the total cost was US$2230 for confined growth, US$1445 for harvest and transportation and US$1124 for dehydration. The analysis elucidated that the proposed compensation was substantially higher than the nitrogen removal at cost of US$2735 t^{-1} in the Stockholm Archipelago region (Söderqvist et al. 2005, Kang and Liu 2015). However, the result from Lake Taihu was reasonable due to water hyacinth being unable to survive over the winter, resulting in relatively higher cost compared with the tropical climate zone. When excluding the cost for seedlings over the winter, the cost in the tropical zone may be less than the cost in Lake Taihu, although the management approach may be different.

12.1.3 Estimating costs of confined growth of water hyacinth

In the tropical and subtropical climate zones where water hyacinth can naturally survive over the winter, there is no need for spending on the growth of water hyacinth. However, for confined growth (i.e., confining water hyacinth to places where pollutants removal is needed), costs of materials and construction of facilities such as posts, nets, floats and labor are necessary (Zhang et al. 2013). The cost-benefit analysis of confined growth of water hyacinth has been reported in the literature only rarely. The average cost was estimated at US\$2866 ha^{-1} including the cost of water hyacinth plantlets, materials (posts, nets and floats), operational spending and labor wages (Kang et al. 2012). At Lake Taihu, one hectare of water hyacinth can yield 891 tons of fresh biomass containing 1.38 tonnes of nitrogen and 0.20 tonnes of phosphorus (Zheng et al. 2008). Under the suggested ecological compensation policy, this amount of nutrient removal would generate an income from compensation of US\$7584, which is more than enough to cover the costs of improving ecosystem services.

12.1.4 Estimating costs of harvest and dehydration of water hyacinth biomass

Phytoremediation using water hyacinth for eutrophic control in aquatic ecosystems and integrating multidimensional strategies in the remediation have well-established principles, but the reality of each individual case is unique due to high risk and huge investment (Malik 2007, Gettys et al. 2009). Wang et al. (2012) reported an ecological engineering project (4.3 km^2) using water hyacinth in phytoremediation of Lake Caohai (24°40'–25°01' N, 102°35'–102°46' E), Yunnan Province, China. This lake was hypereutrophic according to the eutrophic status index (Institute of Environmental Science 1992, MEP-PRC 2002) because a rapid increase in local population and consequently in discharging effluents from the municipal waste treatment plants as well as non-point pollution (Wang et al. 2009, Zhang et al. 2014).

Cost-benefit analysis in phytoremediation case study

The ecological engineering project was designed to integrate multidimensional strategies and to completely harvest and dehydrate water hyacinth biomass (Wang et al. 2012, Wang et al. 2013). During the execution, an area of 430 hectares of water hyacinth was established for removing nitrogen and phosphorus in Lake Caohai from 2010 to 2012 (Wang et al. 2013). The cost-benefit analysis (Table 12.1.4-1) showed that each year the project processed 0.39 million tonnes of fresh biomass of water hyacinth, produced 19,400 tonnes of water hyacinth

Table 12.1.4-1. Cost-benefit analysis of phytoremediation using water hyacinth in Lake Caohai [Equations from (Hronich et al. 2008)].

Water hyacinth growth				
A	Water hyacinth growth area	430 ha	Parameter	
G	Annual dry biomass production	45 tonnes ha^{-1} yr^{-1}	Parameter	
M_r	Total fresh biomass annually	3.87E+08 (kg/year)	$M_r = AG(1000)/(1-R_{W,in})$	Eq.1
ρ_A	Water hyacinth density	20 (kg m^{-2})	Parameter	
T_{yr}	Annual working days	180 (days yr^{-1})	Parameter	
M_D	Daily biomass harvested	2.15E+06 (kg d^{-1})	$M_D = M_r/T_{yr}$	Eq. 2
Harvest				
w_{cut}	Cut width	2.5 (m)	Parameter	
v_{cut}	Cut speed	58 (m min^{-1})	Parameter	
f_H	Re-growth rate required to maintain water hyacinth mat	40 (days)	$f_H = A(10000)\,_A/M_D$	Eq. 3
A_H	Area harvested hourly	8700 (m^2 h^{-1})	$AH = w_{cut}\ v_{cut}\ (60)$	Eq. 4
T_D	Working hours per day	8 (h)	Parameter	
A_{DH}	Daily harvest per harvester	69,600 (m^2 d^{-1})	$A_{DH} = A_H T_D$	Eq. 5
M_H	Plant biomass harvested per harvester	1.39E+06 (kg d^{-1})	$M_H = A_{DH}\rho_A$	Eq. 6
N_H	Whole number of harvesters required	2	$N_H = M_D/M_H$	Eq. 7
A_D	Total hectares per day harvested	13.92 (ha)	$A_D = A_{DH}N_H/10000$	Eq. 8
P_H	Energy requirement per harvester	127 (kW)	Parameter	

Transportation				
C_M	Connectivity of hyacinth mats	100 (Pa)	Parameter	
v_M	Speed of pulling mat in	2 ($m\ s^{-1}$)	Parameter	
ρ	Plant density	167 ($kg\ m^{-3}$)	Parameter (Bagnall 1982)	
l_M	Estimated mat length	719 (m)	$l_M = C_M(2400)/(v_M\rho)$	Eq. 9
M_M	Estimated mat weight	3.59E+04 (kg)	$M_M = l_M w_{cut}\rho_A$	Eq. 10
N_M	Number of mats pulled daily	60 (mats d^{-1})	$N_M = M_D/M_M$	Eq. 11
P_{RB}	Row boat energy requirements	5 (kW)	Parameter	
N_{RB}	Number of operators required	5	$N_{RB} = N_M/[T_D(1.43)]$	Eq. 12
Storage and decomposition				
$R_{W,in}$	Weight fraction of water in biomass	0.95 (no unit)	Parameter	
$M_{W,in}$	Total water in biomass	2.04E+06 ($kg\ d^{-1}$)	$M_{W,in} = R_{W,in}M_D$	Eq. 13
$M_{B,in}$	Total dry biomass	1.08E05 ($kg\ d^{-1}$)	$M_{B,in} = (1-R_{W,in})M_D$	Eq. 14
$R_{W,rem}$	Total water removable	0.97 (no unit)	Parameter	
$M_{P,hr}$	Biomass processed per hour	8.96E+04 ($kg\ h^{-1}$)	$M_{P,hr} = M_D/24$	Eq. 15
$M_{W,rem}$	Water removed	8.26E+04 ($kg\ h^{-1}$)	$M_{W,rem} = M_{W,in}R_{W,rem}/24$	Eq. 16
$M_{T,out}$	Total biomass leaving presses	7.03E+03 ($kg\ h^{-1}$)	$M_{T,out} = M_{P,HR}-M_{W,rem}$	Eq. 17
$M_{W,out}$	Water remaining in biomass	2.55E+03 ($kg\ h^{-1}$)	$M_{W,out} = M_{W,in}/24-M_{W,rem}$	Eq. 18
$M_{B,out}$	Biomass leaving presses	4.48E+03 ($kg\ h^{-1}$)	$M_{B,out} = M_{B,in}/24$	Eq. 19

Table 12.1.3-1. contd....

Table 12.1.3-1. contd.

Storage and decomposition				
$R_{W,out}$	% water leaving presses	36% (mass %)	$R_{W,out} = M_{W,out}/M_{T,out}$	Eq. 20
$R_{B,out}$	% biomass leaving presses	64% (mass %)	$R_{B,out} = M_{B,out}/M_{T,out}$	Eq. 21
N_P	Number of presses required	6	Parameter	
P_P	Energy used by each press	60 (kW)	$P_P = (13.5\ \text{kW--hr/ton})(M_{P,hr}/1000)(1-R_{W,in})$	Eq. 22
P_{PT}	Total energy used	360 (kW)	$P_{PT} = N_P P_P$	Eq. 23
Capital				
C_S	Site	US$4,600,000	Parameter	
C_E	Equipment	US$1,476,900	Parameter	
C_{fix}	Fixed capital costs	US$6,076,900	$C_{fix} = C_S + C_E$	
C_W	Working capital	US$607,690	$C_W = 0.1C_{fix}$	
C_T	Total capital costs	US$6,684,600	$C_T = C_W + C_{fix}$	
Manpower				
MH_H	Harvesting	48 (Manhours d^{-1})	$MH_H = T_D N_H$	
MH_T	Transporting	40 (Manhours d^{-1})	$MH_T = T_D N_{RB}$	
MH_P	Pressing/digestion	8 (Manhours d^{-1})	Parameter	
C_{wage}	Wage + benefits	US$5.00 (Manhour^{-1})	Parameter	
$C_{wage,T}$	Total, per year	US$86,400 (yr^{-1})	$C_{wage,T} = (MH_H + MH_T + MH_P)C_{wage}T_{yr}$	
$C_{wage,S}$	Supervisory labor, per year	US$8640 (yr^{-1})	$C_{wage,S} = 0.1C_{wage,T}$	

Maintenance and operation			
$C_{fuel,H}$	Fuel for harvesters	US\$32,919 (yr^{-1})	$C_{fuel,H} = T_{yr}T_D[3(\$/gal)]([N_H P_H 1{,}000(W/kW)3600(s/hr)]/[43E06(J/kg)])[264.172(gal/m^3)/737.22(kg/m^3)]$
$C_{fuel,RB}$	Transport power required	US\$3240 (yr^{-1})	$C_{fuel,RB} = T_{yr}T_D[3(\$/gal)]([N_{RB}P_{RB}1000(W/kW)3600(s/hr)]/[43E06(J/kg)])[264.172(gal/m^3/737.22(kg/m^3)]$
C_P	Mill press power	US\$233,280 (yr^{-1})	$C_P = [N_P P_P]0.15(\$/kWh)24(hr/day)T_{yr}$
C_{MR}	Maintenance and repairs	US\$147,692 (yr^{-1})	$C_{MR} = 0.1C_E$
C_{OS}	Operating supplies	US\$14,769 (yr^{-1})	$C_{OS} = 0.1C_{MR}$
C_O	Overhead	US\$60,683 (yr^{-1})	$C_O = 0.25(C_{wage,T}+C_{wage,S}+C_{MR})$
C_{LT}	Local taxes	US\$60,769 (yr^{-1})	$C_{LT} = 0.01C_{fix}$
C_I	Insurance	US\$121,538 (yr^{-1})	$C_I = 0.02C_{fix}$
C_{Admin}	Administrative costs	US\$15,170 (yr^{-1})	$C_{Admin} = 0.25C_O$
C_{MO}	Total maintenance and operation costs	US\$785,100 (yr^{-1})	$C_{MO} = C_{wage,T}+C_{wage,S}+C_{fuel,H}+C_{fuel,RB}+C_P+C_{MR}+C_{OS}+C_O+C_{LT}+C_I+ C_{Admin}$

Table 12.1.3-1. contd....

Table 12.1.3-1. contd.

Depreciation				
C_D	Straight-line depreciation	US$303,846 ($yr^{-1}$)	$0.05C_{fix}$	
C_{redit}				
C_{cred}	Water hyacinth removal credit	US$0 ($ha^{-1}$)	Parameter	
$C_{total,yr}$	Total annual cost	US$1,088,900 ($yr^{-1}$)	$C_{total,yr} = C_{MO} + C_D-(C_{cred}A)$	
Biomass production				
$M_{biomass}$	Dry biomass produced annually	1.94E+04 (tonne yr^{-1})	$M_{biomass} = [M_{B,out}/1000\ (kg/tonne)]\ [24(hr/day)]T_{yr}$	
C_{Final}	Price per tonne of water hyacinth bagasse	US$56 ($t^{-1}$)	$C_{Final} = C_{total,yr}/M_{biomass}$	

bagasse at the cost of US$56 t^{-1}, and removed 486 tonnes of nitrogen and 34 tonnes of phosphorus, thus lowering total nitrogen concentration to 2.6 mg N L^{-1} and total phosphorus concentration to 0.4 mg P L^{-1}.

The price (US$56 t^{-1}) of the dry biomass obtained was higher than the price (US$40) reported by Hronich et al. (2008) due to shorter annual working days (180 d yr^{-1}) and longer pulling distance (4500 m) of water hyacinth mats that were expected in the northern subtropical zone in real practice. One of the main targets for the project was to remove nitrogen and phosphorus from water, and these targets were successfully achieved. If the ecological compensation were included in the calculation, the income would be about US$1.3 million annually (with US$67 per tonne dry biomass easily covering the cost of harvesting and dehydration). Further analysis of the removed amounts of nitrogen and phosphorus revealed that the nutrients were unbalanced during the operation, i.e., nitrogen: phosphorus ratio at 14, which was much higher than the normal ratio of 6–7. The unbalanced nutrient conditions may reduce the biomass production, resulting in less effective nutrient removal and thus the cost of phytoremediation may increase. When nutrients are in balance, more nitrogen and phosphorus may be removed to yield higher ecological compensations and/or better water quality.

The total cost of harvesting and dehydration was US$1.09 million yr^{-1} yielding US$2535 ha^{-1} yr^{-1} (Table 12.1.4-1: $C_{total,yr}$). Literature reported that the main difficulties in implementing mechanical harvest method were the high cost of purchasing machinery for harvesting. For instance, to allow the harvesting capacity of 1000 tonnes per day, US$ one million was required at the 2007 market value (Wise et al. 2007). Other costs might be more manageable. For example, integrated herbicide and biological control was optimized at US$40 ha^{-1} yr^{-1} at the 2002 market value in South Africa (van Wyk and van Wilgen 2002). Another report valued ecosystem services management at US$166–181 ha^{-1} at the 2000 market value (van Wilgen and Lange 2011). It appears that in a single-target system, the high investment was not generating high benefits, whereas the multidimensional management system can generate good return on the investment. The above case study in Lake Caohai generated 19,350 tonnes of good-quality biomass (with 36% water content) that can be used for silage feed or methane production or compost. The economic value of utilization should be covered by the relative production cycles separately and is not discussed here.

12.2 Further research

Water hyacinth has caused a global concern in aquatic ecosystems, especially in tropical and subtropical countries. The multidimensional management targets involve the subject fields of botany, biology, ecology, socio-economy, environmental science and chemistry. The knowledge and capacity to manage, control and utilize water hyacinth in order to "solve a major weed problem, a pollution problem, an energy problem, a food problem, and a fertilizer problem" are still in the early stages. At present, the knowledge on this species is insufficient for a clear understanding and a unified opinion on solutions. To achieve the targets of management, control and utilization of water hyacinth, the mitigation of its potential spread as an aquatic weed, and the conversion of biomass to bio-energy or organic fertilizers or feed are still big challenges. Further research in mechanisms of invasion, valuation of socio-economic and ecosystem services (Holl and Howarth 2000), mechanisms of pollution remediation, water hyacinth biology (Gao and Li 2004) and ecology is essential to underpin the solutions for management and control. Other subjects, such as water hyacinth population dynamics (Peterson and Ogwang 2007), potential utilization, and designing more efficient equipment for harvesting and dehydration, are also urgent topics for the future research.

12.2.1 Research on the mechanisms of invasion and weed control

Due to predictions that water hyacinth may be continuously spreading across the world, governments are spending funds for the future management and control of this species. For instance, the Australian government has planned nationwide strategies to manage water hyacinth infestation in 2012–2017, including a management framework to quickly respond to existing and new incursions and to coordinate activities related to this species (Australian Weeds Committee 2012). The plan recognized the severe damages caused by water hyacinth infestation and called for full cooperation from all levels of government and local communities to build strong education, research and capacity programs nationwide. The plan also recognized that water hyacinth remains a difficult and expensive invasive alien species to control using chemical, physical and biological methods. However, the plan may only have limited achievements due to missing an important strategy: multidimensional targets for managing and controlling water hyacinth and utilizing its biomass as an active management option. The plan also failed to recognize that the spread of water hyacinth was closely related to the levels of aquatic eutrophication and climate change, suggesting that the potential spending may increase rather than decrease in the future years.

Water hyacinth infestation is a symptom of broader water shed management and pollution problems. It calls for a concise multi-national and trans-boundary policy to deal with the noxious weed in aquatic systems. In October 2010, world leaders adopted the Strategic Plan for Biodiversity (2011–2020) targeting the need for identification of invasive alien species and their invasion pathways, the need to control and eradicate priority invasive species, and to manage invasion pathways in order to prevent proliferation (CBD 2010). Given the complexity of control options and the potential for climate change to assist the spread of water hyacinth, it is critical to develop comprehensive management strategies and action plans. A multidisciplinary approach should be designed, which ensures that the highest political and administrative levels recognize the potential seriousness of the weed. The plans should also state clearly the role of each government department, stakeholders, municipal councils and local community that are involved in the fight against water hyacinth. Awareness should be raised amongst local communities and all stakeholders about the inherent dangers of water hyacinth infestation to mobilize riparian communities towards the control measures. One practical approach is to involve communities in comprehensive control activities, for example, biological and chemical control or effective harvesting, dehydration and utilization of this species. The methods for water hyacinth control should include reduction of nutrient load in the water bodies through treatment of effluent from sewage treatment plants, urban wastes and industry. Changing land-use practices in the riparian communities through watershed management would help reduce agricultural runoff as a mechanism for controlling the proliferation of water hyacinth. This is considered by many as one of the most sustainable long-term management actions (UNEP and GEAS 2013).

12.2.2 Valuation of socio-economic and ecosystem services

The invasion of water hyacinth has resulted in enormous ecological and economic consequences worldwide. However, the real total economic losses were unknown because the methods for assessment and the valuation of socio-economic and ecosystem service were not well defined and the historic pathways and the mechanisms of invasion were not well understood. Although the spread of this weed in Africa, Australia and North America has been well documented, its invasion in South East Asia, especially in China, has yet to be fully documented. Over the last several decades, water hyacinth has invaded many water bodies across about half of China's aquatic ecosystems, causing a decline of native biodiversity, alteration of ecosystem services, deterioration of aquatic environments, and spread of diseases affecting human health. The infestations have also led to enormous economic losses by impeding water flows, paralyzing navigation and damaging irrigation and hydroelectricity facilities.

To effectively control water hyacinth, a sustainable socio-economic and science-based management framework must be implemented to explicitly incorporate principles from water hyacinth biology and ecology in the management of integrated non-point pollution control (e.g., from agricultural farmland) and waste treatment processing. This framework should recognize full socio-economic and ecosystem valuation and emphasize multiple-scale long-term monitoring and research, integration among different control techniques, combination of control with utilization and landscape-level adaptive management (Lu et al. 2007).

In order for policy makers to make informed decisions, much more economic information is required on the costs and benefits of environmental programs. For example, it is frequently stated that there are insufficient resources to control water hyacinth. However, if the costs of achieving improved targets are compared with the costs of (i) decreases in fish catches and recreational values, (ii) degraded water quality requiring expensive treatment, (iii) reduced water resources availability, and (iv) the costs of increased water-borne diseases, it is likely that resources needed for water hyacinth control are modest in comparison to potential losses from its proliferation (UNEP and GEAS 2013). While researchers continue to investigate the perceived potential uses of water hyacinth, the paucity of information on socio-economic and ecosystem services may impede the implementation of the multidimensional management strategies. Although the use of water hyacinth as raw material should not encourage propagation of the weed at present, the harvest and dehydration approach may promote the control of water hyacinth globally.

Efforts must continue to develop methods for valuation of ecosystem services to demonstrate the market values of either intact or remediated/rebuilt/restored ecosystems. Furthermore, the practitioners of phytoremediation or ecosystem rebuilding/restoration should make the accurate cost-benefit analyses publicly available. Without such figures, it is impossible to know the amount of resources that need to be allocated. There is also a need to develop institutional mechanisms for addressing questions of uncertainty and timescale in funding projects in order to bring people with a range of skills and backgrounds (such as biologists, ecologists, economists, aquatic system managers and policy makers, and people trained in risk management) together to design the strategies that would underpin payments for phytoremediation using water hyacinth or harvesting and dehydration of naturally-growing water hyacinth (Holl and Howarth 2000).

12.2.3 Research on the mechanisms of pollution remediation

Over the last several decades, phytoremediation using water hyacinth for removal of organic, heavy metal and metalloid pollutants have shown great advances. However, there were evidences that increasing amounts of Rare Earth Elements (REE) have been released to aquatic environments, and these elements may be accumulated in water hyacinth biomass (Chua 1998). The extent of REE accumulation and the effects in aquatic food chain including the levels of toxicity are not well known, therefore research is urgently needed to elucidate the potential risk of these pollutants.

In many eutrophic aquatic ecosystems, the water hyacinth growth limiting factor is often phosphorus because phosphorus may be incorporated into particle forms and deposited in the sediment (Zhang et al. 2016). The low concentration of phosphorus in overlaying water may not mean insufficient phosphorus supply to water hyacinth, but the equilibria between sediments and water are important in determining the biological behavior of water hyacinth in uptake and distribution of phosphorus in plant parts. Among the factors influencing phosphorus equilibria, phosphatase plays an important role in promoting uptake and utilization by water hyacinth. There are three groups of phosphatases categorized by their chemical properties: acid, neutral and alkaline. The acid and neutral enzymes are mainly inside the cells and are involved in metabolic phosphate cycles. These two groups may not directly promote transformation of particulate phosphorus to soluble reactive phosphorus or biologically available orthophosphate for plant absorption. Alkaline phosphatase can be located in the periplasmic space, external to the plasma membrane and can transform particulate phosphorus to soluble reactive phosphorus (Hu 2013). One hypothesis was that the alkaline phosphatase was produced by organisms only during phosphorus starvation, suggesting that the activity of this enzyme can be used as an indicator of phosphorus availability to plants (Quisel et al. 1996). In case of phytoremediation using water hyacinth, the activity of alkaline phosphatase in supplying phosphorus to plants is important. However, there has been little research on the activity of alkaline phosphatase on the root surface or in the root zone of water hyacinth and the effect of factors such as temperature, pH and other pollutants.

12.2.4 Research on water hyacinth biology and ecosystem evolution

Water hyacinth is a persistent species in aquatic environments that affects many ecosystem functions. However, good management and control programs have also been reported. For example, the management team on the world's second largest freshwater lake, Lake Victoria, achieved very good control during the period from late 1995 to 2000, reducing the areas of water hyacinth infestation from 17,300 hectares in 1997 to 1000 hectares in 2000 via integrated management and control (Wilson et al. 2007). The dramatic changes in water hyacinth population may affect habitat biodiversity or local socio-economic development and ecosystem services; however, in the long term, water hyacinth population may bounce back due to seed germination. Although we know that water hyacinth invasions are closely related to human economic activities (ornamental plant introduction and species escaping), species biology (seed surviving for a long time and is easily transported, clone reproduction), ecological adaptation (proliferation in eutrophic aquatic systems) and climate change (global warming and increased levels of carbon dioxide), we do not know to what extent the species growth will be influenced by the eutrophic level of the system versus other factors.

Understanding water hyacinth biology is important in order to predict its effects on the evolution of aquatic ecosystems. A global collaborative research effort would facilitate the understanding of biology and ecology of water hyacinth invading, reproducing and moving on a small scale such as local lakes and river basins as well as on a large scale at a country or continental level. Then, the water hyacinth movement and invasion routes and trends can be predicted so that early responses become possible. For instance, the remote sensing data analysis provided the population changes from 1988 to 2001 in the northern part of Bangalore city in India (Verma et al. 2003). The report revealed that the expansion of water hyacinth population was related to the pollution from agriculture and industrial wastewater discharge. An important question is: if pollution can be controlled and eutrophication mitigated, would water hyacinth population shrink or stop growing? An answer to such and similar questions may come from the models incorporating water hyacinth biology and ecology related to the global environment and climate change models, allowing prediction on where and when the water hyacinth problem may occur and may require a quick response.

12.2.5 Research on high-value products from water hyacinth biomass

The cost-benefit analyses showed that global management and control of water hyacinth demand huge financial resources (Mench et al. 2010), without becoming a profitable business as yet (Gettys et al. 2009). The root cause is that the final products from the water hyacinth biomass have low value, even when used in silage feed for animals. Good quality alfalfa hay was only US\$100–250 t^{-1} on the global market in April 2016 (USDA 2016). The quality of water hyacinth biomass using current harvesting and dehydration technology is not as good as that of alfalfa. Even when roots and leaves of water hyacinth can be harvested separately, the quality of the leaf biomass would be almost comparable to alfalfa hay, and the profitability would be almost as good as in alfalfa business, but using water hyacinth involves the risks in heavy initial investment, while the improved ecosystem services may not be directly recognized as beneficial to investors. To change the present situation, research on making value-added products from water hyacinth biomass is urgent.

One potential opportunity is to extract leaf proteins from the by-product of dehydration: water hyacinth juice. Protein content of water hyacinth juice was as high as 5.2–7.4 g L^{-1} (Du et al. 2012), which suggested a potential utilization as a bio-resource due to proteins moving effectively from tissues to the juice (Du et al. 2010). If food-grade protein was to be prepared, starch flocculants can be used at a dose of 1–6 g m^{-3} juice at prices of US\$0.50–12.00 (US\$500–2000 per tonne of cationic starch flocculants). However, there has been no report on such application at the time of this book going to the press.

Another potential opportunity is to extract phenolics and flavonoids from water hyacinth. In many developing countries, water hyacinth biomass or extracts has been used as herbal medicine. In many cases, water hyacinth has shown to be a useful antibiotic agent against some pathogenic strains of bacteria, fungi and algae. The main activity components are allelopathic substances and antioxidants. Allelopathic substances such as alkaloids and phthalate derivatives (Shanab et al. 2010) may be beneficial in agriculture. The human health-related water hyacinth extracts are phenolics and flavonoids or other antioxidants (Aboul-Enein et al. 2014). Kiruba Daniel et al. (2012) used atomic force microscopy, transmission electron microscopy and Fourier transformation infrared spectrum to identify antimicrobial activity of silver nanoparticles from water hyacinth extracts. The phenolic groups present in the plant extract were found responsible for the transferring silver nitrate into silver nanoparticles and imparting stability. The nanoparticles inhibited pathogens such as *Pseudomonas fluorescens* and *Klebsiela pneumoniae* (Patel 2012). Chantiratikul et al. (2009) reported the complex active compounds isolated from water hyacinth biomass for pharmaceutical uses. Total Phenolic (TP) and Total Flavonoid (TF) extracts from water hyacinth ranged from 40–62 (TP) and 13–15 (TF) mg quercetin equivalent g^{-1} dry biomass with antioxidant activities of IC_{50} 145–179 μg mL^{-1}.

Total phenolics and flavonoids are large groups of biochemicals including gallic acid, rutin, vanillin, epicatechin, catechin, kaempferol, quercetin, myricetin and naringinin. The content of total phenolics and total flavonoids varies with plant growth stage and environmental factors such as nutrient conditions. For example, literature reported that the total flavonoids were higher in water hyacinth (43 mg rutin equivalents g^{-1} dry matter) than the medical herb *Labisia pumila* (0.12–0.78 mg rutin equivalent g^{-1} dry biomass) (Ibrahim et al. 2011) and also 10 times higher than legume flour (0.30–3.4 mg rutin equivalents g^{-1} legume flour) (Ren et al. 2012).

Water hyacinth biomass can also be utilized as a lignocellulosic substrate for production of cellulase by *Trichoderma reesei*. The effects of substrate pretreatment, substrate concentration, initial medium pH, mode of inoculation and incubation temperature on cellulase production were reported by Deshpande et al. (2009). Under optimal conditions, a maximal cellulase activity of 0.22 IU mL^{-1} was detected at the end of a 15-day incubation. Specific activity of the enzyme was 6.25 IU mg^{-1} protein; the saccharification of cellulose from water hyacinth was significantly higher by laboratory-produced cellulase than the commercial blend (Deshpande et al. 2009).

The above examples indicated that the biochemistry of water hyacinth is not well understood and the road to higher value products is still distant. Even in the areas of relatively active research (such as extracting total phenolics and total flavonoids), it remains unclear what the content of these compounds is in root or leaf bagasse, in dried juice biomass and in filtrate of juice; also, the content and activities of individual phenolics and flavonoids are not well understood. Without such information, the market value of the water hyacinth products cannot be assessed.

12.2.6 Research on equipment for mechanical harvesting and dehydration

The ideas of multidimensional targets of water hyacinth management, control and phytoremediation are to link the beneficiaries together. In the operations, the harvest and dehydration are the two critical and costliest steps in the whole plan. Although the previous cost-benefit analyses showed that US$40 to 56 t^{-1} water hyacinth bagasse (with 36% water content) could be obtained, the equipment for harvesting and dehydration are still costly and need to be further improved for efficiency and cost reduction regarding the initial investment and the economic profitability of the operation (Gettys et al.

2009), especially in separately harvesting and dehydrating roots and leaves to enhance the biomass utilization. There are patents and designs published such as Automatic Water Hyacinth Root and Leaf Separator and Volume Reduction Harvester (Zhangjiagang Haifeng Water Environmental Protection Machinery Co. 2012), Root and Leaf Auto-separating Water Hyacinth Harvester (Ni et al. 2012) and Scissor Type Water Hyacinth Mowing and Harvesting device (Dong et al. 2013). However, the real machines and the cost of investment are not made public, and the performances are yet to be assessed.

12.2.7 Research on strategies of water hyacinth management and control

Need for systematic approach

Literature indicated that water hyacinth management and control are very costly in invaded aquatic ecosystems. This is because: (1) there are no private enterprises to work on such projects as a profitable business; (2) aquatic water bodies are mostly public entities subjected to lengthy and inefficient legislative procedures; and (3) the cost-beneficiary relationship is not clear. To change this situation, further research is urgently needed to bring private enterprises into the business by developing a systematic approach in designing projects and starting with conservation funds. McConnachie et al. (2012) suggested that the projects on controlling invasive alien plant species need to assess their cost-benefit outcomes, so that management practices can be improved to use scarce conservation funds more effectively. For instance, the Working for Water project in South Africa spent 3.2 billion (ZAR, or US$432 million) conservation funds, but failed to stop the expansion of invaded areas since 1995 (van Wingen et al. 2012). The program concluded that Working for Water project should modify the strategy by focusing control efforts to the high-priority areas.

Incentive legislation needed

Water reclamation and water quality improvement in eutrophic water bodies are very costly. Although in most regions, the conservation funds are available in small amounts, the better utilization of these funds may be achieved by legislating for incentives linked to pollutant removal to assist the execution of multidimensional targets, including the fund recovery from selling the products derived from the biomass of water hyacinth.

Bilateral co-operation

Large fresh water bodies straddling national boundaries are also quite common. In those cases, bilateral co-operation is essential. For example, water hyacinth control has been included in the bilateral program of the Nigeria-Niger and Nigeria-Benin Joint Commission for Economic Co-operation.

The Commission mandates include: exchange of information and sharing of expertise; formulation and investigation as well as organization of the regional training program for water hyacinth control; facilitation of cross border surveys; harmonization of national regulations for the introduction of biological control agents; and establishment of early warning systems for the appearance and movement of floating water weeds in the region. These bilateral co-operations are expected to increase effectiveness in the execution of joint control programs, data collection and management through cross-fertilization of ideas and expert exchange (Uka et al. 2007). The co-operations would be more effective if multidimensional targets are specified.

Today there is a global recognition among scientists and managers that there is no completely effective method to eradicate water hyacinth, and no best option of integrated management and control of the weed. However, the existence of water hyacinth can be regarded as an opportunity to achieve best results in the cost, environmental consequences, and efficacy in controlling and utilizing water hyacinth by combining expertise and resources as well as utilizing various strategies (Téllez et al. 2008).

Management strategies

Water hyacinth is a global concern as well as opportunity and needs global co-operation. Although chemical, biological and mechanical controls have their inherited advantages and drawbacks, using specific methods in specific regions and at particular times may halt weed spreading. However, the long-term targets should be multidimensional so that information on each goal and methods to achieve it is at the core of the research programs. The chemical control may be the last and the biological control the first option because the latter needs at least 4 years to take effect. For instance, biological control of water hyacinth took 4 years in eastern United States (Center et al. 2002) and 4 years (1997–2001) in the Lake Victoria region with low cost and long-term effectiveness (Njoka 2004). By 2006 (after 10 years), 10 weevils per plant were still found in Winam Gulf, Lake Victoria, Kenya, which imply that the biological control agent was still effective (Peterson and Ogwang 2007).

Chemical and biological controls are not the options in eutrophication remediation and biomass resources utilization; instead, using water hyacinth in phytoremediation and utilizing its biomass is predicated on mechanical harvesting and dehydration that require heavy capital investment. However, in hypereutrophic fresh water bodies, there is little other choice than to implement multidimensional strategies to solve problems. Multidimensional strategies are required in cleaning aquatic ecosystems because water hyacinth has strong persistence and capacity to recover from seed, which can remain viable for 28 years (Sullivan and Wood 2012) and germinate in 4–5 days (Pérez et al. 2011). For example, South Africa controlled water hyacinth population well below acceptable levels in 1994 (Hill and Olckers 2001), but the weed came back and water hyacinth population recovered (mainly from seed germination) after the weevil population dropped (Julien 2001). Management strategies may be summarized as follows: chemical control as an emergency tool, biological control as a long-term tool with low cost and multidimensional target management and control for clean and final solution, especially in heavily polluted fresh water bodies.

References cited

Aboul-Enein, A. M., S. M. Shanab, E. A. Shalaby, M. M. Zahran, D. A. Lightfoot and H. A. El-Shemy. 2014. Cytotoxic and antioxidant properties of active principals isolated from water hyacinth against four cancer cells lines. *BMC Complementary and Alternative Medicine* 14(1): 397–406.

Australian Weeds Committee. 2012. Weeds of national significance, water hyacinth (*Eichhornia crassipes*) strategic plan 2012 to 2017. *Australian Weeds Committee*. Canberra, Australia: Australian Weeds Committee. http://www.weeds.org.au/WoNS/waterhyacinth/docs/National_Water_Hyacinth_Strategic_Plan_(final-June_2013).pdf.

Bagnall, L. O. 1982. Bulk mechanical properties of waterhyacinth. *Journal of Aquatic Plant Management* 20: 49–53.

Binder, C. R. 2007. From material flow analysis to material flow management part I: social sciences modeling approaches coupled to MFA. *Journal of Cleaner Production* 15(17): 1596–1604.

Center, T. D., M. P. Hill, H. Cordo and M. H. Julien. 2002. Waterhyacinth. In: Biological Control of Invasive Plants in the Eastern United States, ed. R. G. Van Driesche, B. Blossey, M. Hoddle, S. Lyon, and R. Reardon, 41–64. Morgantown, West Virginia, USA: United States Department of Agriculture & Forest Service Publication FHTET-2002-04.

Chantiratikul, P., P. Meechai, W. Nakbanpote, K. Campus and K. Campus. 2009. Antioxidant activities and phenolic contents of extracts from *Salvinia molesta* and *Eichornia crassipes*. *Research Journal of Biological Sciences* 4(10): 1113–1117.

Chua, H. 1998. Bio-accumulation of environmental residues of rare earth elements in aquatic flora *Eichhornia crassipes* (Mart.) Solms in Guangdong Province of China. *Science of the Total Environment* 214(1-3): 79–85.

Costanza, R., R. Arge, R. De Groot, S. Farber, M. Grasso, B. Hannon et al. 1997. The value of the world's ecosystem services and natural capital. *Nature* 387(May): 253–260.

Deshpande, P., S. Nair and S. Khedkar. 2009. Water hyacinth as carbon source for the production of cellulase by *Trichoderma reesei*. *Applied Biochemistry and Biotechnology* 158(3): 552–60.

Ding, L., W. Zhao, Y. Huang, S. Cheng and C. Liu. 2015. Research on the coupling coordination relationship between urbanization and the air environment: a case study of the area of Wuhan. *Atmosphere* 6(10): 1539–1558.

Dong, X., X. Chen, C. Gao, W. Li, Z. Xiaopei, Y. Liu et al. 2013. Scissor type water hyacinth mowing and harvesting device. China: SIPO, China Patent #CN203608568U (In Chinese).

Du, J., Z. Chang, X. Ye, Y. Xu and J. Zhang. 2012. Pilot-scale study on dehydration effect of water hyacinth with different pulverization degree. *Transactions of the Chinese Society of Agricultural Engineering* 28(5): 207–212 (In Chinese with English Abstract).

Du, J., Z. Z. Chang, X. Ye and H. Huang. 2010. Losses in nitrogen, phosphorus and potassium of water hyacinth dehydrated by mechanical press. *Fujian Journal of Agricultural Science* 25(1): 104–107 (In Chinese with English Abstract).

Gao, L. and B. Li. 2004. The study of a specious invasive plant, water hyacinth (*Eichhornia crassipes*): achievements and challenges. *Acta Phytoecologica Sinica* 28(6): 735–752 (In Chinese with English Abstract).

Gettys, L. A., W. T. Haller and M. Bellaud. 2009. *Biology and control of aquatic plants: a best management practices handbook*. 2nd ed. Marietta GA, USA: Aquatic Ecosystem Restoration Foundation.

de Groot, R., R. Alkemade, L. Braat, L. Hein and L. Willemen. 2010. Challenges in integrating the concept of ecosystem services and values in landscape planning, management and decision making. *Ecological Complexity* 7(3): 260–272.

de Groot, R., B. Fisher and M. Christie. 2010. Integrating the ecological and economic dimensions in biodiversity and ecosystem service valuation. In: The economics of ecosystems and

biodiversity: the ecological and economic foundations, ed. P. Kumar, 1–40. London, UK: United Nations Environment Programme, Earthscan.
Hill, M. P. and T. Olckers. 2001. Biological control initiatives against water hyacinth in South Africa: constraining factors, success and new courses of action. In: Proceedings of the Second Meeting of the Global Working Group for the Biological and Integrated Control of Water Hyacinth, Beijing, China, 9–12 October 2000, ed. M. H. Julien, M. P. Hill, T. D. Center, and J. Ding, 102: 33–38. Canberra, Australia: Australian Centre for International Agricultural Research.
Holl, K. D. and R. B. Howarth. 2000. Paying for restoration. *Restoration Ecology* 8(3): 260–267.
Hronich, J. E., L. Martin, J. Plawsky and H. R. Bungay. 2008. Potential of *Eichhornia crassipes* for biomass refining. *Journal of Industrial Microbiology and Biotechnology* 35(5): 393–402.
Hu, Z. 2013. Seasonal variation of phosphorus forms and mutual transformation in Grand River watershed, Canada. Ph.D. Thesis, Department of Soil and Environmental Chemistry. Southwest University.
Ibrahim, M. H., H. Z. E. Jaafar, A. Rahmat and Z. A. Rahman. 2011. The relationship between phenolics and flavonoids production with total non structural carbohydrate and photosynthetic rate in *Labisia pumila* Benth. under high CO_2 and nitrogen fertilization. *Molecules* 16(1): 162–174.
Institute of Environmental Science. 1992. *Survey on eutrophication of Lake Dianchi*. 1st ed. Kunming, China: Kunming Science and Technology Press.
Julien, M. H. 2001. Biological control of water hyacinth with arthropods: a review to 2000. In: Proceedings of the Second Meeting of the Global Working Group for the Biological and Integrated Control of Water Hyacinth, Beijing, China, 9–12 October 2000, ed. M. H. Julien, M. P. Hill, T. D. Center and J. Ding, 102: 8–20. Canberra, Australia: Australian Centre for International Agricultural Research.
Kakuru, W., N. Turyahabwe and J. Mugisha. 2013. Total economic value of wetlands products and services in Uganda. *The Scientific World Journal* 2013: 1–13.
Kang, Z. and H. Liu. 2015. Eco-compensation for removal of nitrogen and phosphorus from Lake Taihu using cost-benefit analysis. *Jiangsu Journal of Agricultural Sciences* 31(4): 943–945 (In Chinese with English Abstract).
Kang, Z., J. Tang and H. Liu. 2012. Eco-compensation based on cost-benefit analysis of nitrogen and phosphorus removal in Lake Taihu region. In: Advances in China farming system research 2012, ed. Farming System Committee, Chinese Agricultural Association, 182–187 (In Chinese). Beijing, China: China Agricultural Science and Technology Press.
Kiruba Daniel, S. C. G., K. Nehru and M. Sivakumar. 2012. Rapid biosynthesis of silver nanoparticles using *Eichornia crassipes* and its antibacterial activity. *Current Nanoscience* 8(1): 125–129.
de Lange, W. J. and B. W. van Wilgen. 2010. An economic assessment of the contribution of biological control to the management of invasive alien plants and to the protection of ecosystem services in South Africa. *Biological Invasions* 12(12): 4113–4124.
Liu, H., Z. Kang and H. Chen. 2011. Establishment of ecological compensation mechanism for eutrophied water treatment. *Jiangsu Journal of Agricultural Sciences* 27(4): 899–902 (In Chinese with English Abstract).
Lu, J., J. Wu, Z. Fu and L. Zhu. 2007. Water hyacinth in China: a sustainability science-based management framework. *Environmental Management* 40(6): 823–830.
Malik, A. 2007. Environmental challenge vis a vis opportunity: the case of water hyacinth. *Environment International* 33(1): 122–38.
McConnachie, M. M., R. M. Cowling, B. W. van Wilgen and D. A. McConnachie. 2012. Evaluating the cost-effectiveness of invasive alien plant clearing: a case study from South Africa. *Biological Conservation* 155: 128–135.
Mench, M., N. Lepp, V. Bert, J. -P. Schwitzguébel, S. W. Gawronski, P. Schröder et al. 2010. Successes and limitations of phytotechnologies at field scale: outcomes, assessment and outlook from COST Action 859. *Journal of Soils and Sediments* 10(6): 1039–1070.
MEP-PRC. 2002. Environmental quality standards for surface water (GB3838-2002). Baijing, China: Ministry of Environmental Protection of The Peoples's Republic of China.
Ni, J., P. Wu, Y. Yang, F. Dong and W. Guo. 2012. Root and leaf auto-separating water hyacinth harvester. China: SIPO, China Patent #CN202320731U (In Chinese).

Njoka, S. W. 2004. The biology and impact of *Neochetina* weevils on water hyacinth, *Eichhornia crassipes* in Lake Victoria Basin, Kenya. Ph.D. Thesis, Department of Environmental Studies (Biological Sciences). School of Graduate Studies at Moi University.

Pagiola, S., K. von Ritter and J. Bishop. 2004. *Assessing the economic value of ecosystem conservation. The World Bank Environment Department*. Washington, D.C. 20443, U.S.A.

Pascual, U., R. Muradian, L. Brander, E. Gómez-baggethun, B. Martín-lópez, M. Verma et al. 2010. Chapter 5 The economics of valuing ecosystem services and biodiversity. In: The Economics of Ecosystems and Biodiversity: Ecological and Economic Foundations, ed. P. Kumar, 183–255. London, UK: United Nations Environment Programme, Earthscan.

Patel, S. 2012. Threats, management and envisaged utilizations of aquatic weed *Eichhornia crassipes*: an overview. *Reviews in Environmental Science and Bio/Technology* 11(3): 249–259.

Pérez, E. A., J. A. Coetzee, T. R. Téllez and M. P. Hill. 2011. A first report of water hyacinth (*Eichhornia crassipes*) soil seed banks in South Africa. *South African Journal of Botany* 77(3): 795–800.

Peterson, G. and J. A. Ogwang. 2007. Water hyacinth re-invades Lake Victoria. *Resilience Science*. http://rs.resalliance.org/2007/02/22/water-hyacinth-re-invades-lake-victoria/.

Quisel, J. D., D. D. Wykoff and A. R. Grossman. 1996. Biochemical characterization of the extracellular phosphatases produced by phosphorus-deprived *Chlamydomonas reinhardtii*. *Plant Physiology* 111(3): 839–848.

Ren, S., Z. Liu and P. Wang. 2012. Proximate composition and flavonoids content and *in vitro* antioxidant activity of 10 varieties of legume seeds grown in China. *Journal of Medicinal Plants Research* 6(2): 301–308.

Schiller, F., A. S. Penn and L. Basson. 2014. Analyzing networks in industrial ecology—a review of social-material network analyses. *Journal of Cleaner Production* 76: 1–11.

Shanab, S. M. M., E. A. Shalaby, D. A. Lightfoot and H. A. El-Shemy. 2010. Allelopathic effects of water hyacinth (*Eichhornia crassipes*). *PloS One* 5(10): e13200.

Söderqvist, T., H. Eggert, B. Olsson and A. Soutukorva. 2005. Economic valuation for sustainable development in the Swedish coastal zone. *AMBIO: A Journal of the Human Environment* 34(2): 169–175.

Sullivan, P. and R. Wood. 2012. Water hyacinth [*Eichhornia crassipes* (Mart.) Solms] seed longevity and the implications for management. In: Eighteenth Australasian Weeds Conference, ed. V. Eldershaw, 37–40. Melbourne, Australia: Weed Society of Victoria Inc.

Téllez, T. R., E. López, G. Granado, E. Pérez, R. López and J. Guzmán. 2008. The water hyacinth, *Eichhornia crassipes*: an invasive plant in the Guadiana River Basin (Spain). *Aquatic Invasions* 3(1): 42–53.

Uka, U. N., K. S. Chukwuka and F. Daddy. 2007. Water hyacinth infestation and management in Nigeria inland waters: a review. *Journal of Plant Sciences* 2(5): 480–488.

UNEP and GEAS. 2013. Water hyacinth—can its aggressive invasion be controlled? *Environmental Development* 7: 139–154.

UNEP-WCMC. 2011. *Marine and coastal ecosystem services: Valuation methods and their application*. Cambridge, United Kingdom: UNEP-WCMC Biodiversity Series No. 33. 46 pp.

USDA. 2016. *National hay, feed & seed weekly summary for week ending April 8, 2016*. Moses Lake, WA, USA.

Verma, R., S. P. Singh and K. Ganesha Raj. 2003. Assessment of changes in water-hyacinth coverage of water bodies in northern part of Bangalore city using temporal remote sensing data. *Current Science* 84(6): 795–804.

Wang, F. S., C. Q. Liu, M. H. Wu, Y. X. Yu, F. W. Wu, S. L. Lu et al. 2009. Stable isotopes in sedimentary organic matter from Lake Dianchi and their indication of eutrophication history. *Water, Air, & Soil Pollution* (199): 159–170.

Wang, Z., Z. Zhang, J. Zhang, Y. Zhang, H. Liu and S. Yan. 2012. Large-scale utilization of water hyacinth for nutrient removal in Lake Dianchi in China: the effects on the water quality, macrozoobenthos and zooplankton. *Chemosphere* 89(10): 1255–61.

Wang, Z., Z. Zhang, Y. Zhang, J. Zhang, S. Yan and J. Guo. 2013. Nitrogen removal from Lake Caohai, a typical ultra-eutrophic lake in China with large scale confined growth of *Eichhornia crassipes*. *Chemosphere* 92(2): 177–183.

van Wilgen, B. and Wjd. Lange. 2011. The costs and benefits of biological control of invasive alien plants in South Africa. *African Entomology* 19(2): 504–514.

Wilson, J. R. U., O. Ajuonu, T. D. Center, M. P. Hill, M. H. Julien, F. F. Katagira et al. 2007. The decline of water hyacinth on Lake Victoria was due to biological control by Neochetina spp. *Aquatic Botany* 87(1): 90–93.
van Wingen, B. W., G. G. Forsyth, D. C. Le Maitre, A. Wannenburgh, J. D. F. Kotzé, E. van den Berg et al. 2012. An assessment of the effectiveness of a large, national-scale invasive alien plant control strategy in South Africa. *Biological Conservation* 148(1): 28–38.
Wise, R., B. VanWilgen, M. Hill, F. Schulthess, D. Tweddle, A. Chabi-Olay et al. 2007. *The economic impact and appropriate management of selected invasive alien species on the African Continent*. CSIR Report Number: CSIR/NRE/RBSD/ER/2007/0044/C: Global Invasive Species Programme.
van Wyk, E. and B. W. van Wilgen. 2002. The cost of water hyacinth control in South Africa: a case study of three options. *African Journal of Aquatic Science* 27(2): 141–149.
Zhang, L., P. Zhu, Y. Gao, Z. Zhang and S. Yan. 2013. Design of anti-stormy wave enclosures for confined growth of water hyacinth in lakes. *Jiangsu Journal of Agricultural Sciences* 29(6): 1360–1364 (In Chinese with English Abstract).
Zhang, Y., Z. Zhang, Z. Chen, H. Liu, X. Wen, H. Qin et al. 2016. Phosphorus removal pathways in water hyacinth (*Eichhornia crassipes*) ecological restoration systems and influence on phosphorus release in sediment by the macrophyte. *Journal of Nanjing Agricultural University* 39(1): 106–113 (In Chinese with English Abstract).
Zhang, Z., Y. Gao, J. Guo and S. Yan. 2014. Practice and reflections of remediation of eutrophicated waters: a case study of haptophyte remediation of the ecology of Dianchi. *Journal of Ecology and Rural Environment* 30(1): 15–21 (In Chinese with English Abstract).
Zhangjiagang Haifeng water Environmental Protection Machinery Co., L. 2012. Automatic water hyacinth root and leaf separating and volume reduction harvester. SIPO, China Patent #CN102303688 A.
Zheng, J., Z. Chan, L. Chen, P. Zhu and J. Shen. 2008. Feasibility studies on N and P removal using water hyacinth in Lake Taihu region. *Jiangsu Agricultural Science* 3: 247–250 (In Chinese).

Index